Wet Cake Filtration

Wet Cake Filtration

Fundamentals, Equipment, and Strategies

Harald Anlauf

Author

Dr.-Ing. Harald Anlauf
Karlsruhe Institute of Technology (KIT)
Institute of Mechanical Process
Engineering and Mechanics
Straße am Forum 8
76131 Karlsruhe
Germany

Cover
Courtesy of Bokela, Karlsruhe, Germany
and Dr. Harald Anlauf, KIT, Germany

All books published by **Wiley-VCH** are carefully produced. Nevertheless, authors, editors, and publisher do not warrant the information contained in these books, including this book, to be free of errors. Readers are advised to keep in mind that statements, data, illustrations, procedural details or other items may inadvertently be inaccurate.

Library of Congress Card No.:
applied for

British Library Cataloguing-in-Publication Data
A catalogue record for this book is available from the British Library.

Bibliographic information published by the Deutsche Nationalbibliothek
The Deutsche Nationalbibliothek lists this publication in the Deutsche Nationalbibliografie; detailed bibliographic data are available on the Internet at <http://dnb.d-nb.de>.

© 2019 Wiley-VCH Verlag GmbH & Co. KGaA, Boschstr. 12, 69469 Weinheim, Germany

All rights reserved (including those of translation into other languages). No part of this book may be reproduced in any form – by photoprinting, microfilm, or any other means – nor transmitted or translated into a machine language without written permission from the publishers. Registered names, trademarks, etc. used in this book, even when not specifically marked as such, are not to be considered unprotected by law.

Print ISBN: 978-3-527-34606-6
ePDF ISBN: 978-3-527-82063-4
ePub ISBN: 978-3-527-82065-8
oBook ISBN: 978-3-527-82066-5

Typesetting SPi Global, Chennai, India
Printing and Binding Markono Print Media Pte Ltd, Singapore

Printed on acid-free paper

10 9 8 7 6 5 4 3 2 1

Contents

Preface *ix*

1 Introduction and Overview *1*
1.1 General Aspects of Solid–Liquid Separation in General and Cake Filtration in Detail *1*
References *11*

2 Slurry Characterization *13*
2.1 Introduction *13*
2.2 Liquid Properties *14*
2.3 Particle Properties *14*
2.3.1 General Aspects *14*
2.3.2 Characterization of Single Particles *16*
2.3.3 Characterization of Particle Collectives *20*
2.3.4 Characterization of Particle Collective Fractionation *24*
2.4 Slurry *32*
2.4.1 Solid Concentration *32*
2.4.2 Stability *33*
2.5 Sampling *35*
References *38*

3 Cake Structure Characterization *41*
3.1 Introduction *41*
3.2 Porosity *42*
3.3 Particle Arrangement *49*
3.4 Pore Size *52*
References *54*

4 Characterization of Liquid Flow Through Porous Particle Layers *57*
4.1 Introduction *57*
4.2 Dimension Analytic Approach for the Flow Through Porous Particle Layers *57*
4.3 Empirical Approach for the Flow Through Porous Particle Layers *61*
References *63*

5		**Slurry Pretreatment to Enhance Cake Filtration Conditions** *65*
5.1		Introduction *65*
5.2		Thickening *66*
5.3		Agglomeration *70*
5.4		Fractionation/Classification/Sorting *75*
5.5		Filter Aids – Body Feed Filtration *80*
5.6		Thermal Conditioning *83*
5.7		Chemical Conditioning *83*
		References *84*
6		**Filter Cake Formation** *87*
6.1		Introduction *87*
6.2		Filtration Mechanisms During the Initial Phase of Cake Filtration *88*
6.3		Formation of Incompressible Filter Cakes by Pressure Filtration *94*
6.3.1		Principle Model of Time-Dependent Filter Cake Growth *94*
6.3.2		Experimental Determination of Process Characterizing Parameters *98*
6.3.3		Throughput of Discontinuous Cake Filters *104*
6.3.4		Throughput of Continuous Vacuum and Pressure Filters *108*
6.3.5		Aspects of Filter Design and Operation Regarding Cake Formation and Throughput *113*
6.4		Formation of Compressible Filter Cakes by Pressure Filtration *123*
6.4.1		Fundamental Considerations Regarding Compressible Cake Filtration *123*
6.4.2		Experimental Determination of Process Characterizing Parameters *130*
6.4.3		Optimization of Compressible Cake Filtration *133*
6.4.4		Aspects of Filter Design and Operation Regarding Cake Formation and Throughput *136*
6.5		Formation of Filter Cakes in Centrifuges *146*
6.5.1		Fundamental Considerations Regarding Cake Filtration in Centrifuges *146*
6.5.2		Aspects of Centrifuge Design and Operation Regarding Cake Formation and Throughput *152*
		References *169*
7		**Particle Washing** *175*
7.1		Introduction *175*
7.2		Principles of Particle Washing *176*
7.3		Limits of Particle Washing Processes *178*
7.4		Characterization of Particle Washing Results *180*
7.5		Dilution Washing *182*
7.6		Permeation Washing *186*
		References *201*
8		**Filter Cake Deliquoring** *203*
8.1		Introduction *203*
8.2		Characterization of Deliquoring Results *206*

8.3	Desaturation of Filter Cakes	208
8.3.1	Boundary Surface and Surface Tension	208
8.3.2	Three-Phase Contact Line, Contact Angle, and Wetting	215
8.3.3	Capillary Pressure and Capillary Pressure Distribution	222
8.3.4	Desaturation of Incompressible Filter Cakes by Gas Pressure Difference	231
8.3.4.1	Equilibrium of Cake Desaturation with a Gas Pressure Difference	231
8.3.4.2	Kinetics of Filter Cake Desaturation with Gas Pressure Difference	234
8.3.4.3	Kinetics of Gas Flow through Filter Cakes and Energetic Considerations	240
8.3.4.4	Measurement of Cake Desaturation Equilibrium and Kinetics	246
8.3.4.5	Transfer of Desaturation Results from Bench Scale to Rotary Filters	248
8.3.4.6	Interrelation of Throughput, Cake Moisture, and Gas Consumption for Rotary Filters	251
8.3.5	Desaturation of Incompressible Filter Cakes by Steam Pressure Difference	257
8.3.6	Desaturation of Incompressible Filter Cakes in the Centrifugal Field	261
8.3.6.1	Equilibrium of Filter Cake Desaturation in the Centrifugal Field	261
8.3.6.2	Kinetics of Filter Cake Desaturation in the Centrifugal Field	267
8.3.6.3	Aspects of Centrifuge Design and Operation Regarding Cake Deliquoring	268
8.4	Consolidation of Compressible Filter Cakes by Squeezing	271
8.4.1	Fundamental Considerations Regarding the Consolidation Process	271
8.4.2	Aspects of Filter Design and Operation Regarding Cake Consolidation	274
8.5	Consolidation/Desaturation of Compressible Filter Cakes by Gas Differential Pressure	278
8.5.1	Equilibrium of Filter Cake Consolidation/Desaturation	278
8.5.2	Cake Shrinkage and Shrinkage Cracking	285
8.5.3	Prevention of Shrinkage Cracks by Squeezing and Oscillatory Shear	288
8.6	Electrically Enhanced Press Filtration	292
	References	293
9	**Selected Aspects of Filter Media for Cake Filtration**	**299**
9.1	Introduction and Overview	299
9.2	Woven Filter Media for Cake Filtration	304
9.3	Porometry – Using Capillarity to Analyze Pore Sizes of Filter Media	310
9.3.1	Introduction	310
9.3.2	Methods of Pore Size Determination	312
9.3.3	Theoretical Approach to Correlate Bubble Point and Largest Penetrating Sphere	315
9.3.4	Experimental Validation of the Theoretical Findings	318

9.4 Semipermeable Filter Media – Gas Pressure Filtration Without Gas Flow *321*
9.4.1 Introduction *321*
9.4.2 Concept of Gasless Filtration on Vacuum Drum Filters and Physical Background *322*
9.4.3 Realization of the Process in Lab and Pilot Scale *325*
References *330*

Nomenclature *333*

Index *341*

Preface

Mechanical separation of particles from liquids represents a cross-sectional technology, which touches nearly every industrial process, our personal life, and the environment. It is obvious that various physical separation principles and a huge variety of highly specialized apparatuses are needed to solve all separation problems in such different areas of application effectively and economically. Wet cake filtration, in contrast to dry cake filtration for gas cleaning purposes, represents one of the key technologies for solid–liquid separation. This is particularly the case if deliquoring of the separated solid particles is an important issue.

This book should provide a comprehensive overview of more or less all relevant aspects of wet cake filtration. It represents one selected topic among others of the "Karlsruhe School of Solid–Liquid-Separation," which was founded in 1979 by Dr.-Ing. Werner Stahl, Professor for Mechanical Process Engineering at the Technical University of Karlsruhe and refers to 40 years of research, development and teaching in this field. I myself had in 1979 the privilege to become the first PhD student of Stahl and to be part of this group for solid–liquid separation until today under the present guidance of Prof. Dr.-Ing. habil. Hermann Nirschl. This book is dedicated in great memory to Werner Stahl, who had strongly influenced the mechanical solid–liquid separation technology in general by many innovative developments, which originated in an exceptionally creative atmosphere combined with accurate scientific work. One of his most important tasks was to shorten the gap between academia and industry and for every research project had to be considered not only the scientific attractiveness but also the prospective practical benefit. For this reason, special emphasis is placed in this book on the interdependence of theoretical fundamentals and practical applications.

Besides a detailed presentation of the cake filtration process itself and process-related topics, such as slurry characterization or slurry pretreatment, special developments from the Karlsruhe School, such as "Hyperbaric Filtration" or "Steam Pressure Filtration" are included, which meanwhile worldwide have been established as the state-of-the-art technology. In addition, promising new, but not jet, commercially available processes, such as "Gasless Cake Desaturation" or "Shrinkage Crack Free Cake Desaturation" are discussed to document the still ongoing evolutionary technical progress.

It would give me great pleasure if this book would support students and teachers from academia as well as engineers from the industry to deepen their knowledge about the physical background of the different cake filtration phenomena and to find out how the fundamentals can be used most effectively to solve practical solid–liquid separation problems.

Karlsruhe, 2019 *Harald Anlauf*

1

Introduction and Overview

1.1 General Aspects of Solid–Liquid Separation in General and Cake Filtration in Detail

Cake filtration represents one of the several different mechanical methods to separate particles from liquids. Cake filtration offers, in comparison to other mechanical particle–liquid separation techniques, the advantageous possibility of direct solids posttreatment. This enables particle washing by cake permeation and the comparatively lowest mechanically achievable solid moisture contents. Particularly, if low residual moisture content of the solids is an important issue, cake filtration is the preferred technique. Unfortunately, not in every case, cake filtration can be realized from the technical and/or physical point of view. Figure 1.1 shows the principal steps of a cake filtration cycle.

After feeding the filter apparatus with slurry, a filter cake is formed under the influence of a pressure difference above and beneath the filter medium. If necessary, the cake can be washed in the next step to get rid of soluble substances in the liquid, which are still present in the wet cake. Finally, the filter cake is deliquored to displace further liquid from its porous structure. At the end of the process, the cake is discharged from the filter apparatus, and if necessary, the entire apparatus or the filter medium has to be cleaned.

The residual moisture content of the separated solids is considerably influencing the efficiency of a subsequent thermal drying and thus the energy consumption of the whole separation process. As a rule, the thermal methods are usually quite energy-intensive compared with the mechanical liquid separation. In the literature, guiding numbers of more than factor 100 can be found between the energy demands of mechanical and thermal deliquoring [1, 2]. In comparison to mechanical methods, thermal methods require not only heating of the wet system to the boiling point of the liquid but also a phase transition from the liquid to the gaseous aggregate state. The appropriate vaporization enthalpy must be raised. For this reason and also because of the often-undesirable load of temperature-sensitive products, it is in most cases advantageous to separate as much liquid as possible at low temperatures by mechanical means. For physical reasons, a final rest of liquid remains in any case after the mechanical liquid separation in the particle structure. However, this portion of liquid can only be removed from the solid material by thermal means. If a completely dry powder

Wet Cake Filtration: Fundamentals, Equipment, and Strategies, First Edition. Harald Anlauf.
© 2019 Wiley-VCH Verlag GmbH & Co. KGaA. Published 2019 by Wiley-VCH Verlag GmbH & Co. KGaA.

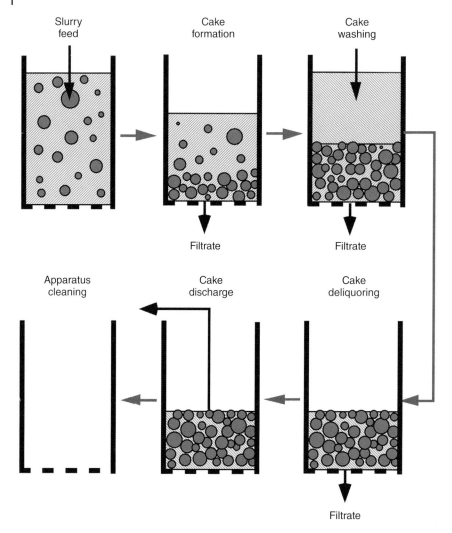

Figure 1.1 Process steps of a cake filtration.

is required as the final product, one of the tasks for the optimization of the whole separation process consists in determining the most favorable point of transfer from the mechanical to the thermal separation step. This interconnection point is very variable and defined by the requirements of the selected thermal drying process. For spray drying, a pumpable and sprayable slurry is still required, whereas the solids should be deliquored to the mechanical limit for a fluidized bed drying because the cake behaves brittle and powdery in that case. At the interface of these two basic processes for solid–liquid separation, combined mechanical–thermal processes have also been developed and established such as centrifuge dryers, nutsche dryers, and in recent times continuously operating steam pressure filters. The advantages of these systems consist in synergies, which result in energy conservation and compact and simplified process design.

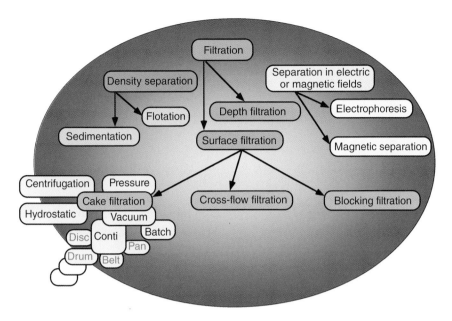

Figure 1.2 Physical principles of solid–liquid separation.

As mentioned before, the application of cake filtration is of course not possible in every case but limited by some boundary conditions. For example, in the case of submicron particles and very low solid concentration in the slurry, cake filtration makes no sense because of very high cake pressure loss of the cake and large quantities of liquid, which must penetrate such a tight cake structure. If the boundary conditions for cake filtration are not fulfilled, alternative separation techniques such as density separation, depth, cross-flow, or blocking filtration must be used. To enhance the filtration performance, electric or magnetic particle properties may be utilized additionally by the realization of an electric or magnetic field in the process room. In the literature, several comprehensive descriptions of more or less the entire technology are published [3–8]. Figure 1.2 gives an overview of the different available physical principles of mechanical particle–liquid separation.

Density separation is based on a difference of density between solid particles and liquid. If the solid density is greater, the particles are settling into the direction of gravity or a centrifugal field and are deposited at a solid wall as sediment, whereby the liquid is displaced to the opposite direction. Static continuously operating thickeners or clarifiers and many types of batch and continuous solid bowl centrifuges are based on this separation principle. If the particles have less density than liquid, they will float on the liquid surface. Also, particles of greater density than liquid float by froth flotation. For this purpose, gas bubbles are generated in the slurry, and if the particles are hydrophobic, they can adhere to the bubbles and float on the surface. This technique is often used to separate hydrophobic from hydrophilic particles of different materials in sorting processes of mineral beneficiation or waste paper recycling. If necessary, the flotation conditions regarding the wetting behavior of particles can be influenced by various surfactants. These are water-soluble molecules with polar (hydrophilic) and nonpolar (hydrophobic/lipophilic) parts. For example, the polar part of a surfactant adheres at the surface of a hydrophilic particle and thus the particle appears hydrophobic from outside. This process runs spontaneously because the systems decrease its free energy. If all particles are hydrophobic as in the case of organic matter, the complete solids can be floated and thus separated very effectively.

Filtration in contrast to density separation is based on the presence of a porous filter medium. Particles and liquid are moving under the influence of a gas pressure difference, a mechanical, hydraulic, hydrostatic, or centrifugal pressure toward the filter medium. The liquid penetrates the filter medium whereby the particles are retained inside the structure or on the surface of the filter medium.

In depth filtration processes highly diluted slurries of very small particles usually in the µm- or sub-µm range are separated. The particles are deposited inside of a three-dimensional network of pores. The pores of the filter medium must be much greater than the particles to be separated in order to minimize the flow resistance for the liquid and to allow the particles to enter the structure and to accumulate inside. The slurry concentration must be very low to prevent the filter from becoming spontaneously blocked by pore bridging at the filter media surface. The filter media can consist of a disperse particle layer from various types of materials such as sand, gravel, activated carbon, diatomaceous earth, and others or premanufactured filter elements made of fibrous materials such as cellulose, carbon, polymers, metal, and others.

Surface filtration can be subdivided principally into blocking, cross-flow, and cake filtration. Blocking or sieve filtration means that single and low concentrated particles are approaching the filter medium and plugging single pores. In the cases of sieve filtration, the pores of the filter medium must be smaller than all particles, which should be separated completely. Not in all cases, a total separation of particles is aimed, but only the retention of oversized particles to protect subsequent separation apparatuses such as hydrocyclones or disc stack separators. These apparatuses are discharging the separated solids highly concentrated through nozzles, which are in danger to become blocked by oversized particles. The particle spectrum, which can be handled by different types of blocking filters, is very broad from the cm to the µm range.

In cross-flow filters, a low concentrated slurry of small particles less than about 10 µm is pumped across a microporous membrane and is consecutively concentrated, whereas the filtrate (permeate) is discharged through the membrane. The formation of a highly impermeable particle layer (filter cake) is prevented as completely as possible by the cross-flow, which should wash away the deposited particles permanently. It depends on the force balance around a particle, whether it is sheared off or adheres at the membrane surface. The particle spectrum to be separated by cross-flow filters ranges from some µm down to small molecules such as Na^+ or Cl^- ions. As an example for reverse osmosis, "poreless" membranes are used to separate salt ions from seawater in order to produce drinking water. In such cases, not a convective liquid transport through real pores but a diffusive transport of water molecules through the molecular structure of the membrane takes place. If the cross-flow only by pumping the slurry across the membrane is not sufficient to limit the particle deposition dynamic cross-flow filters can be applied. Here, the shear forces between membrane and liquid are generated by a rotor/stator system (membrane/membrane or membrane/stirrer) or an oscillation of the membrane.

Last but not least, cake filtration is based on the formation of a relatively thick particle layer from a few millimeters to several decimeters on the surface of the filter medium. Normally, woven fabrics of different materials are used as filter media, but also needle felts, wedge wire screens, sintered materials, or sometimes microporous membranes are applied. The particle size and the slurry concentration must be great enough to enable the cake formation in a reasonable time. Otherwise, the process may be physically possible but will be not more economical or cannot be realized technically with the separation apparatus. All continuously operating apparatuses exhibit a limited residence time of the product in the process room. The particle size spectrum ranges usually from several hundred μm down to some few μm. The lower limit can be extended in many cases by particle agglomeration as one of the several slurry pretreatment techniques (cf. Chapter 5). To prevent pore blockage of the filter medium in most cases, relatively open fabrics are used (cf. Chapter 9). In contrast to the depth filtration, a sufficiently high slurry concentration must enable a more or less spontaneous pore bridging to seal the filter medium against particle penetration into the filtrate.

In Figure 1.2, the physical principles of solid–liquid separation processes are represented. As indicated before, there is hiding a great number of different processes and individual apparatuses behind the shown physical principles. This should be demonstrated exemplarily for the cake filtration principle. As can be seen in Figure 1.2, cake filtration can be carried out by using different driving forces, which are generated by centrifugation, vacuum behind the filter medium, gas overpressure above the slurry surface, and hydraulic or mechanical pressure. Each technique can be realized as a batch or continuously operating process. Continuous vacuum cake filters can be designed as drum, disc, belt, or pan filters, and each of these filters can be subdivided into several specialized modifications.

Figure 1.3 gives a rough overview of basic cake filter types, which are differing in the driving force, mode of operation, and design.

Comprehensive detailed descriptions of many apparatuses can be found in the literature [9].

From each of these filters, several special variants exist to adapt the basic design as optimal as possible to a specific separation problem. Taking the principle of a vacuum drum filter as an example, beside other modifications, different possibilities of cake discharge are possible as depicted in Figure 1.4.

Scraper discharge is well suited for well-desaturated brittle cakes, roller discharge for hard to filter slurries with thin and sticky cakes, leaving belt discharge and separate cloth washing zone for slurries, which extremely tend to clog the filter media, and last but not least precoat discharge to handle highly diluted slurries of very small particles. Here, in a first step, a precoat layer height of several cm and made of diatomaceous earth, cellulose, starch, or other materials is built up on the filter media. In a second step, the slurry to be separated is filtered and the

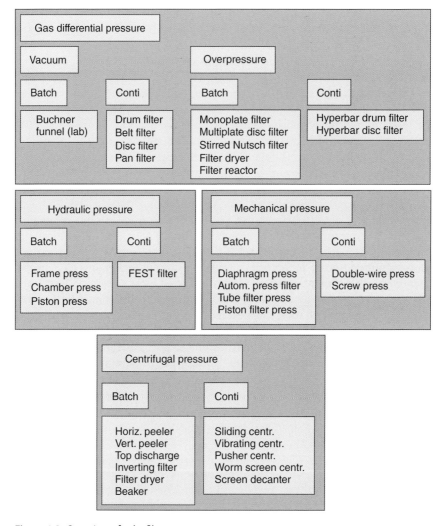

Figure 1.3 Overview of cake filter apparatuses.

precoat itself acts as a filter medium. During one cycle, the surface of the precoat becomes blocked and a very thin cake of often about 1 mm height or less is formed. A sharp knife cuts off this blocked precoat layer and again permeable precoat surface immerses into the slurry. The knife has to proceed permanently to renew the precoat surface until it is nearly consumed. Normally, the knife proceeds about 0.1 mm per revolution of the drum.

Figure 1.5 gives a rough overview of the whole range of cake filter applications and examples for apparatuses in correspondence to particle sizes and pressure differences.

The smaller the particles become, the greater the necessary forces are to separate them. The more difficult the filtration becomes, the longer the filtration periods are necessary, and the greater the pressure must be. These operation

1.1 General Aspects of Solid–Liquid Separation in General and Cake Filtration in Detail

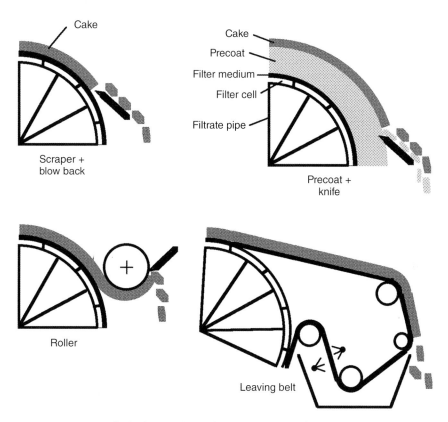

Figure 1.4 Variants of cake filter discharge from vacuum drum filters.

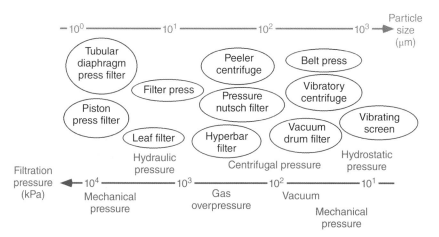

Figure 1.5 Cake filtration equipment depending on particle size and filtration pressure.

conditions are difficult to realize for continuously operating apparatuses, and thus, batchwise operating apparatuses are preferred here. Similar overviews of the ranges of apparatus application had been elaborated on the basis of empirical experience such as Gasper for filters in general [10], Trawinski for centrifuges [11], and Gösele for cake filters [12]. The boundaries of the ranges for apparatus application in such generalizing representations must not be seen as sharp and exact. The meaning is that within these ranges, often applications of the respective apparatuses can be found. The limits of application can be shifted remarkably, for example, by slurry pretreatment measures such as agglomeration.

Each physical principle in Figure 1.2 can be itemized in the described way to arrange all the different existing separation apparatuses in a systematic way. This large variety of physical separation principles and apparatuses is necessary to solve the extremely different problems of solid–liquid separation as cross-sectional technology. As can be seen in Figure 1.6, the separation of particles from liquids touches nearly every industrial process, our personal life, and the environment.

Solid–liquid separation can be focused on very different tasks such as thickening, purification, fractionation, sorting, extraction, or deliquoring. The separation has to be mastered for wide ranges of particle size and shape, specific solid and liquid weight, slurry concentration, chemical composition and rheology, flow rate, process and technical boundary conditions, and last but not least demands on the separation results. Because of the very complex boundary conditions of mechanical solid–liquid separation problems, the analysis of the specific separation problem is essential to find the best solution for getting a desired result [13]. However, in principle, there is nearly no particle–liquid system today, which cannot be separated anyhow. However, the final question is nearly in every case, how much does it cost. If the separation is not economical, the resulting product may not be saleable on the market. Thus, the challenge is to find an effective, economical, and sustainable solution. This is the motivation not only to look after the best suited state-of-the-art solution to get the desired separation result but to also search for the most cost-efficient separation processes. This is one driving force for permanent technical improvement of the equipment and most effective operating conditions. This evolutionary process of mutation and selection exhibits some similarities to biological processes. Driven by the need to improve a process, an ingenious (engineer!) idea leads to a technical mutation. If this mutation is more successful than its competitors, it will be implemented in the market. Otherwise, it will disappear again [14]. At present, more than 3000 particle separation apparatuses are on the market and consistently new developments can be recognized [15].

In most cases, not one single separation apparatus represents the optimal solution to find the best solution for a special separation task but an appropriate combination of apparatuses often in combination with slurry pretreatment. One example may be the combination of a static thickener and a pusher centrifuge (cf. Section 5.2). Although the pusher centrifuge may represent for a special case the best-suited apparatus to get lowest cake moisture, the centrifuge must be fed with adequately concentrated sludge to enable its function. If the slurry to be separated is too diluted, a prethickening is necessary. Otherwise, no complete cake

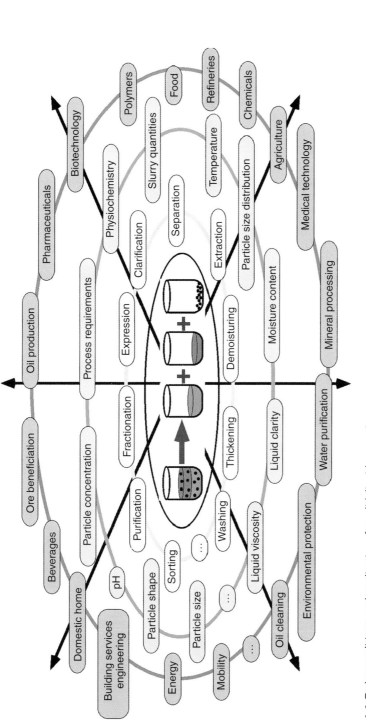

Figure 1.6 Tasks, conditions, and applications for solid–liquid separation processes.

is built in the short residence time of some seconds in the process room of the centrifuge and only sludge will be discharged.

The choice of a discontinuous or a continuous process can have great influence on the separation results. Discontinuous apparatuses enable to choose each process step independently from the other, but only during a part of the batch time cake is produced. This restricts the throughput. In continuous apparatuses, all process steps are linked together and the transport velocity and the geometrical length of the respective process zone are limiting the process time for each process step. On the other hand, cake is produced all the time and the throughput is remarkably higher than in the case of batchwise operating apparatuses.

Last but not least, other boundary conditions of the process can have an impact on the freedom to choose the apparatus, which seems from the view of pure physics to be best suited for solving a specific separation problem. If for an example a peeler centrifuge may be the best choice, in the case of a pharmaceutical application with very high demands on hygienic standards, a highly efficient siphon peeler centrifuge (cf. Section 6.5.2) cannot be applied because of cleaning problems and a conventional and less effective version must be used.

To find out the best suited solution for a separation problem in every case, the first step consists of the careful analysis of the entire process as well as the properties of the slurry to be separated (cf. Chapter 2) and the specification of the requirements regarding the separation results has to be formulated. This leads to first ideas for eventually well-suited processes. A cake filtration may be the result. The second step generally includes bench scale separation experiments to verify the principle feasibility of the hypothesis and to get the necessary database for scaling up considered types of apparatuses. On the basis of the laboratory results and already including economic and other superordinate aspects in most cases, a pilot scale test must be carried out. Now, specific apparatus parameters can be investigated, which cannot be captured by the simple bench scale test. From the successful realized pilot test, the final quantitative scale up to the size of the industrial equipment can be done. Special concepts for scaling up cake filtration processes under different boundary conditions will be given later on and can be found in the literature also for other solid–liquid separation processes [16]. The combination of model equations and filtration experiments guarantees realistic forecast data for the technical application, reduces experimental effort, and extends the inter- and extrapolation of the included parameters.

However, all these considerations about the optimal solution of solid–liquid separation problems are based on a fundamental knowledge of the physical phenomena and the key parameters, which are influencing the separation results. To characterize the whole cake filtration process, several experimental methods and model equations from different authors are available, which are all based on a similar physical fundament [17].

As mentioned before, the procedures and apparatuses for the mechanical solid–liquid separation in general and in particular for cake filtration are under permanent research and development in view of new challenges and requirements to the separation results. Actual results from the research are used permanently for the apparatus design. New materials and production engineering are used. Modern sensor and data transfer technology allows the remote monitoring of separation processes and a result-dependent control and regulation. One trend goes toward the "intelligent" machine, which reacts to changes of the feed conditions automatically.

Another trend toward numerical simulation tools will become in future a more and more powerful measure in these processes, although particle technology is always faced with distributed parameters, which makes things much more complicated than in the case of uniform phases. For example, the theoretical research to simulate the capillarity in porous particle systems has made notable progress and helps to understand the phenomena, but unfortunately cannot yet give quantitative data for the practical apparatus design until today [18]. Numerical simulation of cake filtration will not be discussed here in more detail. However, on the way to approach that target, a certain number of today's still unsolved questions of basic research in this field have to be answered. One of these questions is the quantitative correlation between separation results and real particle collective characteristics. Respecting the fact that each physically different particle size measurement method leads for nonspherical particles to different equivalent particle diameters, one has to answer the nontrivial question, which particle diameter is the relevant diameter for any further calculation with particle sizes (cf. Section 4.3). The solution of such kind of problems is not only of academic interest but relevant for the technical practice to make a safe forecast of the cake filtration results. If, for example, upstream in crystallization, milling, or fractionation process, the operation conditions are changing with the consequence of a change in the resulting particles size distribution or the particle shape, the filtration results are influenced seriously.

At the end, a digital twin of filtration apparatuses is desired not even as a three-dimensional design or training tool but also to study process conditions with different material parameters or critical conditions. This requires in general a simulation of the apparatus behavior not only for the stationary but also for the dynamic case. The main challenge for an extensive digitalization is the combination of the complex material behavior, which, as mentioned above, is very difficult to describe in a simulative environment, the complex geometries of the apparatuses, and combined with the complex process behavior. For the material behavior, characterization devices have to be developed, which help to implement the material properties directly into the simulation procedure. This is necessary because a direct numerical simulation on a microscale level, where each particle is approximated, is far too time-consuming. With multiscale simulations, it is necessary to derive shortcut models, which allow simulating the whole process within seconds or even faster. This allows simulating not just process parameters but also raw material or energy consumption relations. Those relations could be directly used to optimize the whole process according to inverse simulations.

References

1 Couturier, S., Valat, M., Vaxelaire, J., and Puiggali, J.R. (2007). Enhanced expression of filter cakes using a local thermal supply. *Separation and*

Purification Technology 57: 321–328. https://doi.org/10.1016/j.seppur.2007.04.020.

2 Reichmann, B. and Tomas, J. (2001). Expression behaviour of fine particle suspensions and the consolidated cake strength. *Powder Technology* 121: 182–189. https://doi.org/10.1016/S0032-5910(01)00336-9.

3 Anlauf, H. (2012). Mechanical solid-liquid separation – processes and techniques. In: *Modern Drying Technology*, vol. 4 (ed. E. Tsotsas and A.S. Mujumdar), 47–97. Weinheim: Wiley-VCH. ISBN: 978-3-527-31559-8.

4 Wakeman, R. and Tarleton, S. (2005). *Solid/Liquid Separation: Principles of Industrial Filtration*. Oxford: Elsevier.

5 Sutherland, K.S. (2005). *Solid/Liquid Separation Equipment*. Weinheim: Wiley-VCH. ISBN: 3-527-29600.

6 Svarovsky, L. (2000). *Solid–Liquid Separation*. Oxford: Butterworth Heinemann. ISBN: 0750645687.

7 Rushton, A.S., Ward, A.S., and Holdich, R.G. (1996). *Solid–Liquid Filtration and Separation Technology*. Weinheim: Wiley-VCH.

8 Luckert, K. (ed.) (2004). *Handbuch der mechanischen Fest-Flüssig-Trennung*. Essen: Vulkan. ISBN: 3-8027-2196-9.

9 Dickenson, T.C. (1997). *Filters and Filtration Handbook*. Oxford: Elsevier Advanced Technology. ISBN: 1-85617-322-4.

10 Gasper, H. (2000). Filterbauarten und ihre Einsatzbereiche – Übersicht. In: *Handbuch der industriellen Fest/Flüssig-Filtration*, 2e (ed. H. Gasper, D. Oechsle and E. Pongratz), 35–47. Weinheim: Wiley-VCH. ISBN: 3-527-29796-0.

11 Trawinski, H. (1983). Entwicklungstrends der Verfahren für mechanische Phasentrennung mit Zentrifugalkräften. *Chemie-Technik* 12 (6): 15–21.

12 Gösele, W., Leibnitz, R., Pongratz, E., and Tichy, J. (2004). Filterapparate – Physikalische Einflüsse auf den Filterwiderstand. In: *Handbuch der mechanischen Fest-Flüssig-Trennung* (ed. K. Luckert), 176–181. Essen: Vulkan. ISBN: 3-8027-2196-9.

13 Tarleton, S. and Wakeman, R. (2007). *Solid/Liquid Separation: Equipment Selection and Process Design*. Oxford: Butterworth–Heinemann. ISBN: 1-85617-421-2.

14 Anlauf, H. (2017). Towards mitigation of particle/liquid separation problems by evolutionary technological progress. *Journal of the Taiwan Institute of Chemical Engineers* 58 (2): https://doi.org/10.1016/j.jtice.2017.09.043.

15 Anlauf, H. (2006). Recent developments in research and machinery of solid–liquid separation processes. *Drying Technology* 24: 1235–1241. https://doi.org/10.1080/07373930600838066.

16 Wakeman, R. and Tarleton, S. (2005). *Solid/Liquid Separation: Scale-Up of Industrial Equipment*. Oxford: Elsevier Advanced Technology. ISBN: 1-85617-420-4.

17 Wakeman, R. and Tarleton, S. (2005). *Principles of Industrial Filtration*. Oxford: Elsevier Advanced Technology. ISBN: 1-85617-419-0.

18 Xu, J. and Louge, M.Y. (2015). Statistical mechanics of unsaturated porous media. *Physical Review E: Statistical, Nonlinear, and Soft Matter Physics* 92, 062405 92 (6): 1–17. https://doi.org/10.1103/PhysRevE.92.062405.

2

Slurry Characterization

2.1 Introduction

The knowledge about the composition and properties of slurry to be separated contains essential key information to select and to design a separation process in general and a cake filtration process in particular. The slurry properties can be subdivided into properties of the liquid, properties of the solids, and properties of the entire slurry, where liquid and particles are interacting and influencing each other. On the one hand, the material of the apparatus and the filter medium must not be affected negatively in contact with the slurry, and on the other hand, the particles should not be negatively affected by the apparatus operation. This means that, for example, the pH of liquid or abrasiveness of particles should not damage parts of the apparatus or the filter medium. If the particles exhibit a fragile crystalline structure, they should not be destroyed during separation in the apparatus. If the particles have to be washed for removing soluble substances, it is interesting to know whether the particles themselves are porous or not. If the slurry needs to be handled sanitary, the technical environment must be designed adequately according to the hygienic requirements. If the liquid is inflammable, no electrostatic charging of apparatus parts in contact with the product is allowed and explosion protection must be guaranteed. The particle size determines to a great extent the choice of a well-suited separation technology in general and appropriate filter media in the case of filtration processes particularly. Depending on the standard deviation of the particle size distribution and/or the concentration of the slurry, segregation effects of the particles according to their size should be considered during separation. Normally, segregation is connected with negative consequences for the separation results. In cases of particle segregation, either the filter media are in danger to be blinded or a particle layer of the smallest particles forms on top of a filter cake and hinders its permeation during washing and/or its desaturation. Cake filtration needs, in most cases, a certain minimum of solid concentration in the slurry to initiate fast particle bridging across the pores of the filter medium. Additionally, greater solid concentrations avoid particle segregation in the slurry because of hindered settling effects and lead last but not least to increased cake production per time. Fast filter cake build up is sometimes not only an economical issue but also decides about the functional capability of the separation apparatus. For example, in continuously operating pusher centrifuges, the residence time of the product in the process room comprises not more than several seconds. In this short time, the cake must be formed, eventually washed, and desaturated. Otherwise, the overhung centrifuge basket becomes flooded and the slurry is shot without separation into the solid chute. Increased slurry concentration reduces the cake formation time.

The mentioned aspects and examples make on the one hand no claim to be complete and are on the other hand not all relevant for each separation problem. In the following, a rough overview about the most important slurry parameters for cake filtration will be discussed. Which of them are really of interest depends on the analysis of the specific separation problem.

2.2 Liquid Properties

There do exist several liquid characterizing parameters such as chemical composition, concentration of specific dissolved substances, conductivity, density, pH, temperature, surface tension, or viscosity (rheology). Some of these properties are partly correlated with each other. The concentration of ions is correlated with the conductivity. The surface tension and viscosity are depending on the temperature. Normally, Newtonian flow behavior can be supposed for the liquid flow in the pores of filter cakes, which means that the viscosity is independent from the shear gradient and thus constant. In contrast to that, the flow behavior of slurries often show a shear thinning effect, especially if the particles are not spherical, but this normally does not affect the cake filtration directly because only the flow of the pure liquid in the cake structure is of relevance here. However, if the flow behavior of the liquid is not Newtonian, its specific rheological properties have to be measured and considered [1].

Depending on the specific separation problem, more than the mentioned parameters such as fugacity, toxicity, reactivity, deleteriousness, or others can play a role and have to be characterized to design the technical separation process properly.

2.3 Particle Properties

2.3.1 General Aspects

The physical (dispersity) properties of particles to be separated are not only sensitively influencing the final properties of the solid product but also the demands on the separation process to be designed. Similar to the liquid properties, a huge variety of particle properties exists and has to be considered specifically for each case of application. Such properties are the size and size distribution, shape, porosity, stability, specific surface, surface roughness, density, magnetism, and others. Some particle properties are generated only in interaction with the surrounding liquid, such as the electric surface charge. Particles not only can have very different size and shape. As exemplified in Figure 2.1, it is often not easy to define what the real particle size is, and the question arises, what does it mean if the particle analyzer gives as a result an exact particle diameter.

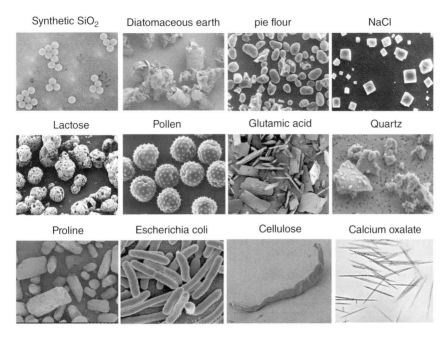

Figure 2.1 Different types of particles.

For particles such as spheres or cubes, it is easy to determine the particle size. They can be described by one characteristic length. The situation becomes more difficult if particles have an irregular shape or consist of needles, fibers, or a fibrous structure, as in the case of many filamentous microorganisms (*Aspergillus niger*). In addition, particles must not be compact but can be porous like diatomaceous earth or agglomerates of primary particles.

The particle size and particle size distribution influences to a great extent the cake filtration conditions [2]. The pore size and porosity of the cake directly depend on these parameters, and as a consequence, its flow resistance, washing behavior, and capillarity. Problems can evolve in the case of very broad particle size distributions and low concentrated slurries because they tend to segregate, although slurry should be as homogeneous as possible for optimal filtration conditions.

The particle shape influences the structure of a filter cake sensitively. If particles according to Figure 2.2 are flat and flaky, as kaolin (china clay), they can form a nearly impermeable layer in a roof tile arrangement. If they are arranged like a house of cards, they will form a highly permeable cake structure.

Especially for separation processes, it is interesting that which part of particles is separated and which part gets lost with the separated liquid. For cake filtration processes especially, the start of filtration is relevant because during this time, particle bridges must be formed across the meshes of the filter fabric, and a certain undesired amount of particles is able to get into the filtrate. As a rule, the mean particle size and the mean pore size of the filter media should not be the same to avoid irreversible pore blinding. Thus, normally pore sizes of woven filter

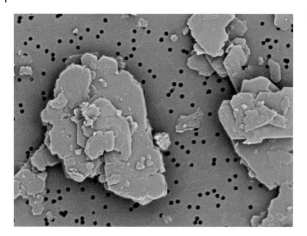

Figure 2.2 Kaolin particles.

fabrics are chosen greater than the particles. In the case of a particle fractionation, as a pretreatment measure for a cake filtration, a cut size has to be defined. Besides the sharpness of cut, the amount and size distribution of coarse and fine fraction should be known.

2.3.2 Characterization of Single Particles

Particles can be described by dispersity characteristics (index i). These are physical parameters, which must be objectively definable, assignable to the single disperse element, and allow an arrangement of the disperse elements according to the respective dispersity characteristics. Such dispersity characteristics consist of geometrical parameters, such as volume, surface, cross section, and projection area, or physical parameters, such as inertia of mass, settling or diffusion velocity, impulse, energy, influence on electric or magnetic fields, and interdependency with electromagnetic radiation in the form of scattering, diffraction, or absorption.

On the basis of these characterizing parameters, a great variety of different particle size analysis techniques has been developed to describe single particles and particle collectives with a remarkable extension during the last decades toward nanoscale particles [3–7].

The simplest possibility to describe the particle size consists in the direct measurement by image analysis and calculation of the size from the geometry. Unfortunately, an exact description of the geometry makes only sense for simple particle geometries as can be shown in Figure 2.3.

A cube is described clearly by the edge length a, a sphere by the diameter x and a cylinder by the diameter d, and the length l. If particles do not have such a simple geometry and an image analysis is preferred, statistical lengths can be measured, if the sample comprises a sufficient great number of particles to meet statistical requirements. As demonstrated in Figure 2.4, a measuring direction must be defined and then the particle size is determined according to this direction.

One out of several and frequently used definitions is the Feret diameter x_F. Here, the distance between two particle-inclosing tangents vertical to the

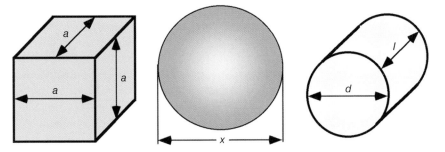

Figure 2.3 Cube-, sphere-, and cylinder-shaped particle.

Figure 2.4 Statistical lengths.

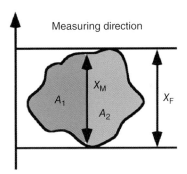

measuring direction is determined. One other possibility is to separate the particle projection area in two equal halves A_1 and A_2 and take the length of the separation line as Martin diameter x_M.

Alternatively to direct measurement, particle sizes can be determined indirectly by using a physical property (index i) and a correlation between the physical property and the particle size. The result is an equivalent diameter x_i. This should be explained in Figure 2.5 on the example of a sedimentation analysis.

The sedimentation velocity in the laminar Stokes range ($Re < 0.25$), for Newtonean liquid behavior, and single particle sedimentation w_{St} is in Eq. (2.1) correlated with the particle diameter x

$$w_{St} = \frac{(\rho_s - \rho_L) \cdot g \cdot x_{St}^2}{18 \cdot \eta_L} \tag{2.1}$$

If in a sedimentation experiment the settling velocity w_{St} is measured, the density of solids ρ_s and liquid ρ_L, the gravity g, and the dynamic viscosity of the liquid η_L is known, the equivalent diameter of a sphere x_{St} can be calculated. The

Figure 2.5 Equivalent diameter.

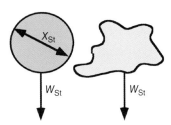

equivalent particle diameter x_{St} represents the diameter of a sphere, which settles with the same velocity, as the analyzed particle of irregular shape. According to the respective measurement principle, the same particle results in different equivalent diameters, if it is not spherical. A laser diffraction measurement leads to a different particle diameter than a sieve analysis for the same nonspherical particle because the physical basis of the measurements is different. This is important to realize for the interpretation and comparison of particle sizes, which are measured using different physical principles. If a formula contains as a parameter the particle size, one has to decide which is the relevant particle size. This problem will be discussed later on in connection with the filter cake formation process.

Besides physical properties, equivalent particle diameters can also be derived from geometrical properties, as can be demonstrated on the following examples:

Equation (2.2) defines the diameter of a sphere x_p with the same projection area A_p:

$$A_p = \frac{\pi}{4} \cdot x_p^2 \tag{2.2}$$

Equation (2.3) defines the diameter of a sphere x_S with the same surface area S:

$$S = \pi \cdot x_S^2 \tag{2.3}$$

Equation (2.4) defines the diameter of a sphere x_V with the same volume V:

$$V = \frac{\pi}{6} \cdot x_V^3 \tag{2.4}$$

In addition to the size, the shape of particles is often of interest. In most cases, a simple qualitative description such as "spherical," "fibrous," "needle shape," or "flaky" is sufficient information for a first impression and estimation of the eventual influence on the process. This information can be achieved simply by looking through the microscope. However, also quantitative methods are available [8, 9]. A relation between different equivalent particle diameters and thus the possibility of their comparison is given by shape factors according to Eq. (2.5)

$$\psi_{i,j} = \left[\frac{x_i}{x_j}\right] \tag{2.5}$$

The shape factor $\psi_{i,j}$ allows to compare the particle diameter x_i with the dispersity characteristic i and the particle diameter x_j with the dispersity characteristic j. As an example, the shape factor $\psi_{S,V}$ compares according to Eq. (2.6) the surface-related particle diameter and the volume-related particle diameter

$$\psi_{S,V} = \left[\frac{x_S}{x_V}\right] \tag{2.6}$$

A further possibility is given by the sphericity of Wadell ψ_W. In Eq. (2.7), the surface of a sphere of the same volume as the analyzed particle is related to the real surface of the analyzed particle

$$\psi_W = \psi_{V,S}^2 = \frac{\pi \cdot x_V^2}{\pi \cdot x_S^2} = \frac{x_V^2}{x_S^2} \tag{2.7}$$

Figure 2.6 illustrates the sphericities for cylinders, ellipsoids, and pyramids.

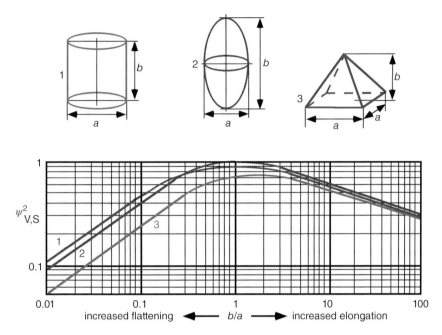

Figure 2.6 Sphericity for different particle shape (cf. [7], p. 24).

The sphericity describes the "roundness" of a particle as deviation from the ideal shape of a sphere. For a given volume, the sphere exhibits the smallest possible surface area, and as a consequence, every nonspherical particle of the same volume results in a sphericity <1. The aspects discussed above should be sufficient in most cases to describe the geometric properties of single particles for the purpose of cake filtration. However, one could go much more into detail to describe the microstructure of the particle surfaces, for example, by determination of the fractal dimension of a particle contour. This could be relevant for particles as can be shown in Figure 2.7 for precipitated and grinded calcium carbonate, which exhibit an extremely different surface structure, although both consist of calcium carbonate.

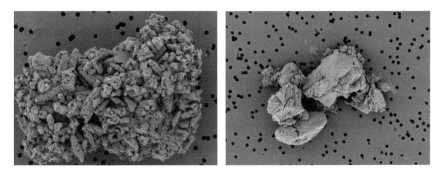

Figure 2.7 Precipitated and grinded calcium carbonate.

2.3.3 Characterization of Particle Collectives

After fractionation of a particle collective into size classes according to a specific dispersity characteristic, the respective amount of particles for each size class must be determined. The kind of quantity depends on the measuring principle, as is explained in Table 2.1.

$Q_0(x_i)$ represents the cumulative number distribution of particles, which have been analyzed according to the dispersity characteristic i (furthermore, the index i is mostly disclaimed for simplification). Particle size analysis techniques can be grouped according to the dispersity characteristic and the used kind of quantity. For example, counting of single particles leads to a number distribution ($r = 0$) and different dispersity characteristics are available such as statistical lengths ($i = 1$), light extinction ($i = 2$), laser scattering ($i = 3$), interference with an electric field ($i = 3$), and others.

In general, the cumulative particle size distribution $Q_r(x)$ characterizes the share of particles smaller than a certain particle size related to the total amount of particles. As can be seen from Figure 2.8, $Q_r(x_{min}) = 0$ for the smallest particle x_{min} and $Q_r(x_{max}) = 1$ for the greatest particle x_{max}.

A particle size interval $\Delta x = (x_2 - x_1)$ represents a particle quantity $\Delta Q_r(x_2, x_1)$. The median value $x_{50,r}$ characterizes the particle size, which divides the total particle amount in two halves of particles smaller and greater than $x_{50,r}$.

The cumulative particle size distribution does not show directly which particle size is most frequently represented in the particle collective.

To get this information, the particle size frequency distribution $q_r(x)$, shown in Figure 2.9, is the appropriate function.

Table 2.1 Definition of particle quantities.

Method	Quantity	Dimension	Index	Denotation
Counting	Number	L^0	$r = 0$	$Q_1(x_i)$
Extinction	Area	L^2	$r = 2$	$Q_2(x_i)$
Weighing	Mass volume	L^3	$r = 3$	$Q_3(x_i)$

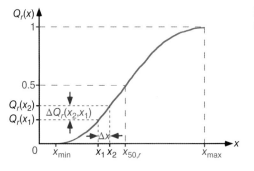

Figure 2.8 Cumulative particle size distribution.

Figure 2.9 Particle size frequency distribution.

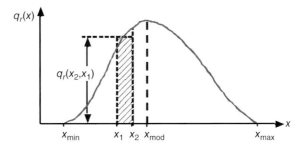

Here, according to Eq. (2.8), the share of quantity in a certain particle size interval $x_2 - x_1$ is related to the interval width $x_2 - x_1$

$$q_r(x_{av}) = \frac{Q_r(x_2) - Q_r(x_1)}{x_2 - x_1} = \frac{\Delta Q_r|x_1, x_2|}{\Delta x} \tag{2.8}$$

If the number of particles is large, $Q_r(x)$ can be continuously differentiated (Eq. (2.9))

$$q_r(x) = \frac{dQ_r(x)}{dx} \tag{2.9}$$

As can be seen from Eq. (2.10), the area below the function $q_r(x)$ between x_2 and x_1 correlates with the share of particle quantity between x_2 and x_1

$$\int_{x_1}^{x_2} q_r(x) \cdot dx = \int_{x_1}^{x_2} dQ_r(x) = Q_r(x_2) - Q_r(x_1) \tag{2.10}$$

The integration between x_{min} and x_{max} gives the standardization condition (Eq. (2.11))

$$\int_{x_{min}}^{x_{max}} q_r(x) \cdot dx = \int_{x_{min}}^{x_{max}} dQ_r(x) = 1 \tag{2.11}$$

The maximum of $q_r(x)$ is named modal value x_{mod}.

It is important to realize that particle size distributions of different quantity characteristics must not be compared directly. This should be explained in Table 2.2 on the example of four spheres and a doubling diameter from particle to particle.

Table 2.2 Comparison of quantity characteristics.

	Relative Ø	Number (%)	Volume (%)
	1	25	1
	2	25	8
	3	25	27
	4	25	64
	Sum	100	100

Each particle, independent of its size, represents 25% of the total quantity, if the number is the quantity characteristic. If the volume characterizes the quantity, the particle size plays an important role because the volume is calculated with the third power of x. The smallest particle represents only 1% of the total particle volume, whereby the largest one represents 64%.

To compare particle size distributions $q_r(x)$ and $q_l(x)$ with different quantity characteristics r and l, one distribution has to be converted into the other using Eq. (2.12)

$$q_r(x) = \frac{x^{r-l} \cdot q_l(x)}{\int_{x_{min}}^{x_{max}} x^{r-l} \cdot q_l(x) \cdot dx} \tag{2.12}$$

Often, it is not necessary to know the particle size distribution in detail. Especially, if particle size information should be used in process calculations, this information should be reduced to a few parameters, which are able to represent the relevant particle size influence on the respective process. In Section 4.3, these aspects will be discussed in more detail on the example of filter cake permeation. In those cases, the entire distribution can be characterized by a location parameter or average particle size and a spreading parameter (span) for the width of the particle size distribution.

There do exist several possibilities to define a location parameter. The frequently used characteristic particle diameters x_{mod} and $x_{50,r}$ are shown in Figure 2.10 (cf. Figures 2.8 and 2.9).

As can be seen, both these particle diameters are normally not identical. Only in the case of an absolutely symmetric distribution (Gaussian normal distribution), x_{mod} and $x_{50,r}$ are identical. The modal and median particle diameter can be obtained directly from the particle size distribution curve and no further calculations are necessary.

The modal value can be identified easily as the maximum of the frequency distribution. In Figure 2.10, a monomodal size distribution is shown. In cases of multimodal distributions, several maxima are present.

The other way to characterize the position of a particle size distribution is the easy-to-find percentiles. Percentiles represent the size for which a given percentage of the entire particle collective is reached. For example, the quartiles represent the 25% and 75% values. If 50% of the particles are smaller than 100 µm, the x_{50} of the collective is 100 µm. This special value is named median value and most

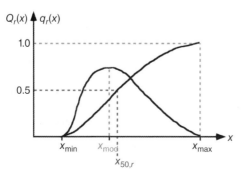

Figure 2.10 Characteristic particle diameters.

frequently used. However, it has to be respected on the one hand that a median size is not a real average size because it does not weigh the relative contribution of each single size present in the distribution. On the other hand, the median size depends on the kind of distribution used. Number, area, or volume distribution of the same particle collective leads to completely different modal values. By this reason, it is not enough to denominate the median value as x_{50}, but the more precise notation $x_{50,r}$ is necessary. Weighed average values can be defined in a different way such as arithmetic, quadratic, cubic, and others. The concept is worked out extensively in the German standard DIN 66141. In particle liquid separation generally and in cake filtration particularly, often the Sauter average diameter $x_{av,SV}$ is used because it describes the particle collective at best for this operation. A reason for that may be that the flow of liquid through a filter cake takes place across the particle surfaces. The parameter $x_{av,SV}$ represents the diameter of a sphere with the same specific surface S_V as the whole particle collective.

According to Eq. (2.13), $x_{av,SV}$ can be determined without the shape factor from two measuring methods, which use particle volume and particle surface as dispersity characteristic

$$x_{av,SV} = \frac{x_{av,V}^3}{x_{av,S}^2} = \frac{6}{S_V} \qquad (2.13)$$

The general calculation from the results of physically different measurement methods needs to implement a specific shape factor $\psi_{SV,i}$ as Eq. (2.14) demonstrates

$$\begin{aligned} x_{av,SV} &= \psi_{SV,i} \cdot \frac{\int_{x_{min}}^{x_{max}} x_i^3 \cdot q_0(x_i) \cdot dx_i}{\int_{x_{min}}^{x_{max}} x_i^2 \cdot q_0(x_i) \cdot dx_i} = \psi_{SV,i} \cdot D_{3,2,i} \\ &= \psi_{SV,i} \cdot \int_{x_{min}}^{x_{max}} x_i \cdot q_2(x_i) \cdot dx_i \\ &= \psi_{SV,i} \cdot \frac{1}{\int_{x_{min}}^{x_{max}} \frac{1}{x_i} \cdot q_3(x_i) \cdot dx_i} \end{aligned} \qquad (2.14)$$

Besides an averaging parameter, it is necessary to get information about the width of the distribution in the form of a spreading parameter or span. It is obvious that two particle size distributions with the same average particle size can be produced by monosized particles or a size distribution of any width and these different distributions will have of course a different impact on the filtration conditions and filter cake properties.

The width of a particle size distribution can be characterized according to Figure 2.11, for example, by the difference between the largest particle diameter and the smallest particle diameter $|x_{max} - x_{min}|$.

Generally, it is not possible or extremely difficult to determine exactly the largest and smallest particle diameter in a particle collective because of the limited resolution of the measurement instruments. By that reason, often the difference between the particle diameters, which represent 10% and 90% of the total amount of particles $|x_{90,r} - x_{10,r}|$, is used. A further method to express the width of the particle size distribution is given in Eq. (2.15) in form of

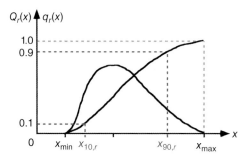

Figure 2.11 Width of a particle size distribution.

the variance σ^2 or standard deviation σ

$$\sigma^2 = \int_{x_{min}}^{x_{max}} (x - x_{av})^2 \cdot q(x) \cdot dx \tag{2.15}$$

The general formulation in Eq. (2.15) must be specified by the index r because the calculation depends on the kind of distribution used. Additionally, because the variance or standard deviation is related to an average value, the kind of mean (arithmetic, quadratic, cubic, geometric, etc.) must be indicated. As an example, the arithmetic standard deviation $\sigma_{ar,r}$ is formulated in Eq. (2.16)

$$\sigma_{ar,r} = \sqrt{\int_{x_{min}}^{x_{max}} (x - x_{av,ar,r})^2 \cdot q_r(x) \cdot dx} \tag{2.16}$$

The above discussed aspects should be sufficient here for the description of the particle collective regarding its influence on the cake filtration process. However, one could go much more in detail to discuss multimodal distributions or the asymmetry of distributions. Furthermore, characteristic types of particle size distributions can be approximated by well-defined mathematical functions, such as logarithmic normal distribution, power distribution, or Rosin-Rammler-Sperling-Bennet (RRSB) distribution (cf. [7], 36–50).

2.3.4 Characterization of Particle Collective Fractionation

Technical solid–liquid separation processes and especially cake filtration are normally not able to guarantee 100% separation of the particles. For cake filtration in most cases, relatively open woven filter media are chosen to prevent the cloth from blockage by particles. To hinder particles from penetrating the filter medium, particle bridges must be formed in the very first moment of the cake filtration process. The faster the bridge formation, the fewer particles are getting into the filtrate. Prethickening of the slurry to be separated or agglomeration promotes fast bridge formation. In other cases, slurry has to be fractionated before separation, as a pretreatment measure, in order to optimize the whole process. Options are here to cut off a small amount of the finest particles (desliming), to cut off the coarsest particles (degritting), or to fractionate the slurry in the middle at a certain cut size, to assign optimally suited separation techniques to each fraction.

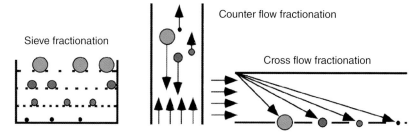

Figure 2.12 Sieving, counterflow, and cross-flow fractionation.

Separation means the separation according to different phases, such as solid particles and liquid. In general, all three aggregate states can be separated from each other. In each case, one phase is defined as the disperse phase (e.g. particles), which is distributed in a continuous phase (e.g. liquid). During the cake filtration process, the solid and liquid phases are changing their character. Initially, when the cake is formed, the particles are dispersed in the liquid. After the end of cake formation, the solid particles represent the continuous phase in the form of an interconnected porous network and the liquid to be removed from this cake structure is distributed as a disperse phase in the cakes pores.

Sorting means the subdividing of a particle collective according to other physical characteristics, such as a kind of material, surface properties, magnetic or electric properties, color, shape, density, or others. If density is the criterion, the particle collective is divided in a heavy component and a light component.

Fractionation means the subdividing of a particle collective according to the particle size into partial fractions smaller and greater than a cut size x_t, as demonstrated in Figure 2.12 on the examples of sieving, counterflow fractionation, and cross-flow fractionation.

Sieving separates particles directly according to their size. Counter and cross flow fractionation is based on different particle settling velocities in gas or liquid because of different particle mass and thus size. The density of all particles is presupposed as the same, and between solids and fluid, there must exist a density difference. Constant density of all solids is essential because different solid densities would lead to the situation that a great and light particle would settle with the same velocity than a small and heavy particle.

A further example for an excellent fractionation device, which is based on centrifugal sedimentation, is given by the hydrocyclone in Figure 2.13.

The tangential feed of the fixed cyclone first leads to a potential vortex of the slurry with increasing centrifugal acceleration toward the center. This is due to the condition of continuity and the tangential velocity of the spiral flow and thus the centrifugal acceleration toward the center must increase. The maximal acceleration is reached at the radius of the vortex finder. At this position, the cut size x_t is determined. The coarse particle fraction, which is separated at the cyclones wall, is discharged in the concentrated form downward through the underflow nozzle. The main part of the liquid including the fine particle fraction is not able to leave the cyclone through the throttled underflow nozzle but must change its

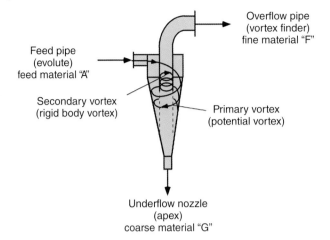

Figure 2.13 Hydrocyclone.

flow direction and leaves the cyclone upward through the overflow pipe. Here, the liquid rotates in the form of a rigid body.

The coarse fraction in principle contains all particles greater than a cut size x_t and the fine fraction contains all particles smaller than x_t. Any particle size between x_{min} and x_{max} can be defined as cut size. One can distinguish between statistically defined cut sizes from particle size distributions and physically defined cut sizes from physical separation conditions. To characterize the results of a fractionation or a more or less complete separation, the information is needed, which share of the particles in the feed is found after the process in the coarse and fine fraction. A first answer to this question is given by the formulation of an integral material balance as depicted in Figure 2.14.

A feed material M_A (index A) with a cumulative particle distribution $Q_A(x)$ and a frequency function $q_A(x)$ enters a separator with a certain separation function $T(x)$, which leads to a split into a coarse fraction (index G) and a fine fraction (index F) as indicated in Eq. (2.17)

$$M_A = M_G + M_F \tag{2.17}$$

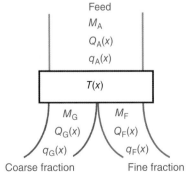

Figure 2.14 Integral material balance.

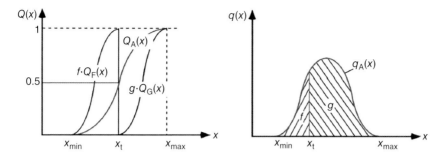

Figure 2.15 Ideal sharp fractionation.

Normally, not the absolute amount of material is interesting, but according to Eq. (2.18), the percentage of fines f and the percentage of coarse material g in relation to 100% of the feed material

$$\frac{M_G}{M_A} = g \quad \frac{M_F}{M_A} = f \tag{2.18}$$

The sum of the relative portions g and f of course add up to 1 or 100% (Eq. (2.19))

$$g + f = 1 \tag{2.19}$$

Equation (2.17) describes the integral material balance for the fractionation process. In many cases, it is interesting to know the separation efficiency for each particle size. For this purpose, in Eq. (2.20), not an integral but a differential material balance is formulated

$$q_A(x) = g \cdot q_G(x) + f \cdot q_F(x) \tag{2.20}$$

Equation (2.20) can also be written in the integral form

$$Q_A(x) = g \cdot Q_G(x) + f \cdot Q_F(x) \tag{2.21}$$

An ideal sharp fractionation at the cut size x_t is shown in Figure 2.15.

In this case, no particle smaller than x_t gets into the coarse fraction and no particle greater than x_t gets into the fine fraction. The amount of misplaced particles amounts zero. The share of coarse and fine particles is calculated in Eq. (2.22) by integration of the frequency function

$$\int_{x_t}^{x_{max}} q_A(x) \cdot dx = g \quad \int_{x_{min}}^{x_t} q_A(x) \cdot dx = f \tag{2.22}$$

As depicted in Figure 2.16, in the technical reality, the fractionation will never be ideally sharp, but more or less misplaced particles can be found in both fractions.

The particle size distributions of the fine and the coarse fraction are overlapping now. In the fine fraction, particles of $x > x_t$ can be found, and in the coarse fraction, particles of $x < x_t$. These portions of misled particles can be calculated according to Eq. (2.23)

$$\int_{x_{min}}^{x_{max}} q_A(x) \cdot dx = f \cdot \int_{x_{min}}^{x_{F,max}} q_F(x) \cdot dx + g \cdot \int_{x_{G,min}}^{x_{max}} q_G(x) \cdot dx \tag{2.23}$$

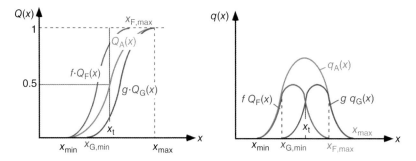

Figure 2.16 Real technical fractionation.

The cut size in Figure 2.16 is defined as particle size, which can be found with the same frequency in the fine and the coarse fraction.

There are several reasons for the imperfection of fractionation. If, for an example, a defect is present in a screen, greater particles than the nominal screen opening size will get as misplaced particles into the screen underflow. If small particles are adhering to particles, which are greater than the screen opening size, they will be found as misplaced particles in the sieve residue.

To characterize for each particle size in the feed the percentage of particles, which is separated in the coarse fraction, the grade efficiency function $T(x)$ or Tromp curve as depicted in Figure 2.17 for ideal sharp and technical fractionation can be used.

The grade efficiency of a certain particle size $x_{G,min}$ down the smallest particle x_{min} is zero, and from a certain particle size, $x_{F,max}$ up to the greatest particle size x_{max}, all particles are found to 100% in the coarse material. The grade efficiency

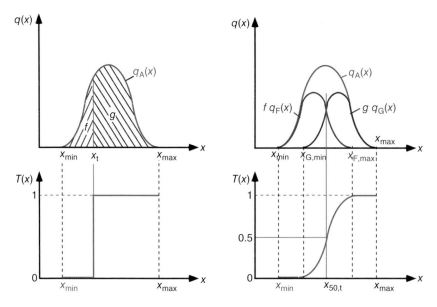

Figure 2.17 Grade efficiency function.

Figure 2.18 Analytical cut size.

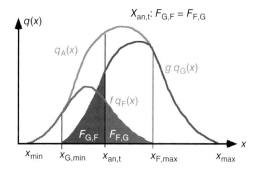

curve can be calculated from two of the three particle size distributions

$$T(x) = g \cdot \frac{q_G(x)}{q_A(x)} = 1 - f \cdot \frac{q_F(x)}{q_A(x)} = \frac{g \cdot q_G(x)}{g \cdot q_G(x) + f \cdot q_F(x)} \tag{2.24}$$

The special statistical cut size $x_{50,t}$ characterizes the particle size, which is present to 50% in the fine as well as in the coarse fraction (cf. Figure 2.17). This cut size is known as the median cut size or preparative cut size. According to Eq. (2.25), $T(x)$ becomes 0.5 in this case

$$T(x_{50,t}) = 0.5 \tag{2.25}$$

From this follows Eq. (2.26)

$$g \cdot q_G(x_{50,t}) = f \cdot q_F(x_{50,t}) \tag{2.26}$$

A different defined cut size is the analytical cut size $x_{an,t}$, as shown in Figure 2.18. Here, the quantity of misplaced particles in the fine $F_{G,F}$ fraction and in the coarse fraction $F_{F,G}$ are equal, as formulated in Eqs. (2.27) and (2.28)

$$g \cdot \int_{x_{G,min}}^{x_{an,t}} q_G(x) \cdot dx = f \cdot \int_{x_{an,t}}^{x_{F,max}} q_F(x) \cdot dx \tag{2.27}$$

$$g \cdot Q_G(x_{an,t}) = f \cdot (1 - Q_F(x_{an,t})) \tag{2.28}$$

The analytical cut size can be determined easily from the cumulative particle size distribution of the feed and the fine fraction as becomes clear from Figure 2.19.

This can be mathematically formulated as demonstrated in Eqs. (2.29)–(2.31)

$$Q_A(x_{an,t}) = g \cdot Q_G(x_{an,t}) + f \cdot Q_F(x_{an,t}) \tag{2.29}$$

Figure 2.19 Determination of the analytical cut size.

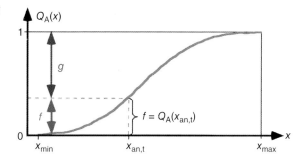

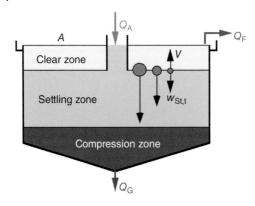

Figure 2.20 Physical cut size of a static circular thickener/clarifier.

$$Q_A(x_{an,t}) = f \cdot (1 - Q_F(x_{an,t})) + f \cdot Q_F(x_{an,t}) \tag{2.30}$$

$$Q_A(x_{an,t}) = f \tag{2.31}$$

An example for a physical cut size may be the cut size $x_{St,t}$ in a static thickener/clarifier, as shown schematically in Figure 2.20, which is often applied as a slurry pretreatment apparatus for a subsequent cake filtration.

The cut size $x_{St,t}$ depends on the separation condition that the settling velocity $w_{St,t}$ is equal to the upstream directed velocity of the clear liquid v. To guarantee particle-free liquid overflow, the cut size should represent here the smallest particle present in the slurry. For the settling velocity $w_{St,t}$, it is presupposed here that settling of single and spherical particles takes place in a liquid of Newtonean flow behavior and in the laminar flow regime, which means particle Re-numbers are smaller than 0.25. In that case, the Stokes equation is valid (cf. Section 5.2.). The liquid underflow v_G of the concentrated sludge can be normally neglected in comparison to the clear liquid overflow v_F. This means that the feed liquid flow velocity v_A is approximately equal to the clear liquid flow velocity v_F and thus v. Equation (2.32) formulates the separation condition

$$w_{St,t} = \frac{\Delta \rho \cdot g \cdot x_{St,t}^2}{18 \cdot \eta_L} = v = v_F \approx v_A = \frac{Q_A}{A} \tag{2.32}$$

From this follows Eq. (2.33) for the cut size $x_{St,t}$

$$x_{St,t} = \sqrt{\frac{18 \cdot \eta_L}{(\rho_s - \rho_L) \cdot g} \cdot \frac{Q_A}{A}} \tag{2.33}$$

For a given feed volume flow, the critical upstream flow velocity to hinder the smallest particles from getting into the clear liquid normally causes a very large separation area and thus a great diameter of the thickener. To avoid this, very often the slurry is flocculated before fed into the thickener. Normally, the whole particle size distribution is bound in the flocks, which settle because of their size much faster than the smallest particle and thus need much less separation area.

Another physical cut size is the cut size $x_{sieve,t}$, which is determined according to Figure 2.21 by the largest pore of a screen or filter media such as woven fabrics, which are mainly used for cake filtration.

Figure 2.21 Physical cut size of a filter medium.

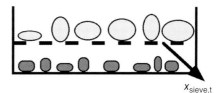

Figure 2.22 Sharpness of cut.

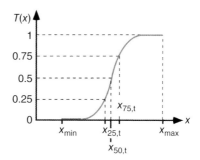

The pores of a filter medium are normally not circles and the particles are not spheres. However, the largest sphere, which is able to penetrate the pore, can define the largest pore. The largest particle can be defined for a convex geometry by a circle around its smallest cross section (cf. Section 9.3.1).

Besides the cut size, the sharpness of cut κ is important for the quality of a separation. It depends on the inclination of the grade efficiency function depicted in Figure 2.22.

One standard definition given in Eq. (2.34) uses the ratio of $x_{25,t}$ and $x_{75,t}$ to describe the sharpness of cut

$$\kappa = \frac{x_{25,t}}{x_{75,t}} \tag{2.34}$$

For different conditions of a fractionation process, the sharpness of cut exhibits different values.

- Ideal fractionation: $\kappa = 1$
- Technical fractionations: $0.3 \leq \kappa \leq 0.6$
- Sharp technical fractionations: $0.6 \leq \kappa \leq 0.8$ and even
- Analytical fractionations: $0.8 \leq \kappa \leq 0.9$.

Even for analytical fractionations, the sharpness of cut is not perfect because of a certain probability of failures. If, in the case of a coincidence failure, two particles instead of one are present in a measuring cell, both smaller particles are handled as one greater particle.

However, other definitions such as the imperfection value I for hydrocyclones in Eq. (2.35) are in use

$$I = \frac{x_{75,t} - x_{25,t}}{2 \cdot x_{50,t}} \tag{2.35}$$

There is no real physical reason for the different definitions, but only historical traditions in different industrial sectors.

2.4 Slurry

2.4.1 Solid Concentration

The solid concentration of slurry has under different aspects strong impact on the cake filtration conditions. First of all, the particle concentration obviously influences the cake formation velocity. The greater the concentration, the faster the filter cake will grow. As a consequence, the filter throughput increases, or for constant feed volume rate, the filter area and thus the apparatus can become smaller. Secondly, an increase of concentration above a critical value leads to improved slurry homogeneity, if the particle size distribution is quite broad. Any slurry in a filter apparatus is exposed to the gravity or centrifugal field and thus sedimentation of the particles takes place. In some cases, such as drum, disc, or stirred nutsche filters, a stirrer in the slurry can reduce or even overcome this problem. In other cases such as belt or pan filters, unhindered sedimentation takes place in the slurry above the horizontal filter area. Unfavorable segregation of the slurry can be reduced or even avoided at greater concentrations because single particle sedimentation switches to swarm or zone sedimentation and the average settling velocity decreases. Swarm or zone sedimentation means that the particles are hindering each other. Size, shape, and density of the particles lose their individual influence on the settling velocity and a sharp front of settling particles is forming. The critical concentration for the formation of such a sharp settling front depends on the particles size, the mass forces acting on the particles, and the stability of the slurry. The smaller the particles, the lower the forces on the particles are. The more agglomerated the particles, the lower the critical concentration becomes.

The handling properties of identically concentrated slurries can vary remarkably depending on the particle size. Slurry of nanoscale particles with 10 vol% solids seems to be a highly concentrated sludge or paste, whereby slurry of particles in the 100 µm range behaves as diluted slurry. Last but not least, the formation time of particle bridges over the meshes of the filter medium is decreased for greater concentrations and thus less filtrate pollution will occur.

The slurry solid concentration can be expressed by different parameters, which are often expressed in percent. To avoid serious mistakes, it is important in any case to know how those percentages are calculated exactly. The information about only percent is meaningless, if the definition of the respective concentration is not known.

The solid mass concentration c_m in Eq. (2.36) relates the solid mass to the total mass of the slurry

$$c_\mathrm{m} = \frac{m_\mathrm{s}}{m_\mathrm{s} + m_\mathrm{L}} \tag{2.36}$$

The solid volume concentration c_v in Eq. (2.37) relates the solid volume to the total volume of the slurry

$$c_\mathrm{v} = \frac{\frac{m_\mathrm{s}}{\rho_\mathrm{s}}}{\frac{m_\mathrm{s}}{\rho_\mathrm{s}} + \frac{m_\mathrm{L}}{\rho_\mathrm{L}}} = \frac{V_\mathrm{s}}{V_\mathrm{tot}} \tag{2.37}$$

Figure 2.23 Determination of the concentration parameter κ in a lab tube.

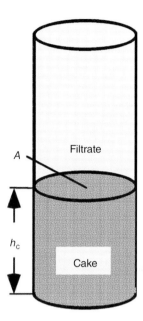

The slurry density $\rho_{s,L}$ in Eq. (2.38) relates the slurry mass to the slurry volume

$$\rho_{s,L} = \frac{m_s + m_L}{\frac{m_s}{\rho_s} + \frac{m_L}{\rho_L}} \tag{2.38}$$

The concentration parameter κ in Eq. (2.39) relates the cake volume to the separated liquid volume (derivation in Section 6.3)

$$\kappa = \frac{h_c \cdot A}{V_L} = \frac{c_v}{1 - c_v - \varepsilon} = \frac{c_v}{1 - c_v - \frac{V_L}{V_L + V_s}} \tag{2.39}$$

This parameter is correlated with the solid volume concentration c_v and the cake porosity ε, which represents the relative void volume in the cake and thus the relative amount of liquid, if all voids are completely filled with liquid. This especially for cake filtration used parameter is very easy to measure in a lab tube, as shown schematically in Figure 2.23 or in a lab nutsche filter and correlates cake thickness and filtrate volume.

All mentioned definitions of the slurry solid concentration could be transferred according to Table 2.3 to each other.

2.4.2 Stability

The interaction between particles and dissolved ions in the liquid decides according to Figure 2.24, whether the particles are existing separately from each other (stable slurry) or agglomerate (instable slurry) by van der Waals adhesion forces [10] (cf. Section 5.3).

Table 2.3 Conversion of slurry concentration describing parameters.

	(κ)	(c_v)	(c_g)	$(\rho_{s,L})$
$\kappa =$	$\dfrac{m_s \rho_L}{m_L \rho_s (1-\varepsilon) - \varepsilon m_s \rho_L}$	$\dfrac{c_v}{1-\varepsilon-c_v}$	$\dfrac{(1-c_g)\rho_L}{c_g(1-\varepsilon)(\rho_s-\rho_L)}$	$\dfrac{\rho_{sL}-\rho_L}{\rho_s-\rho_{sL}+\varepsilon(\rho_L-\rho_s)}$
$c_v =$	$\dfrac{\kappa(1-\varepsilon)}{1+\kappa}$		$\dfrac{m_s \rho_L}{m_s \rho_L + m_L \rho_s}$	$\dfrac{c_g \rho_L}{c_g \rho_L + (1-c_g)\rho_s}$	$\dfrac{\rho_{sL}-\rho_L}{\rho_s-\rho_L}$
$c_g =$	$\dfrac{\rho_s(1-\varepsilon)\kappa}{\rho_s(1-\varepsilon)\kappa + \rho_L(1+\varepsilon\kappa)}$	$\dfrac{c_v \rho_s}{c_v \rho_s + (1-c_v)\rho_L}$		$\dfrac{m_s}{m_s + m_L}$	$\dfrac{\rho_s(\rho_{sL}+\rho_L)}{\rho_{sL}(\rho_s-\rho_L)}$
$\rho_{s,L} =$	$\dfrac{\rho_s(\rho_{sL}+\rho_L)}{\rho_{sL}(\rho_s-\rho_L)}$	$= c_v \rho_s + (1-c_v)\rho_L$	$\dfrac{\rho_s \rho_L}{c_g \rho_L + (1-c_g)\rho_s}$	$\dfrac{\rho_s \rho_L(m_s+m_L)}{m_s \rho_L + m_L \rho_s}$

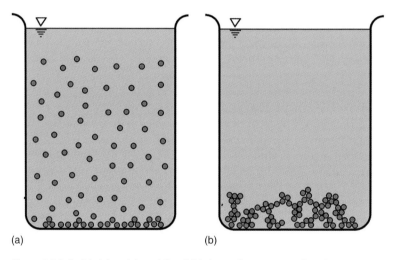

Figure 2.24 Stable (a) and destabilized (b) slurry after some settling time.

This naturally influences the separation conditions significantly. These effects are significant for particle diameters of less than about 100 μm and of increasing importance, the smaller the particles become. The reason for that is the competition between van der Waals force and weight of the particles. The van der Waals force behaves directly proportional to the particle diameter x and the weight behaves proportional to the third power of x. In addition, it has to be considered that particles are normally carrying electrical charges on their surfaces. Uniformly negative or positive charged particles are repelling each other and van der Waals forces cannot act because they are sensitively dependent on the particle distance. To bring the particles near enough together for adhesion, the repulsive forces must be reduced and the particles must be moved relatively to each other with sufficient kinetic energy. The repulsive forces at the particle surface can be reduced directly by shifting the pH or indirectly by the concentration and valence of ions in the surrounding liquid. Some of the ions are bound immobile at the particle surface and the resulting charge can be characterized by the zeta potential ς, which is measured in mV. There are different methods and instruments available to measure the zeta potential [11, 12]. Countercharged ions, which are present in the surrounding liquid, are accumulating in the form of an electrical double layer more and more toward the charged particle surface and are shielding the surface charges. In the literature, extensive information about the fundamentals of electrostatic surface forces and the phenomenon of the electrical double layer can be found [13–15].

The destabilization of slurries is frequently used as a slurry pretreatment measure to improve the separation conditions. The settling velocity of the solids is increasing, an eventually existing tendency for particle segregation in the slurry is prohibited, the filtrate pollution in the initial stage of the cake formation becomes less because of the prebuilt bridges, the cake formation time is reduced because of the more permeable cake structure, and the capillary forces of the larger pores of the cake are reduced.

2.5 Sampling

Normally, the total amount of a particle system, slurry, paste, or cake is much too large to be characterized or analyzed completely. Thus, a sample of appropriate quantity for the respective measurement device is needed. In many cases, the sample has to be in the dimension of a few grams or less. In many cases, the challenge is to extract such a small but nevertheless representative sample out of a huge amount of material in the dimension of tons. If this does not work correctly, the result of the analysis may be absolutely precise but meaningless with respect to the properties of the real material [16]. According to each situation, an individual sampling strategy has to be defined (cf. [7], 85–89).

If one has to analyze a batch of a bulk material, as indicated schematically in Figure 2.25, which is well accessible, several small samples from different locations have to be taken by chance.

In the case of a bulk material stream, the samples have to be taken according to Figure 2.26 periodically out of the stream.

Depending on the kind of transport process, for example, a swiveling shovel in the case of a conveyor belt or a flap in the case of a chute can be used. To get a representative sample from a moving belt, it is necessary to take material from the whole width of the stream because an identical material composition across

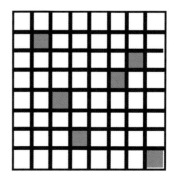

Figure 2.25 Sampling from a batch of bulk material.

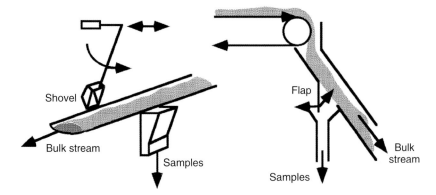

Figure 2.26 Periodical sampling from a continuous bulk material stream.

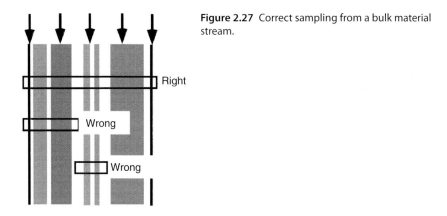

Figure 2.27 Correct sampling from a bulk material stream.

the belt width is not guaranteed. In Figure 2.27, it is shown how the sample is taken right or wrong.

For a periodic sampling from a continuous material stream, it must be considered that the bulk composition could vary periodically. Thus, the samples must be taken at a well-suited frequency as explained in Figure 2.28.

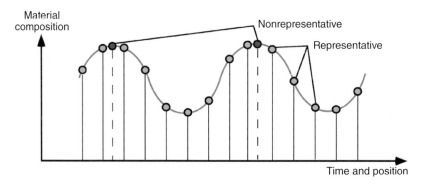

Figure 2.28 Choice of the right sampling frequency from a continuous material stream.

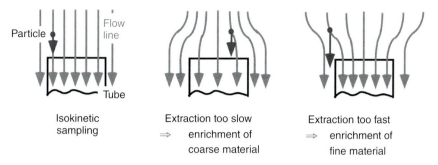

Figure 2.29 Sample extraction from a slurry stream.

As can be clearly seen, the sample would not be representative if the sampling frequency coincides with the frequency of the varying material composition. Thus, the right sampling frequency must be higher.

If the sample has to be taken out of a fluid stream such as slurry, the extraction must be realized under isokinetic conditions, as shown in Figure 2.29.

The flow velocity in the tube must be the same as the flow velocity around the tube. If the extraction velocity is too slow, liquid is hindered to enter the tube and flow lines are deviating from the tube inlet aperture. Greater particles, which cannot follow the bent flow line, are settling into the tube and an enrichment of coarse particles in the sample results. If the extraction velocity is too high, flow

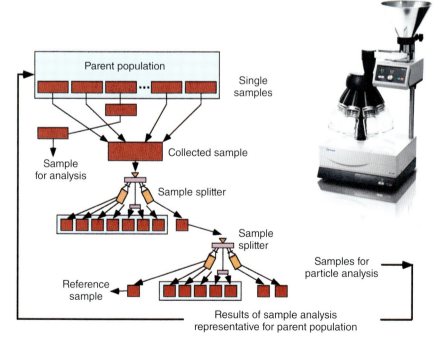

Figure 2.30 Sample splitting strategy and rotating sample splitter. Source: Courtesy of Verder Scientific/Retsch GmbH & Co. KG.

Figure 2.31 Sample splitting by quartering.

lines from outside the tube are sucked in and coarser particles are settling beside the tube. As a consequence, the sample contains an increased amount of fines.

Besides the isokinetic extraction, the coverage of the whole slurry stream and the consideration of eventual periodic variations of the slurry flow is again an important issue. In most cases, the total amount of a sample, which consists of several small random samples, is too large to handle in a particle size analyzer. However, at this point, a sub-sample can be taken to determine, for example, the density of the material. The rest of the collected sample must be subdivided into smaller portions by a correct procedure of sample splitting. As explained in Figure 2.30, this splitting process has to be repeated until the sample has the right size for the analysis (cf. [7], p. 86).

If finally the sample has the right size, several but at least two of these small samples should be analyzed. Sometimes, it is advantageous to take one sample as a reference sample. If later in the process something is going wrong, it is easier to decide whether the apparatus or the product has changed meanwhile.

Besides other sample splitting devices in Figure 2.30, a rotating sample splitter was chosen. The particles of the sample are transported via a vibrating chute into a funnel and from the funnel evenly into a number of rotating flasks.

The easiest way to split a sample can be seen in Figure 2.31.

The material is given on the crossing point of two blades, which are crossing each other in a 90° angle. This procedure quarters the sample into four equal parts.

References

1 Shirato, M., Aragaki, T., Iritani, E. et al. (1975). Constant pressure filtration of power-law non-Newtonian fluids. *Journal of Chemical Engineering of Japan* 10 (1): 54–60.

2 Scarlett, B. and Ward, A.S. (1986). Particle size analysis. In: *Solid/Liquid Separation Equipment Scale-Up*, 2e (ed. D.B. Purchas and R. Wakeman), 17–46. London: Uplands Press & Filtration Specialists Ltd.

3 Hess, W. (2005). *Partikelmesstechnik*. Weinheim: Wiley-VCH. ISBN: 3-527-30311-1.

4 Merkus, H.G. (2009). *Particle Size Measurements – Fundamentals, Practice, Quality*. Dordrecht: Springer https://doi.org/10.1007/978-1-4020-9016-5.

5 Polke, R., Schäfer, M., and Scholz, N. (2003). Charakterisierung disperser Systeme. In: *Handbuch der Mechanischen Verfahrenstechnik* (ed. H. Schubert), 7–100. Weinheim: Wiley-VCH. ISBN: 3-527-30577-7.
6 Kaye, B.H. (1999). *Characterization of Powders and Aerosols*. Weinheim: Wiley-VCH. ISBN: 3-527-28853-8.
7 Stieß, M. (1995). *Mechanische Verfahrenstechnik 1*. Heidelberg: Springer. ISBN: 3-540-59413-2.
8 Blott, S.J. and Pye, K. (2008). Particle shape: a review and new methods of characterization and classification. *Sedimentology* 55 (1): 31–63. https://doi.org/10.1111/j.1365-3091.2007.00892.x.
9 Rodriguez, J.M., Edeskär, T., and Knutsson, S. (2013). Particle shape quantities and measurement techniques – a review. *Electronic Journal of Geotechnical Engineering* 18: 169–198. ISSN: 1089-3032.
10 London, F. (1937). The general theory of molecular forces. *Transactions of the Faraday Society* 33: 8–26, ISSN 0014-7672, doi:https://doi.org/10.1039/TF937330008B.
11 Hunter, R.J. (1989). *Zeta Potential in Colloid Science, Principles and Applications*. New York: Academic Press. ISBN: 10: 0123619610.
12 Delgado, A.V., González-Caballero, F., Hunter, R.J. et al. (2007). Measurement and interpretation of electrokinetic phenomena. *Journal of Colloid and Interface Science* 309: 194–224. https://doi.org/10.1016/j.jcis.2006.12.075.
13 Dukhin, S.S. (1995). Electrochemical characterization of the surface of a small particle and nonequilibrium electric surface phenomena. *Advances in Colloid and Interface Science* 61: 17–49.
14 Hunter, R.J. (2001). *Foundations of Colloid Science*, 2e. Oxford: Oxford University Press. ISBN: 9780198505020.
15 Lyklema, J. (ed.) (1995). *Fundamentals of Interfaces and Colloid Science*. New York: Academic Press. ISBN: 9780124605237.
16 Sommer, K. (1979). *Probenahme von Pulvern und körnigen Massengütern, Grundlagen, Verfahren, Geräte*. Berlin-Heidelberg-New York: Springer.

3

Cake Structure Characterization

3.1 Introduction

The originating cake structure during the cake formation process is not only dependent on the above described properties of the slurry but also on process parameters such as filtration pressure, the kind of filter apparatus, such as centrifuge, or gas pressure filter, and design parameters such as horizontal or vertical arrangement of the filter area and more or less effective possibilities for slurry homogenization. The quantity, size, and geometry of the voids between the particles of the cake are determining the flow conditions of the filtrate, bulk density, liquid content, and subsequent processes, such as cake washing and cake deliquoring. However, the real geometry including all aspects, such as particle size distribution, particle shape, surface roughness of the particles, isotropy or anisotropy of the particle arrangement, tortuosity of the hydraulically connected pore channels, and others are not measurable with appropriate effort for real particle systems to describe the pore geometry exactly. As a consequence from the practical point of view and the necessity to design cake filtration apparatuses quantitatively, one needs to simplify the problem as much as possible and to complicate it as much as necessary. There are several different pore models available from the simple parallel-arranged cylindrical tubes to complex three-dimensional network models. A deeper insight and the more knowledge about specific influencing parameters is required, the more detailed the model must become, the more effort for measurements is needed. Here, not all existing models will be described systematically, but those models should be picked out which have been proven to describe the discussed cake filtration phenomena properly for practical purposes. In general, the structure of porous layers, such as filter cakes, sediments, or filter media can be characterized by porosimetric and porometric methods. Porosimetry is defined as the analysis of the relative void volume and porometry as the analysis of pore sizes and geometry. As examples for modern measurement methods to analyze the structure of porous systems, such as filter cake, mercury intrusion porometry [1], capillary flow porometry [2], or micro-computer-tomography (μ-CT) [3, 4] should be mentioned. The cake structure is given by the spatial arrangement of the particles. In principle, as two boundary cases, the regular arrangement and the uniform random arrangement can be distinguished. In regular packing, the particles are arranged

Wet Cake Filtration: Fundamentals, Equipment, and Strategies, First Edition. Harald Anlauf.
© 2019 Wiley-VCH Verlag GmbH & Co. KGaA. Published 2019 by Wiley-VCH Verlag GmbH & Co. KGaA.

in periodically repeating elementary structures as in the case of the molecules in a crystal. This type of arrangement needs organizing forces and is generally not found in packed particle beds and particularly in filter cakes but is helpful as simple model in several cases to understand fundamental phenomena. In practice, it is normally calculated with the stochastic homogeneity or the homogeneous random arrangement as best technically achievable mixing state. In that case, the probability to find the center of mass of an optional particle at an optional location in the cake volume is identical for each location and time in the cake. In addition, the orientation of any particle in any spatial direction exhibits the same probability. However, the stochastic homogeneity is a boundary case, but the best possible particle arrangement for a real technical process like a cake filtration. Deviations from the stochastic homogeneity are normally leading to deteriorate filtration results and thus the cake homogeneity is aimed as perfect as possible. If lab-scale tests, in which homogeneous filter cakes under nearly ideal conditions can be formed, are compared with imperfect results from technical filters, often an optimization potential can be identified, which can be traced back to the cake structure.

3.2 Porosity

The hollow space or void volume between the particles can be quantified according to Figure 3.1 in the simplest case as an integral absolute volume V_v.

As can be seen from Eq. (3.1), the total volume V_{tot} of the porous filter cake consists of the solid volume V_s and the void volume V_{void}, which is defined here at first as voids between compact particles

$$V_{tot} = V_s + V_{void} \tag{3.1}$$

The void volume is normally not given in absolute numbers but, according to Eq. (3.2), as relative void volume or porosity ε

$$\varepsilon = \frac{V_{void}}{V_{tot}} = \frac{V_{void}}{V_s + V_{void}} = \frac{V_{tot} - V_s}{A \cdot h_c} = \frac{(A \cdot h_c) - \frac{m_s}{\rho_s}}{A \cdot h_c} = 1 - \frac{m_s}{A \cdot h_c \cdot \rho_s} \tag{3.2}$$

The void volume V_{void} is related to the total volume of the porous cake V_{tot}. In Eq. (3.2), the information is included, how to determine the cake porosity by experiment. In a laboratory filter cell (cf. Section 6.3.2) with a circular and known filter area A, a cylindrical cake of measurable height h_c has to be formed. Its mass m_s can be determined after drying the moist filter cake by weighing, and with

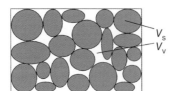

Figure 3.1 Solid and void volume of a filter cake.

the known solid density ρ_s, the porosity can be calculated. In some cases, the geometry of the cake is not defined as clear as in the case described before, but only a cake fragment of irregular shape, for example, from the cake discharge of a vacuum drum filter, is available. One possibility to get a geometrically defined piece of cake is to take out a sample of the cake with a circular tube. Alternatively, a pycnometric method according to Figure 3.2 can be applied.

The dry piece of cake is weighed first and then given into a nonwetting liquid. The replaced liquid volume corresponds to the total cake volume and again the porosity can be calculated.

In a similar way, the irregular-shaped cake sample can be immersed into a wetting fluid, as demonstrated in Figure 3.3.

In that case, the liquid fills the pores and the displaced liquid corresponds to the solid volume of the agglomerate.

A further alternative to measure the porosity of such irregular-shaped porous bodies is the mercury intrusion. Here, mercury is pressed as the nonwetting phase under high pressures into the porous sample and the intruded mercury volume is registered. If the pressure is increased step by step, in addition, a pore size distribution can be measured because pressure from outside, capillary pressure, and pore size are correlated with each other. This method cannot be applied for pressure-sensitive structures because of the danger of deformation or destruction of the sample.

Typical values for cake porosities are $0.3 \leq \varepsilon \leq 0.7$. For first estimations, a porosity of $\varepsilon = 0.4$ may be meaningful. Beside the porosity frequently, the solidosity $(1-\varepsilon)$ is used as indicated in Eq. (3.3)

$$1 - \varepsilon = \frac{V_s}{V_{tot}} = c_v. \tag{3.3}$$

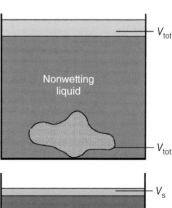

Figure 3.2 Porosity determination for irregular-shaped cakes using nonwetting liquid.

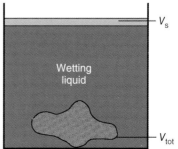

Figure 3.3 Porosity determination for irregular-shaped cakes using wetting liquid.

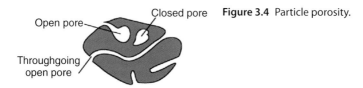

Figure 3.4 Particle porosity.

Comparing with Eq. (2.37), the solidosity is a synonym for the solid volume concentration, which characterizes the solid content in a slurry.

A further definition to characterize the relative void and solid volume is given by the pore number e in Eq. (3.4), which is often used in soil mechanics and for the description of filter cake shrinkage during deliquoring

$$e = \frac{V_{void}}{V_s} \quad (3.4)$$

The cake porosity is depending on various parameters such as particle size distribution, particle shape, and interdependencies between particles in the form of van der Waals or electrostatic forces and cake formation conditions.

As Figure 3.4 clarifies, the particles themselves are sometimes porous and are containing closed ($V_{void,cl}$) and/or open ($V_{void,op}$) micropores.

This particle porosity ε_p can be calculated, according to Eq. (3.5), analogous to the definition in Eq. (3.2)

$$\varepsilon_p = \frac{V_{void,p}}{V_p} = \frac{V_{void,op} + V_{void,cl}}{V_p} \quad (3.5)$$

If the closed pores are not interesting or not measurable, this part of the particles' void volume can be ignored. One possibility to get information about closed pores is given by the microsection technique. However, this method is very work intensive and not applicable in any case. The particles must be embedded in a solid polymeric matrix and then grinded down slice by slice. Each new surface has to be analyzed by image analysis with regard to the area porosity ε_A, which is defined in Eq. (3.6) as the relation of void cut face A_{void} to total area A_{tot}

$$\varepsilon_A = \frac{A_{void}}{A_{tot}} \quad (3.6)$$

The area porosity is varying depending on the random location of the cut through the porous system. As Eq. (3.7) indicates, the values of $\varepsilon_{A,i}$ are fluctuating around the expected value $\varepsilon_{av,A}$, which corresponds to the previously defined volume porosity ε (cf. Eq. (3.2))

$$\varepsilon_{av,A} = \lim_{n\to\infty} \frac{1}{n} \cdot \sum_{1}^{n} \varepsilon_{A,i} = \varepsilon \quad (3.7)$$

If particles are forming an agglomerate as can be seen in Figure 3.5, in Eq. (3.8), an agglomerate porosity ε_{ag} can be determined

$$\varepsilon_{ag} = \frac{V_{void,ag}}{V_{ag}} = \frac{V_{void,ag}}{V_p + V_{void,ag}} \quad (3.8)$$

Figure 3.5 Agglomerate porosity.

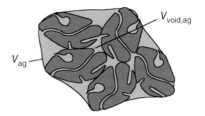

If porous particles are forming an agglomerate during a slurry pretreatment process such as flocculation or coagulation, the porosity $\varepsilon_{ag,p}$ can be calculated, according to Eqs. (3.9)–(3.11), with consideration of the micro- and macropore system

$$\varepsilon_{ag,p} = \frac{V_{void,ag} + V_{void,p}}{V_{ag}} = \frac{V_{void,ag}}{V_{ag}} + \frac{V_{void,p}}{V_{ag}} = \varepsilon_{ag} + \frac{V_{void,p}}{V_{ag}} \tag{3.9}$$

$$\frac{V_{void,p}}{V_{ag}} = \frac{V_{void,p}}{V_p} \cdot \frac{V_p}{V_{ag}} = \varepsilon_p \cdot (1 - \varepsilon_{ag}) \tag{3.10}$$

$$\varepsilon_{ag,p} = \varepsilon_{ag} + \varepsilon_p \cdot (1 - \varepsilon_{ag}) \tag{3.11}$$

After the agglomeration process, the slurry is filtered, and in the resulting cake originate, according to Figure 3.6, the macropores between the agglomerates.

The cake porosity ε_c under the assumption of compact particles is calculated in Eq. (3.12) from the relation of void volume $V_{void,c}$ and the total cake volume V_{tot}

$$\varepsilon_{tot} = \frac{V_{void,p} + V_{void,ag} + V_{void,c}}{V_{tot}} \tag{3.12}$$

Here, the particles are agglomerates, which consist of porous primary particles and a total porosity ε_{tot} is formulated in Eq. (3.13) in the dependency of the single porosity portions

$$1 - \varepsilon_{tot} = \frac{V_s}{V_{tot}} = \frac{V_s}{V_p} \cdot \frac{V_p}{V_{ag}} \cdot \frac{V_{ag}}{V_{tot}} = (1 - \varepsilon_p) \cdot (1 - \varepsilon_{ag}) \cdot (1 - \varepsilon_c) \tag{3.13}$$

In most cases, it is sufficient to simply calculate the integral cake porosity according to Eq. (3.2), but sometimes, more detailed information is desirable. In the case of particle washing, it is interesting to know whether there is one more

Figure 3.6 Filter cake porosity.

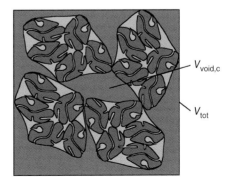

or less uniform pore size system present or a multimodal pore size distribution. The less uniform the pore sizes, the less piston like the flow of wash liquid becomes. The flow velocity in the coarse pores around the agglomerates is greater than in the smaller pores inside the agglomerates and much greater than in the very small pores inside the particles. This increases washing times and the amount of necessary wash liquid until the desired purity of the cake is achieved. Especially porous particles are normally not purified inside by permeation, but the contaminants leave the particle pores by time-intensive diffusion.

A further example may be the cake desaturation. In the case of multimodal pore size systems, hydraulic breaks of the liquid flow can occur with the consequence of remaining isolated wet islands in the cake. The inner liquid in the micropores of particles is not possible to deliquor mechanically and the information about the quantity of liquid bound in micropores can give a hint to explain high residual cake moisture contents after mechanical deliquoring processes.

Until now, porosity has been discussed without consideration of the particle size, size distribution, shape, or particle interaction. However, these parameters are influencing the porosity as well as the pore sizes and the cake structure and thus the cake filtration conditions sensitively.

Looking first to the influence of particle size on porosity, although it is not realistic for the technical case, one can get valuable information from idealized models as shown in Figure 3.7 for a cubic and a rhombohedral structure of mono-sized spheres.

A regular structured particle arrangement consists of periodically repeating elements. Each particle has a defined number of contact points (coordination number k) to its neighbors. In the case of a cubic arrangement, the coordination number k amounts to 6 and in the case of the most densely packed rhombohedral structure, the coordination number k amounts to 12. The elementary structure of the cubic arrangement is a cube with an edge length x containing a sphere with a diameter x and the porosity can be easily calculated in Eq. (3.14) analogous to the definition in Eq. (3.2)

$$(1-\varepsilon) = \frac{V_s}{V_{tot}} = \frac{\frac{\pi}{6} \cdot x^3}{x^3} = \frac{\pi}{6} = 0.526 \qquad (3.14)$$

The rhombohedral arrangement can be treated in the same way in Eq. (3.15)

$$(1-\varepsilon) = \frac{V_s}{V_{tot}} = \frac{\frac{\pi}{6} \cdot x^3}{\frac{1}{2} \cdot \sqrt{2} \cdot x^3} = 0.741 \qquad (3.15)$$

Table 3.1 shows further regular sphere arrangements.

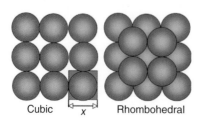

Figure 3.7 Cubic and rhombohedral structure of spheres.

Table 3.1 Packing density of different regular sphere arrangements.

	k	$(1 - \varepsilon)$	ε
Cubic	6	$\pi/6$	0.477
Orthorhombic	8	$(\pi/6)(2/\sqrt{3})$	0.395
Tetragonal	10	$(\pi/6)(2/\sqrt{3})^2$	0.302
Rhombohedral	12	$(\pi/6)(2/\sqrt{2})$	0.259

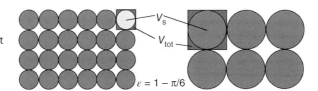

Figure 3.8 Elementary cell volume of regular particle arrangements with different particle size.

The packing density depends on the number of contact points between the particles, which is expressed by the coordination number k. As a consequence of this examination, the porosity of regular particle arrangements must be independent of the particle size because the elementary structure is, according to Figure 3.8, the same for any sphere diameter.

This finding can be transferred to randomly arranged porous layers from cake filtration or sedimentation processes. Figure 3.9 illustrates the findings of Sorrentino [5] to validate the theoretical consideration.

For these experiments, monosized particles of different diameters have been used. Related to the specific kind of separation process, a constant and particle size independent basic porosity value ε_0 results for particle sizes beyond a critical value toward greater particles. The porosity level depends sensitively on the specific conditions under which the particle layer is formed. Here, a static sedimentation was compared with a vacuum filtration. If the vacuum filtration would be compared with a centrifugal filtration of the same filtration pressure, the centrifuge cake would often show a slightly decreased porosity because of the energy input by motor vibrations.

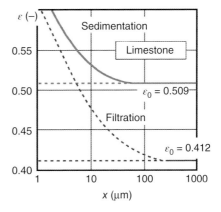

Figure 3.9 Porosity as a function of particle size for randomly packed porous layers.

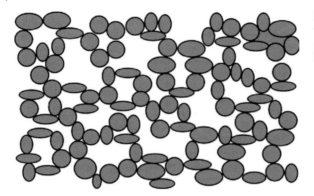

Figure 3.10 Influence of van der Waals forces on the porosity of fine particle layers.

Yu and Standish have described the porosity functions in Figure 3.9 by Eq. (3.16) [6–8].

$$\varepsilon = \varepsilon_0 + (1 - \varepsilon_0) \cdot e^{-a \cdot x^b} \tag{3.16}$$

The particle interaction parameters a, b as well as the basic porosity ε_0 must be determined experimentally. This leads to the first conclusion that the porosity of even monosized particle layers cannot be calculated theoretically, but some material parameters have to be measured by practical experiments, and these must be orientated as close as possible to the planned technical filtration process.

If the particle size undergoes a certain critical value and under the precondition of absent repelling forces, the porosity increases the smaller the particles become because van der Waals forces are getting more and more influence. As can be seen in Figure 3.10, an increasingly loose structure is formed.

If the particles are becoming smaller and the mass forces are decreasing, adhering van der Waals and repulsive electrostatic forces are deciding about the stability of the cake forming slurry. If no repulsive forces act, nanoscale particles also form nearly incompressible and densely packed filter cakes. Figure 3.11

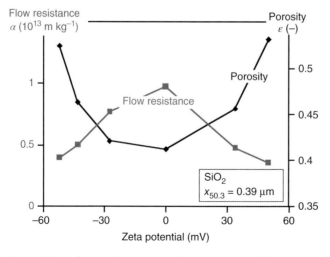

Figure 3.11 Influence of zeta potential on porosity and flow resistance of filter cakes.

illustrates, according to the measurements of Heuser [9], the influence of pH on the porosity and flow resistance of cakes from particles in the submicrometer range.

3.3 Particle Arrangement

Looking more into details of a real filter cakes, the homogeneous structure indicated in Figure 3.10 is a too simple model for compressible filter cakes. Only from stable slurries, homogeneous and incompressible structures with a relatively narrow pore size distribution can be expected as depicted in Figure 3.12.

Porosity ε and specific cake flow resistance r_c are constant over the cake height, and thus, the liquid pressure p_L decreases and the pressure on the solid p_s increases linearly toward the filter medium during cake formation.

In the case of destabilized slurry, a compressible and inhomogeneous structure with a broad pore size distribution will form. If adhesion forces are acting on the particles, the compressible cake behavior leads to a porosity gradient over the cake height as can be seen in Figure 3.13 with a densely packed structure near the filter medium and a more loose packed structure at the cake surface.

As a consequence, p_L and p_s are no longer linear across the cake height. In addition, further mechanisms can lead to a compressibility of a filter cake. In Figure 3.14, different reasons for the consolidation of deformable cake structures under the impact of higher pressures are collocated, which has been described in more detail by Alles et al. [10].

If particles are loosely packed at low filtration pressure, the originated structure can be condensed afterward at greater pressure to certain content by overcoming of the friction between the particles and their rearrangement. If the filter

Figure 3.12 Incompressible filter cake.

Figure 3.13 Compressible filter cake.

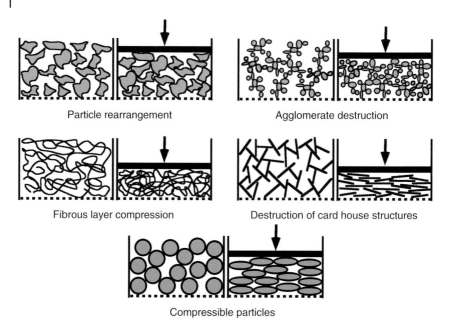

Figure 3.14 Reasons for cake compressibility.

cake consists of agglomerated particles, the cake can be compacted by agglomerate destruction. Fibers are forming highly compressible structures, which can be deformed at comparably low pressures. In the case of flaky particles under certain conditions, a very permeable structure like a house of cards can be formed, which breaks down at greater pressures and changes to a very tight structure of high flow resistance. Last but not least, a reason for compressibility is also given, if the particles themselves are deformable as in the case of many biological microparticles.

Until now, particles of the same size have been discussed. In a further step, the question has to be answered how the particle size distribution influences porosity. As Figure 3.15 illustrates, the broader the particle size distribution becomes, the smaller the porosity will be because smaller particles are filling the voids between the greater particles (2), (3).

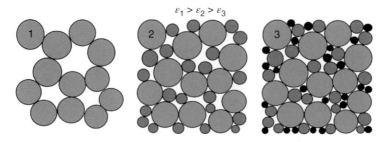

Figure 3.15 Influence of particle size distribution on porosity.

Figure 3.16 Porosity of mixtures from fine (*i*) and coarse (*j*) particles.

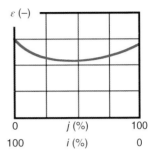

The smaller particles are not fitting exclusively and perfectly into the voids of the greater particles without changing the structure of the original coarse-grained cake (1). In that case, one would have only to reduce the void volume of the coarse-grained cake by the volume of the smaller particles. In reality, the particles are arranged randomly, which leads to a certain expansion of the cake structure, which is formed only by the large particles. Besides the different size of the particles, the quantity of each particle size fraction influences the porosity. This may be demonstrated qualitatively on the example of a binary mixture. In Figure 3.16, the porosity is plotted against the percentage of the coarse particles *j* in the fine fraction *i* and vice versa.

The porosity function exhibits a characteristic minimum of porosity. Toward the pure fine or pure coarse fraction, the porosity increases successively. The pure fine fraction may result in a little higher porosity because of an increasing influence of adhesive forces (cf. Figure 3.9). By measuring for a given particle size distribution, the cake porosity function $\varepsilon = f(x)$ for single particle sizes and $\varepsilon = f(x_j/x_i)$ for several binary mixtures, the cake porosity for any particle size distribution in principal can be predicted [11]. Of course, this model needs a huge effort to measure the required data for getting reasonable results. In addition, the results are depending on the cake formation procedure (cf. Figure 3.9).

If the filter cake is formed by particles of different size, no guarantee is given that a homogeneous structure, as can be seen in Figure 3.15, originates. A segregation of the particles by sedimentation during the cake formation process can occur. In that case, according to the right-hand side (2) of Figure 3.17, the greatest particles will be found near the filter media and the finest particles form a layer of comparatively high flow resistance on top of the filter cake.

Figure 3.17 Homogeneous and segregated filter cake structure.

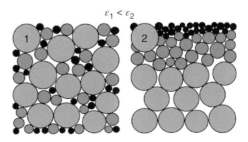

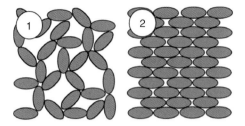

Figure 3.18 Influence of nonspherical particle shape on the filter cake structure.

The danger of particle segregation exists especially in the case of broad particle size distributions and low solid concentration in the slurry. Low slurry concentration leads to single-particle sedimentation with high sensitivity of particle size, whereby higher concentrations show a swarm or zone sedimentation behavior of the slurry with a sharp sedimentation front, which separates the slurry from clear liquid above. Here, the particles are hindering each other and settle nearly independent of their size and shape with the same velocity. The segregation effect becomes much more dramatic in filter centrifuges than in pressure filters. In a centrifuge with an acceleration of $1000 \times g$, the settling velocity is nearly 1000 times greater than in a pressure filter with only $1 \times g$. Because centrifugal forces are mass forces and the mass force is depending on the third power of the particle diameter, the segregation tendency increases with an increase of the centrifugal acceleration and width of the particle size distribution. If particles are agglomerated, segregation is suppressed because more or less, the whole particle size distribution is represented in the agglomerates. Especially if the solid concentration of the slurry is quite low and no homogenization of the slurry can be realized during the cake formation process, agglomeration is an effective measure to get homogeneous filter cakes.

The inhomogeneous cake structure in Figure 3.17 (2) is under nearly all circumstances of negative impact in comparison to a homogeneous structure (1). Because of the fine particle layer, demixed filter cakes exhibit a high flow resistance and a worse washing efficiency because of the broad pore size distribution. Furthermore, demixed filter cakes contain more liquid in comparison to homogeneous filter cakes because of the higher porosity and are more difficult to desaturate because of an increased capillary entry pressure in the fine particle layer.

Last but not least, the particle shape can have remarkable influence on the cake structure (cf. Figure 3.14). As an example of flaky or lentil-shaped particles in Figure 3.18, it is demonstrated that either very permeable (1) or impermeable (2) cakes can be formed depending on their arrangement in the cake structure.

Which modification of the structure will be formed depends on physical chemistry, solid concentration in the slurry, and operational conditions such as shear forces or pressure.

3.4 Pore Size

Besides the porosity, the pore size and pore size distribution determine the flow resistance of a porous particle layer. The pores are forming not a discrete disperse

Figure 3.19 Tubular pore model.

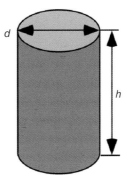

collective such as the particles but a hydraulically interconnected channel system of different cross sections and lengths. Similar to a characteristic average particle size such as the Sauter average diameter (cf. Eq. (2.13)), a hydraulic mean pore diameter $d_{av,h}$ can be calculated by a rough estimation. Assuming, according to Figure 3.19, that all pores are cylindrical tubes of diameter d and length l, the lateral area of the tubes should represent the pore surface and the cylinder volume should represent the pore volume.

The mean hydraulic pore diameter $d_{av,h}$ is calculated in Eq. (3.17) from the ratio of pore volume V_p and pore surface S_p

$$\frac{V_p}{S_p} = \frac{\frac{\pi}{4} \cdot d^2 \cdot h}{\pi \cdot d \cdot h} = \frac{d}{4} \Rightarrow d_{av,h} = 4 \cdot \frac{V_p}{S_p} \tag{3.17}$$

According to Eq. (3.18), the correlation with the porous cake structure can be achieved by extension of Eq. (3.17)

$$d_{av,h} = 4 \cdot \frac{V_{void}}{V_{tot}} \cdot \frac{V_{tot}}{V_s} \cdot \frac{V_s}{S} = 4 \cdot \varepsilon \cdot (1-\varepsilon)^{-1} \cdot S_V^{-1} \tag{3.18}$$

The void volume V_{void} correlates with the pore volume V_p and the particle surface S correlates with the pore surface S_p. Now, the correlation between mean hydraulic pore diameter and Sauter average diameter can be determined in Eq. (3.19)

$$S_V = \frac{6}{x_{av,SV}} \Rightarrow d_{av,h} = \frac{2}{3} \cdot \frac{\varepsilon}{(1-\varepsilon)} \cdot x_{av,SV} \tag{3.19}$$

The hydraulic mean diameter strongly depends on the porosity. If for a first assumption, the porosity is set to $\varepsilon = 0.4$ (cf. Section 3.2), the hydraulic mean diameter becomes 44% of $x_{av,SV}$. If the porosity is increased, by agglomeration, to $\varepsilon = 0.6$, d_h equals $x_{av,SV}$.

This approach of course gives only a first and very rough but not completely unrealistic impression about the relevant pore size for the filtration process. In reality, the pores are of course not cylindrical. The particles are enclosing a certain void volume, which is characterized by the diameter of the sphere, which fits into this void, but the flow resistance of the porous system or the desaturation of pores is depending more on the hydraulic diameter of the pore throat, which

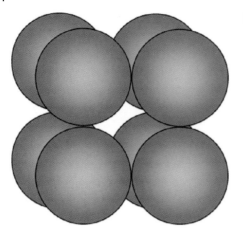

Figure 3.20 Pore void and pore throat.

is considerably smaller than the void diameter, as depicted in Figure 3.20 on the example of a cubic arrangement of spheres.

Normally, the pore size distribution and details of the pore geometry of filter cakes are not required for the determination of parameters such as flow resistance, if experimental data are available. If such data are required for numerical flow simulations, three-dimensional structure reconstruction of magnetic resonance tomography (MRT) or µ-CT can be used. Pore size analysis is very interesting for filter media to check which particles are able to penetrate the media and which particles are retained or which standard deviation the nominal pore size exhibits, respectively. Besides direct microscopic analysis of open filter media with transmitted light, sieving with glass beads and capillary flow porometry are usual methods. Especially, the measuring of the bubble point to characterize the largest pore will be discussed later on in Chapter 9 after discussion of fundamental capillary phenomena.

References

1 Abell, A.B., Willis, K.L., and Lange, D.A. (1999). Mercury intrusion porosimetry and image analysis of cement-based materials. *Journal of Colloid and Interface Science* 211: 39–44.
2 Agarwal, C., Pandey, A.K., Das, S. et al. (2012). Neck-size distributions of through-pores in polymer membranes. *Journal of Membrane Science* 415–416: 608–615.
3 Landis, E.N. and Keane, D.T. (2010). X-ray microtomography. *Materials Characterization* 61 (12): 1305–1316. https://doi.org/10.1016/j.matchar.2010.09.012.
4 Taina, I.A., Heck, R.J., and Elliot, T.R. (2008). Application of X-ray computed tomography to soil science: a literature review. *Canadian Journal of Soil Science* 88: 1–20.
5 Sorrentino, J.A. (2002). *Advances in Correlating Filter Cake Properties with Particle Collective Characteristics*. Aachen: Shaker. ISBN: 3-8322-0972-7.

6 Yu, A.B. and Standish, N. (1987). Porosity calculations of multicomponent mixtures of spherical particles. *Powder Technology* 52: 233–241.
7 Yu, A.B. and Standish, N. (1991). Estimation of the porosity of particle mixtures by a linear mixture packing model. *Industrial and Engineering Chemistry Research* 30 (6): 113–124.
8 Yu, A.B. and Zou, R.P. (1996). Evaluation of the packing characteristics of mono-sized non-spherical particles. *Powder Technology* 88: 71–79.
9 Heuser, J. (2003). *Filterkuchenwaschprozesse unter besonderer Berücksichtigung physikalisch-chemischer Einflüsse*. Aachen: Shaker.
10 Alles, C., Anlauf, H., and Stahl, W. (1999). Fine particles in compressible filter cakes. *Advances in Filtration and Separation Technology* 13: 898–905.
11 Yu, A.B., Bridgewater, J., and Burbidge, A. (1997). On the modelling of the packing of fine particles. *Powder Technology* 92: 185–194.

4

Characterization of Liquid Flow Through Porous Particle Layers

4.1 Introduction

Cake formation, washing, and deliquoring are based on the flow of at least one liquid through particle layers. In the case of washing, sometimes the original liquid is displaced by a different liquid (flushing), and in the case of cake desaturation, a parallel liquid and gas flow takes place. The slurry properties and the process conditions determine the filter cake structure. The filter cake structure determines its flow resistance or permeability as a characteristic parameter. To design a cake filtration process, the correlation between the energetic expenditure in the form of a pressure difference, the process results in the form of volume flow, and the relevant properties of the cake structure, and the fluid must be known. In most cases of practical applications, the filter cake is supposed as a black box, and the permeability is measured by a simple but precise and practice-relevant experiment. If a deeper insight is needed, the influence of cake structure parameters on the permeability, such as porosity, particle size, or tortuosity can be described as well by several models, but the experimental effort to get the necessary data of these parameters increases. Generally, postulated here is a stationary and isothermal flow of an incompressible Newtonean fluid through a homogeneous and isotropic porous layer. The particles and thus the pore sizes should be much smaller than the dimensions of the layer to be allowed, to neglect inlet or outlet as well as wall effects.

4.2 Dimension Analytic Approach for the Flow Through Porous Particle Layers

In general, the dimension analysis represents a well-suited tool to formulate a correct mathematical relation between several parameters of different dimensions by description of a physical/technical problem in a dimensionless form. Advantages are the reduction of parameters to describe the problem, a safe up- or down-scaling, a more deep insight into the process, and a greater variability in the choice of test parameters [1, 2].

On the example of a cylindrical pipe, as most simple pore model for a filter cake, the procedure should be demonstrated. Afterward, the result will be transferred to a porous particle layer and compared with other models.

Wet Cake Filtration: Fundamentals, Equipment, and Strategies, First Edition. Harald Anlauf.
© 2019 Wiley-VCH Verlag GmbH & Co. KGaA. Published 2019 by Wiley-VCH Verlag GmbH & Co. KGaA.

4 Characterization of Liquid Flow Through Porous Particle Layers

Table 4.1 Dimension matrix.

	Core matrix			Rest matrix		
	ρ_L	d	v_L	Δp	$Q_{v,L}$	l
Mass (M)	1	0	0	1	0	0
Length (L)	−3	1	2	−1	3	1
Time (T)	0	0	−1	−2	−1	0

Table 4.2 First steps of matrix transformation.

	Core matrix			Rest matrix		
	ρ_L	d	v_L	Δp	$Q_{v,L}$	l
$X_1 = M$	1	0	0	1	0	0
$X_2 = 3M + L$	0	1	2	2	3	1
$X_3 = -T$	0	0	1	2	1	0

In a first step, a complete list of relevant influencing parameters has to be prepared. These parameters have to be independent from each other.

Here, the target variable should be the pressure loss Δp. The influencing parameters can be subdivided into geometry parameters (pipe diameter d, pipe length l), material parameters (liquid density ρ_L, kinematic liquid viscosity v_L), and process parameters (liquid volume flow $Q_{v,L}$). Thus, the relevance list contains six parameters: $\Delta p, d, l, \rho_L, v_L$, and $Q_{v,L}$.

In the next step, a dimension matrix, as shown in Table 4.1, has to be formed.

The matrix consists of a quadratic core matrix and a rest matrix. The rows contain the basic dimensions of the parameters (mass M, length L, time T), and the columns contain the parameters themselves inclusive of the target parameter (liquid density ρ_L, diameter d, kinematic liquid viscosity v_L, pressure loss, liquid volume flow $Q_{v,L}$, and length l). The target parameters should be put into the rest matrix because they are forming at the end of the numerator of the resulting dimensionless characteristic numbers. Apart from that, the sequence of the parameters is in principle regardless, but should be arranged in the way to make the next step as easy as possible. In the next step, the core matrix has to be converted into a unit matrix, which needs to have only 1 on the diagonal and apart from that at every place 0. The closer the arrangement to the unit matrix from the beginning, the less mathematical operations for conversion are necessary at the end.

At first, for the second row, a linear transformation has to be carried out in the following form: $X_1 = M$; $X_2 = (3M + L)$; $X_3 = -T$. The third row has to be multiplied with -1. The result of this calculation can be seen in Table 4.2.

To get the unit matrix, finally a second linear transformation has to be carried out for the second row ($Y_1 = X_1, Y_2 = X_2 - 2X_3, Y_3 = X_3$) as shown in Table 4.3.

Now, three dimensionless characteristic numbers Π can be formulated because the rest matrix consists of three parameters. The numerators are formed by

4.2 Dimension Analytic Approach for the Flow Through Porous Particle Layers

Table 4.3 Second matrix transformation – unit matrix.

	Core matrix			Rest matrix		
	ρ_L	d	v_L	Δp	$Q_{v,L}$	l
$Y_1 = X_1$	1	0	0	1	0	0
$Y_2 = X_2 - 2X_3$	0	1	0	-2	1	1
$Y_3 = X_3$	0	0	1	2	1	0

the parameters of the rest matrix. The denominators are formed according to Eqs. (4.1)–(4.3) by the parameters of the core matrix to the power, they are given in the rest matrix

$$\Pi_1 = \frac{\Delta p}{\rho_L^1 \cdot d^{-2} \cdot v_L^2} \tag{4.1}$$

$$\Pi_2 = \frac{Q_{v,L}}{\rho_L^0 \cdot d^1 \cdot v_L^1} = \frac{Q_{v,L}}{d \cdot v_L} \tag{4.2}$$

$$\Pi_3 = \frac{l}{\rho_L^0 \cdot d^1 \cdot v_L^0} = \frac{l}{d} \tag{4.3}$$

All Π numbers of a relevance list are equivalent to each other and can be converted by products of any power into others. This leads to the resistance function of flow through pipes in Eq. (4.4)

$$\Pi_1 \cdot \Pi_2^{-2} \cdot \Pi_3^{-1} = Eu \cdot \frac{d}{l} = \zeta = \frac{\Delta p \cdot d^4}{\rho_L \cdot Q_{v,L}^2} \cdot \frac{d}{l} \tag{4.4}$$

The parameter ζ denotes the resistance coefficient of the pipe flow and Eu the Euler number, which represents the ratio of pressure and inertia forces.

Π_2 can be converted according to Eq. (4.5) into the Reynolds number Re, which is the ratio of inertia to viscosity forces and describes the flow regime

$$\Pi_2 = \frac{Q_{v,L}}{d \cdot v_L} = \frac{\frac{V_L}{t}}{d \cdot \frac{\eta_L}{\rho_L}} = \frac{\frac{d^2 \cdot l}{t}}{d \cdot \frac{\eta_L}{\rho_L}} = \frac{\rho_L \cdot d \cdot w}{\eta_L} = Re \tag{4.5}$$

Now, these results should be transferred to the permeation of porous layers or filter cakes, respectively. The characteristic parameters for the relevance list can be seen in Figure 4.1.

Now, seven parameters are relevant to describe the problem: Δp, w, ρ_L, η_L, ε, $x_{av,SV}$, h_c. From the dimension analysis, four dimensionless characteristic numbers are expected according to a core matrix of the three basic parameters: length,

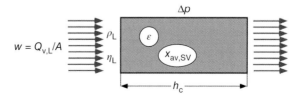

Figure 4.1 Permeation of a porous filter cake.

mass, and time (rows), three influencing parameters, and four parameters in the rest matrix (columns). These are expressed in Eqs. (4.6)–(4.9)

$$\Pi_1 = \frac{\Delta p}{\rho_L \cdot w^2} = Eu \tag{4.6}$$

$$\Pi_2 = \frac{w \cdot x_{av,SV} \cdot \rho_L}{\eta_L} = Re \tag{4.7}$$

$$\Pi_3 = \frac{h_c}{x_{av,SV}} \tag{4.8}$$

$$\Pi_4 = \varepsilon \tag{4.9}$$

From these dimensionless numbers in Eq. (4.10), the general resistance function for the permeation of porous layers (filter cakes) can be formulated

$$Eu \cdot \frac{x_{av,SV}}{h_c} = f(Re, \varepsilon) \tag{4.10}$$

The function $f(Re, \varepsilon)$ correlates with the pipe friction number, the drag coefficient for the flow around particles c_w, or the Newton number Ne for stirrers. The function must be measured by experiments and exhibits for double logarithmic representation the general trend, shown in Figure 4.2. Here, curves for two different porosities ε_1 and ε_2 are plotted.

For small Re-numbers, in the zone of laminar flow regime, friction forces are dominating. In nearly all cases of wet cake filtration, this condition is adequately fulfilled and will be presupposed for all further considerations. The subsequent transition zone for greater Re-numbers and the turbulent flow regime are disregarded furthermore.

From the general resistance function for laminar flow, the Darcy equation can be derived as a basic equation for the description of all processes, where fluids are permeating porous structures [3]. Filter cake formation, washing, and deliquoring can be described on the basis of the Darcy equation, which must be adjusted in each case to the special boundary conditions. Pressure loss Δp and cake thickness h_c should be proportional to each other according to the presupposition of a homogeneous and isotropic porous layer. Transforming of the results of the

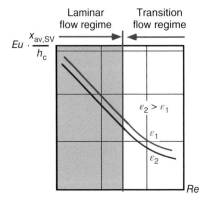

Figure 4.2 Resistance function for permeation of porous layers.

previous dimension analysis, the Darcy equation can be expressed according to Eqs. (4.11)–(4.13)

$$Eu \cdot \frac{x_{av,SV}}{h_c} = \frac{\Delta p}{\rho_L \cdot w^2} \cdot \frac{x_{av,SV}}{h_c} = f(Re, \varepsilon) = \frac{\text{const.}(\varepsilon)}{Re} \quad (4.11)$$

$$Re = \frac{w \cdot x_{av,SV} \cdot \rho_L}{\eta_L} \quad (4.12)$$

$$\frac{\Delta p}{h_c} = \frac{\text{const.}(\varepsilon)}{x_{av,SV}^2} \cdot \eta_L \cdot w = \frac{\eta_L \cdot w}{p_c} \quad (4.13)$$

The specific permeability of the filter cake p_c (m²) characterizes the influence of the pore structure on the permeation. Finally, the integral form of the Darcy equation in Eq. (4.14) describes the one-dimensional flow velocity w in dependence of Δp, η_L, p_c, and h_c

$$w = \frac{Q_{v,L}}{A} = \frac{p_c \cdot \Delta p}{\eta_L \cdot h_c} \quad (4.14)$$

The flow velocity w represents the empty pipe velocity of the fluid $Q_{v,L}/A$.

4.3 Empirical Approach for the Flow Through Porous Particle Layers

To get more detailed information about the influence of the cake structure on the flow among several others [4], the Carman/Kozeny approach in Eqs. (4.15) and (4.16) can be used, which is based on theoretical considerations and experimental data [5, 6]

$$p_c = \frac{1}{180} \cdot \frac{\varepsilon^3}{(1-\varepsilon)^2} \cdot x_{av,SV}^2 = \frac{1}{36 \cdot \lambda \cdot \beta} \cdot \frac{\varepsilon^3}{(1-\varepsilon)^2} \cdot x_{av,SV}^2 \quad (4.15)$$

$$p_c = \frac{1}{C \cdot \phi(\varepsilon)} \cdot x_{av,SV}^2 = \frac{1}{C \cdot \phi(\varepsilon)} \cdot \psi_{SV,i}^2 \cdot D_{3,2,i}^2 \quad (4.16)$$

The constant $C = 180$ presumes that the total particle surface is available as a pore surface and the flow takes place through a bundle of straight cylindrical capillaries. Surface reduction due to partial particle surface coverage (λ) and elongation of the fluid pathway due to tortuosity (β) are resulting in a factor of 5. Figure 4.3 shows these two structure effects in a schematic sketch.

Depending on the particle shape and arrangement, parts of their surface can be covered, which reduces the accessible pore surface.

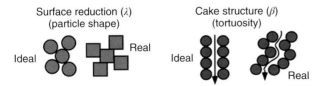

Figure 4.3 Ideal and real cake structure.

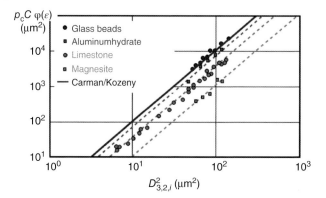

Figure 4.4 Determination of particle shape factors by filtration experiments.

To predict the cake permeability for any particle size distribution, at first, the resulting cake porosity must be known. This is a work-intensive procedure as has been discussed in Section 3.3. Secondly, the true Sauter average diameter $x_{av,SV}$ is needed. Normally, a particle size analyzer is measuring not directly the specific surface of the particles and the Sauter average diameter $D_{3,2,i}$ must be calculated from a particle size analyzer, which works on a certain physical basis i (laser diffraction, sedimentation, image analysis, etc.). To obtain the true Sauter average diameter, a shape factor $\psi_{SV,i}$ must be determined. For this purpose, Sorrentino proposed a procedure, which needs permeation experiments, porosity measurements, and particle size analysis. These data have to be evaluated according to Figure 4.4 [7].

The shape factor can be obtained by shifting the individual straight line of any particle system to the ideal Carman/Kozeny function. Depending on the particle shape, the shift factor is different for each product. For spherical glass beads, the shift factor is obviously 1 and the data for glass beads can be found directly on the Carman/Kozeny function.

To confirm this finding, an independent experiment had to be carried out to determine the shape factor. From Figure 4.4, for limestone, a shape factor of $\psi_{SV,i} = 0.6$ could be determined. Particle sizes had been measured here with laser diffraction. A photometric direct measurement of the specific surface for limestone and glass beads has led to the results, given in Figure 4.5.

The correlation between specific surface S_V and Sauter mean diameter $D_{3,2,i}$ is given by the shape factor $\psi_{SV,i}$

$$S_V = \frac{6}{\psi_{SV,i}} \cdot \frac{1}{D_{3,2,i}} \qquad (4.17)$$

The discussed correlations are evaluated since now for products that are forming nearly incompressible filter cakes. Because of the great necessary effort to obtain the needed data, the permeability is measured for practical use normally only by a simple cake filtration experiment, as described and discussed later on (cf. Section 6.3.2) in connection with the filter cake formation. However, the simple measurement allows no forecast for changing particle size distributions or different products but provides reliable results for safe scale-up calculations.

Figure 4.5 Photometric determination of particle shape factors.

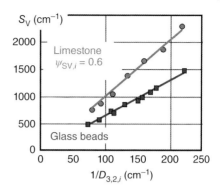

References

1 Szirtes, T. (1998). *Applied Dimensional Analysis and Modelling*. New York: McGraw-Hill.
2 Zlokarnik, M. (2000). *Scale-up-Modellübertragung in der Verfahrenstechnik*. Weinheim: Wiley-VCH. ISBN: 3-527-29864-9.
3 Darcy, H.P.G. (1856). *Les fontaines publiques de la ville de Dijon: Exposition at application des principes à suivre et des formules à employer dans les questions de distribution d'eau*. Paris: Victor Delamont.
4 Rumpf, H. and Gupte, A.R. (1971). Einflüsse der Porosität und Korngrößenverteilung im Widerstandsgesetz der Porenströmung. *Chemie Ingenieur Technik* 43 (6): 367–375.
5 Carman, P.C. (1937). Fluid flow through granular beds. *Transactions of the Institution of Chemical Engineers* 15: 150–166.
6 Carman, P.C. (1939). Fundamental principles of industrial filtration. *Transactions of the Institution of Chemical Engineers* 17: 167–188.
7 Sorrentino, J.A. and Anlauf, H. The use of particle size distribution for estimating cake permeability. *Advances in Filtration and Separation Technology* 14: 595–600.

5

Slurry Pretreatment to Enhance Cake Filtration Conditions

5.1 Introduction

The slurry composition, the operational parameters, and the design of the filter apparatus are sensitively influencing the cake formation conditions. The smaller the particles, the more heterogeneous and the less concentrated the slurry becomes, the more difficult the filtration will be, and the more filtration problems are originating. Smaller particles result in smaller filter cake pores and thus lower cake permeability and higher capillary pressure. As consequences, the solid throughput decreases and the cake residual moisture content rises up. If the slurry contains a very broad particle size distribution, the filter process can be disturbed in different ways. The danger of particle segregation in the slurry and thus inhomogeneous filter cake structure increases. In addition, very small particles could blind the filter media or very large particles could cause sediment formation in a filter trough, which no longer can be successfully dispersed by a stirrer. In the case of drum or disc filters, a rising sediment in the filter trough can block the rotating filter drum or disc. If the slurry concentration is low, the danger of particle segregation increases because of single-particle sedimentation, delayed particle bridge formation across the filter media pores can lead to increased filtrate turbidity, or the liquid cannot be removed fast enough to maintain the filter operation. For an example, a pusher centrifuge with very short residence times of the product in the process room would fail in such a case and not separated slurry would be discharged (flooding). To avoid the described deficiencies and not to accept disadvantageous slurry properties, several slurry pretreatment measures are available to enhance the cake filtration conditions, which should be discussed briefly in this chapter.

A special case of slurry pretreatment is given, if molecules, which are dissolved in the liquid or adhering to the particle surfaces, should be separated from the particles. Normally, the filter cake after formation can be washed more or less comfortable directly on the cake filter apparatus by permeation with wash liquid, but in some cases, it makes sense to release the cake filter apparatus from this operation and to use its full capacity for cake formation and deliquoring. In such cases, special "washing machines" remove dissolved substances before the purified slurry is separated afterward. Although this procedure can be defined as a pretreatment step, it will be discussed in more detail later on together with particle washing in Chapter 7, which deals with particle purification in general.

5.2 Thickening

Slurry thickening or concentration, respectively, before a cake filtration process can be helpful under several circumstances. First of all, an economical slurry prethickening generally relieves the subsequent separation step. Pressure filters or filter centrifuges can be designed accordingly smaller and cheaper because of the reduced amount of liquid, which has to be removed.

As an example, in Figure 5.1, a milling circuit with the following concentration of the diluted hydrocyclone overflow and separation of the concentrate by using a vacuum disc filter is illustrated.

The task for the hydrocyclone is to fractionate the feed stream for the mill in particles, which are still too large and have to be given back to the mill and particles, which are small enough to be separated from the liquid in the next process step. The characteristic of a hydrocyclone is the discharge of the main part of the liquid, including the fine particle fraction, via the overflow nozzle and thus the production of quite diluted slurry. A previous thickening of this diluted slurry supports the function of the disc filter because the danger of particle segregation and increased filtrate pollution is reduced. Secondly, the necessary filter area and thus the filter size can be reduced because less slurry has to be separated after a preconcentration.

A thickening of diluted slurry with a certain particle size distribution changes the sedimentation behavior of the particles from disadvantageous single particle to advantageous swarm sedimentation, if a critical concentration is exceeded.

In diluted slurries, a segregation of particles according to their size takes place as can be seen in Figure 5.2.

Filter cakes with a layer of the smallest particles on top are the consequence. The particles are settling independently from each other and their velocity w_{St} depends very sensitively on their size, as can be seen from the Stokes law of particle sedimentation for single spherical particles, laminar flow, and Newtonian

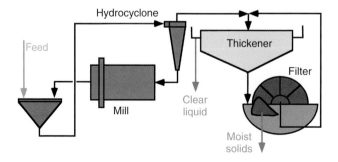

Figure 5.1 Hydrocyclone, static thickener, and vacuum disc filter.

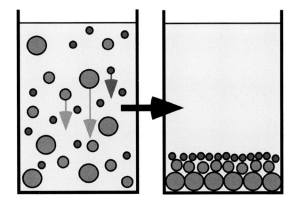

Figure 5.2 Single-particle sedimentation and segregation.

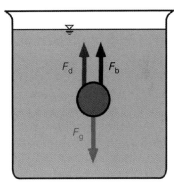

Figure 5.3 Force balance around a settling particle.

fluids [1] in Eq. (5.1)

$$w_{St} = \frac{(\rho_s - \rho_L) \cdot g \cdot x^2}{18 \cdot \eta_L} \tag{5.1}$$

The Stokes law represents the basic equation for laminar sedimentation and should be derived shortly. Basis is a force balance around a single particle as sketched in Figure 5.3.

The force of gravity F_g (Eq. (5.2)) is equal to the sum of buoyancy force F_b (Eq. (5.3)) and drag force F_d (Eq. (5.4))

$$F_g = m_s \cdot g = \rho_s \cdot V_s \cdot g \tag{5.2}$$

$$F_b = V_s \cdot \rho_L \cdot g \tag{5.3}$$

$$F_d = \frac{\rho_L}{2} \cdot w^2 \cdot A \cdot c_w(Re) \tag{5.4}$$

The drag force depends on the projection area A of the particle and a drag coefficient c_w, which depends on the Re-number (cf. Section 4.2). In Figure 5.4, the drag coefficient is plotted against the Re-number for a spherical particle.

In the Stokes range, the drag coefficient is, according to Eq. (5.5), inversely proportional to the Re number

$$c_w = \frac{24}{Re} = 24 \cdot \frac{\eta_L}{w \cdot x \cdot \rho_L} \tag{5.5}$$

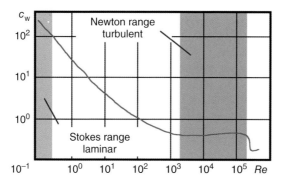

Figure 5.4 Drag coefficient c_w as a function of Re-number.

Now, in Eq. (5.6), the drag force can be formulated for a spherical particle and from the entire force balance the settling velocity (cf. Eq. (5.1)) can be calculated

$$F_g = F_b + F_d \Rightarrow V_s \cdot \rho_s \cdot g = V_s \cdot \rho_L \cdot g + 3 \cdot \pi \cdot x \cdot \eta_L \cdot w \tag{5.6}$$

In the case that intensive stirring of the slurry during cake formation is possible, as for stirred pressure nutsche filters and vacuum drum or disc filters, segregation of the particles because of sedimentation can be suppressed or at least reduced. If that is not possible, as for discontinuously operating peeler and inverting filter centrifuges or continuously operating vacuum belt and pan filters, a segregation of the particles can be rather avoided. Especially in the case of centrifuges, a clear liquid zone is formed on top of the filter cake, and this liquid must penetrate the fine particle layer of high flow resistance. Thus, the cake formation time will be increased remarkably. This problem is particularly relevant for centrifuges because the sedimentation velocity is here, by the centrifugal factor C, greater than in the earth field. The formation of a fine particle layer on top of a filter cake is also very unfavorable for the deliquoring conditions because of its high capillary entry pressure. This affects filter centrifuges as well as vacuum and pressure filters.

If the particle concentration in the slurry is increased up to a critical value by a thickening process, the particles are not touching but hindering each other during sedimentation, and the single-particle sedimentation changes to swarm or zone settling as depicted in Figure 5.5.

If a part of the liquid is separated from diluted slurry and a critical value of the solid concentration is exceeded, the particles are forming a sharp front of settling particles and segregation effects are not completely suppressed, but drastically

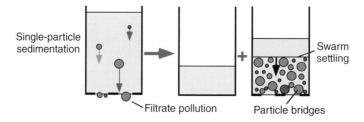

Figure 5.5 Change from single to swarm settling by concentration.

reduced because of the increased particle interaction. If pure zone settling is reached, no more segregation of particles can be observed. The critical concentration to get a sharp settling-front depends on the particle size, size distribution, Froud Fr number, or C value as a ratio of centrifugal and gravitational acceleration and the stability of the slurry.

According to the findings of Richardson and Zaki, the settling velocity w_{sf} of the particle front in Eq. (5.7) is mainly depending on the slurry concentration [2, 3]

$$w_{sf} = w_{St} \cdot (1 - c_v)^m \tag{5.7}$$

The exponent m is a function of the Re number and exhibits the value 4.65 for laminar flow.

If flocs are involved in the case of slurry destabilization, the critical solid concentration for forming a sharp settling front decreases. Equation (5.7) has to be modified according to Eq. (5.8) to express the settling velocity w_{sf}^* of the flocculated particle system

$$w_{sf}^* = w_{St}^* \cdot (1 - k \cdot c_v)^m \tag{5.8}$$

Not the particle but the floc concentration is relevant now. This floc concentration is incorporated by a floc-factor k.

The parameters w_{St} and w_{St}^*, respectively, must be determined by experiments because the settling velocity for a single particle or floc, which represents the entire particle size distribution of the slurry, is not known. To get the necessary data, several settling experiments with different solid volume concentrations have to be carried out, and the settling velocity of the particle front has to be measured. To evaluate the data, Eq. (5.7) for laminar flow has to be rewritten in the form of Eq. (5.9) that a linear function results

$$w_{sf}^{1/4.65} = w_{St}^{\frac{1}{4.65}} - w_{St}^{\frac{1}{4.65}} \cdot c_v \tag{5.9}$$

Now, the experimental results can be plotted according to Eq. (5.9) and are shown qualitatively in Figure 5.6.

As long as the settling of the particle front follows the law of Richardson and Zaki, a declining straight line can be observed. When the free settling ends and the gel point is reached, where all particles now have direct contact with each other, a further movement of the front downward follows the completely different kinetics of sediment compression, and the measured data are deviating from the straight line.

The characteristic parameter w_{St} can be determined by extrapolation of the straight line to the ordinate. Here, the slurry concentration approaches zero, which means sedimentation of single particles of a characteristic size. The floc factor k can be determined by extrapolation of the straight line to the abscissa.

Preconcentration of slurries leads, in principal, to the described positive effects regarding cake filtration, but one should be aware of the problem that in the case of continuously operating cake filters, an increase of slurry concentration would lead to an increase of the residual cake moisture content. An increase of slurry concentration would lead to a thicker cake, which would need more time

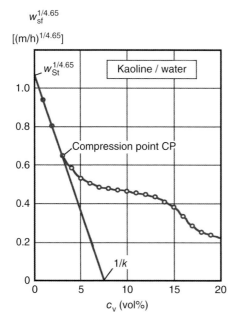

Figure 5.6 Representation of settling data to characterize the swarm setting behavior.

to reach the same cake moisture as the thinner cake before, but the deliquoring time remains constant. These effects will be discussed more in detail in Section 8.3.4.6 for rotary filters.

Slurry thickening can be realized by various sedimentation processes such as static circular or longitudinal settling basins, lamella clarifiers, hydrocyclones, decanter centrifuges, or disc stack separators, filtration processes such as cross-flow membrane filters, fully submerged discontinuous leaf, and candle filters or froth flotation processes.

5.3 Agglomeration

Very often, thickening by sedimentation is combined with agglomeration because it makes the process much more effective. To get clear liquid overflow, the thickener must be calculated to separate the smallest particles. In the case of a circular static thickener, the sedimentation velocity of the particles must be greater than the upward-directed velocity of the liquid. For a constant feed rate, the necessary liquid flow velocity for small particles leads to accordingly large clarification areas and thus large dimensions of the thickeners. If the single small particles are bound into agglomerates, their settling velocity increases and the necessary clarification area decreases. In addition, the sedimentation behavior can change from single particle to swarm sedimentation, and a sharp settling front of the particles makes it more comfortable to operate the continuously operating apparatus. The clear liquid zone, the swarm settling zone, and the compression zone are clearly separated from each other during steady-state operation and can be easily controlled regarding their position.

Figure 5.7 Avoiding particle segregation and filtrate pollution by agglomeration.

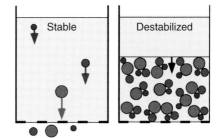

For the cake filtration process itself, agglomeration of the particles results in several advantageous effects. First of all, segregation effects in diluted slurry can be avoided by agglomeration. According to Figure 5.7, normally the whole particle size distribution is included in the agglomerates and thus no segregation can take place.

Additionally, the "prebuilt bridges" in the form of agglomerates reduce the filtrate pollution in the first moment of particle bridge formation.

Besides a homogeneous cake structure without particle segregation, agglomerates form, according to Figure 5.8, a much more porous cake with larger pore diameters.

Precondition is of course stability of the agglomerates against the pressure they are faced within the cake structure during filtration. There are special treatment procedures especially for the agglomeration on the basis of polymers available to modify the floc properties by well-defined shear stress [4]. In the case of a successful agglomeration, the permeability of the more open pore structure of the filter cake increases and thus the specific throughput of the apparatus rises up. In extreme cases of very poor filtration properties of the slurry, a cake filtration could be made economical only by agglomeration of the particles.

Agglomeration also facilitates or enables the desaturation of a filter cake. The capillary pressure distribution and especially the capillary entry pressure can be shifted to lower values. If in the case of a vacuum filter the physically limited pressure difference is not able to desaturate the filter cake from a stable slurry at all and fully saturated mud would be discharged from the filter apparatus, an agglomeration of the particles can lead to a remarkable desaturation of the cake and a change from sticky to brittle behavior.

Particle agglomeration in slurries can be realized by coagulation or flocculation. In the case of coagulation, van der Waals forces are responsible for the adhesion between very small particles in the micrometer range. As can be seen from

Figure 5.8 Structure of filter cakes without and with particle agglomeration.

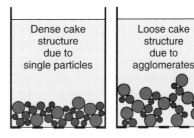

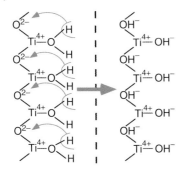

Figure 5.9 Formation of hydroxyl groups on a TiO_2 surface.

Eq. (5.10), the van der Waals force F_{vdW} is proportional to the particle diameter x, the material characterizing Hamaker constant H, and inverse proportional to the square of the distance a between the particles

$$F_{vdW} \propto \frac{x}{a^2} \cdot H \tag{5.10}$$

On the other hand, according to Eq. (5.11), the particle weight is proportional to the third power of x

$$F_g \propto x^3 \tag{5.11}$$

This means that the weight is increasing much faster with increasing particle diameter than the van der Waals force, and particles become unable to stick together, if the weight is greater than the van der Waals force. The situation becomes much more unfavorable for agglomeration if particles in a slurry are carrying electric charges, which is normally the case. There do exist different reasons for electric charges on a particle surface, such as defects in crystal lattices by substituted atoms, ion adsorption, dissociation, or formation of hydroxyl groups. As an example in Figure 5.9, the hydroxyl group formation on titanium dioxide surfaces is demonstrated.

In a first step, polar water molecules are binding with the negatively charged oxygen atom at the positively charged Ti atoms. In second step, a positively charged hydrogen atom leaves the H_2O molecule and binds at a negatively charged oxygen atom of the TiO_2 matrix with the result of an overall negatively charged surface.

Because the electrically charged particles show the same polarity, they are repelling each other. If these repelling forces are strong enough, the particles are not able to come near enough together for adhesion despite the kinetic energy they may exhibit. If ions are present in the slurry, as depicted in Figure 5.10, the surface charge can be compensated by countercharged ions, and thus, the repelling forces are reduced.

In the case of a negatively charged particle surface with a surface potential ψ_o, small negative and bigger positive ions are bound by van der Waals and electrostatic forces very strong and immobile at the surface. This layer is called Stern layer and consists of the inner and outer Helmholtz layer. The remaining and thus relevant electrostatic potential ψ_s is called zeta potential and can be measured by different methods, which all are based on the possibility to shear off the diffuse

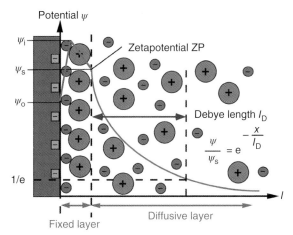

Figure 5.10 Electric repulsion potential of particles in a slurry.

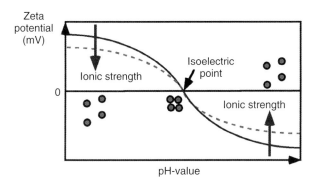

Figure 5.11 Influence of the pH value on the zeta potential.

ion double layer around the particles [5–7]. One possibility to directly influence the surface charge and thus the zeta potential are changes of the pH value in the slurry, as can be seen in Figure 5.11.

For a certain pH value, the zeta potential becomes zero at the so-called "isoelectric point." This means that no electrostatic repulsion takes place, and maximal agglomeration probability by van der Waals forces exists. Electrostatic repulsion and van der Waals adhesion are competing against each other and can be described by the energy balance according to the DLVO theory (Derjaguin, Landau, Verwey, and Overbeek) [8, 9], which is represented in Figure 5.12.

The electrostatic repulsion energy at the surface of the particle ($l = 0$) is given by ψ_s and is approaching asymptotically zero at an infinite distance from the surface ($l \to \infty$). At the isoelectric point, the electrostatic repulsion becomes completely zero. The van der Waals attraction exhibits a specific value at the minimal contact distance of 4 Å to other particles, which depends on the material of the particle, the medium between the particles, and the particle geometry. The van der Waals attraction is a result of electron interaction between particles and thus approaches to zero already after very short distances from the particle surface.

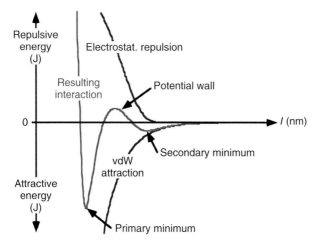

Figure 5.12 Particle/particle interaction energies.

The resulting interaction decides about adhesion or repulsion. The "energy wall" can be overcome by kinetic energy of the particles, which they get from Brownian motion, settling, stirring, or in extreme cases ultrasonic sound.

If the energy wall is too strong, electrostatic repulsion can be reduced by reduction of the Debye length l_D [10] in Eq. (5.12)

$$l_D = \sqrt{\frac{\varepsilon_r \cdot \varepsilon_0 \cdot k \cdot T}{e^2 \cdot N_A \cdot \sum c_i \cdot z_i^2}} \tag{5.12}$$

In this equation, ε_r means relative dielectric constant of slurry (–), ε_0 means absolute dielectric constant (= 8.8542×10^{-12} A s V^{-1} m^{-1}), k means Boltzmann constant (= 1.3806×10^{-23} J K^{-1}), T means absolute temperature (K), e means elementary charge (= 1.6022×10^{-19} C), N_A means Avogadro number (1 mol = 6.0222×10^{23} molecules), c means concentration of ions, and z means ionic valence.

The Debye length is defined, according to Eq. (5.13), as a distance from the particle surface, if the electric repulsion potential is decreased by the factor $1/e$ (cf. Figure 5.9)

$$\frac{\psi}{\psi_s} = e^{-\frac{x}{l_D}} \tag{5.13}$$

An increase of ion concentration c and an increase of ionic strength z reduce the Debye length l_D and facilitates coagulation. Often $FeCl_3$ or $AlCl_3$ is added to the slurry for such purposes.

An alternative or additional measure is the use of polymeric flocculents, which are adhering to the particle surfaces and lead to agglomeration by bridging, as can be seen in Figure 5.13.

In the most simple case, natural polysaccharides, such as starch, can be used, but normally much more effective synthetic polymers are applied. These water-soluble macromolecules are normally based on amides of acrylic acid with a polymerization grade of 40.000–200.000. For optimization of their

Figure 5.13 Bridging agglomeration by polymeric flocculants.

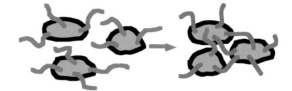

effectiveness, one can add cationic or anionic reactive groups. Depending on the chemical structure, nonionic, cationic, or anionic flocculents are available [11–13]. The choice of the most effective flocculent for a specific particle system needs experience of the supplier and the user with similar applications and filter experiments. After identification of a suitable flocculent, the design of an optimal flocculation process is essential for a successful operation. At first, the flocculent solution has to be prepared for activating the polymeric molecules. This means that the molecules at the beginning exhibit a ball structure and must be stretched. In a second step, the flocculent solution must be mixed under highly turbulent conditions with the slurry to provide each particle with polymeric molecules and to avoid inhomogeneous distribution of the polymers in the slurry. In that case, some particles are overload with polymers, which leads to static repulsion. Other particles get no polymers at all. In the following step, the flocs must form in a flow regime, which is calmed down. Especially for improved deliquoring conditions in cake filtration applications, the flocs must be stable enough to withstand the solid pressure in the filter cake structure of pressure filters or the shear stress in filter centrifuges. One effective measure for this purpose is a cylindrical stirrer in a circular housing, which are forming a coaxial gap (Couette-flow). If the flocculated slurry is pumped slowly from the bottom to the top of this apparatus, flocs are formed and condensed in a very well-defined shear field, which can be adjusted by the rotation speed of the stirrer. Higher speed means greater shear forces, and smaller and more stable flocs. After flocculation, the slurry should be handled carefully. For example, no centrifugal pumps should be used, but volumetric pumps, such as eccentric screw pumps, diaphragm pumps, or peristaltic pumps, can be used. In addition, the homogenization of the slurry in the feed tank of the filter should be carried out in a gentle way. This means that the energy dissipation in the vessel should be as homogeneous as possible, and instead of a propeller mixer in the turbulent flow regime, a better multistage stirrer at lower Re-numbers and less local shear stress should be used.

5.4 Fractionation/Classification/Sorting

If thickening to avoid segregation or agglomeration to increase pore sizes is not the method of choice, fractionation could be an alternative to improve the filtration conditions.

Particle fractionating of feed slurry can make sense by different reasons. Oversized particles may cause nozzle blockage, segregation, sediment formation in tanks, or wear out of apparatuses by abrasion. Separating coarse particles from the slurry is called degritting.

On the other hand, the fine particles of a particle size distribution can cause problems. As discussed before, a certain portion of fine particles in slurry leads to smaller pores in the cake and thus lowers permeability and higher capillary pressures. In the worst case, fine particles are forming a dense layer on top of a filter cake. Separating the fine fraction from slurry is called desliming.

The improvement for the filtration conditions by elimination of the fine fraction is obviously related to the question, what should happen with the fine fraction. The simplest case is given if a very small amount of solids (mass related) is allowed to handle as waste after separation. Although the finest particles may represent very less mass, they will represent a great number and thus have a significant potential for pore blockage.

If the coarse and the fine fraction should be separated both as a product, a fractionation only makes sense, if the overall efficiency of the process is increased, the average moisture of the separately deliquored fractions is lower, or other reasons make a fractionation necessary.

An example may be the separation of aluminum hydroxide in the bauxite beneficiation process. After crystallization, a very broad particle size distribution from some micrometer up to 1 mm is present in the aluminum hydroxide slurry. This is not beneficially separable with one single type of filter apparatus. In this special case, the slurry is fractionated by hydrocyclones. The coarse fraction of high concentration from the cyclones underflow is given directly to a horizontal vacuum pan filter, where the fast settling coarse particles are settling toward the filter medium and no stirring problem exists. Because of the great settling velocity of the particles, the slurry must be distributed evenly on the horizontal disc to enable a constant cake height. The diluted fine fraction is first concentrated by a static settler and then separated by a vacuum disc filter. Figure 5.14 shows the principal flow sheet of such a process.

Fractionation, thickening, and agglomeration are combined in this process. Agglomeration of the finest particles by coagulation takes place automatically here because the liquid contains a maximum of ions, which means minimal Debye length and maximal agglomeration probability. The coarse particle fraction can be perfectly separated on a horizontal pan filter because the particles are settling very fast and could not be homogenized properly in the suspension trough of the disc filter. In addition, the coarse particles lead to very thick filter cakes in the decimeter range because of the resulting high cake permeability. These thick filter cakes would be in danger not to be held on the vertical filter

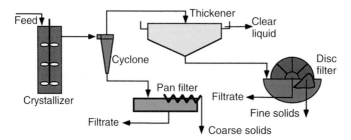

Figure 5.14 Particle fractionation for optimal separation and deliquoring.

area of a disc filter, but to fall backward into the slurry. On the other hand, the fine fraction would decrease the filter performance of the pan filter and a disc filter is much more effective in this case.

Slurry fractionation and removal of fine particles can be carried out by several methods. Hydrocyclones, bended screens, upstream classifiers, spiral classifiers, tube centrifuges, or overloaded apparatuses, such as circular settling basins and decanter centrifuges, can be applied.

Sometimes, fine particles of material, different from the solid product, are hindering the cake deliquoring process. Such an example is given for the separation of starch by peeler centrifuges. Very small colloidal proteins, present in the slurry, would block pores in the filter cake with the consequence of reduced throughput and increased residual cake moisture. The comparatively coarse starch particles are settling very fast in the centrifugal field and are forming a sediment layer, whereby a high amount of the proteins remains in the liquid phase above the sediment and near to the liquid surface. This liquid layer, together with a part of the disturbing finest particles, can be skimmed or decanted by a special skimming pipe, which is immersed into the liquid surface. This procedure is a fractionation and a sorting process. Additionally, this measure accelerates the filter cycle because the skimmed liquid must not go the conventional way and drain through the already formed filter cake.

The situation becomes more difficult if solid particles of different materials and similar particle size distribution are present in the slurry and should be separated independently from each other. Neither a screening process nor sedimentation would be successful. Although the particles may have different density, they cannot be separated by sedimentation. The small and heavy particles would settle down with the same velocity, as the great and light particles. In such cases, other particle properties must be used to separate the different materials.

If the wettability of the particles is differing, froth flotation could be the solution of the problem [14, 15]. Flotation in general means that particles are moving contrary to the gravity toward the surface of the slurry. Flotation is used in particular in the ore, mineral, and coal beneficiation to separate fine-grained particle mixtures of liberated minerals and tailings. In the field of paper recycling, flotation is used for deinking of the pulp. However, flotation is also applied as energetic favorable and low-cost thickening process for total separation of organic particles, which have only small difference in specific weight to the surrounding liquid. This is used not only in the sewage water treatment but also to preconcentrate biological particles, such as microalgae.

If the particles exhibit a greater density than the surrounding liquid phase, they nevertheless can rise up against gravity in a flotation cell as froth, if they adhere to gas bubbles. Figure 5.15 illustrates the general principle of this process.

In order to enable the attachment of gas bubbles to the particle surfaces, these must be hydrophobic. If they are not hydrophobic by nature, they can be made hydrophobic by adding special surfactants, which are called collectors. Surfactants are boundary surface-active molecules, which consist of a hydrophobic nonpolar hydrocarbon chain and a hydrophilic polar group. According to the kind of polarity, one differentiates between anionic, cationic, and nonionic surfactants. Anionic collectors are among other xanthanates, carboxylates, alkyl

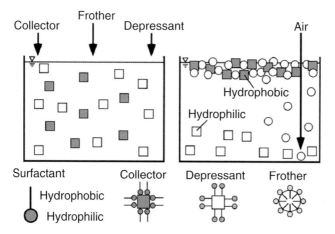

Figure 5.15 Principle of the froth flotation process.

sulfates, or mercaptanes. Cationic collectors are alkyl amines. The particles, which should not be discharged with the froth, must show hydrophilic surface properties and are settling in the direction of gravity to the bottom of the flotation cell. If they are not hydrophilic by nature, they can be made hydrophilic by special surfactants, which are called depressants.

As depressants, alkali cyanide, lime hydrate, zinc sulfate, or water glass are used among other substances. The froth itself can be stabilized against coalescence by frothers. Polypropylene glycol or aliphatic alcohols are used as frothing agents among other substances. The optimally floatable particle diameters are in the range between 40 and 150 µm for solid-specific weights of more than 3000 kg m^{-3}. Upward the flotation process is limited by the particle weight and downward by the reducing selectivity. Because of the ascending particle-loaded bubbles, three-phase froths are originating in a flotation process. Unfortunately, not only the desired particles of the product are adhering at the bubbles but also a small amount of finest hydrophilic particles. This phenomenon is denominated heteroagglomeration and limits the selectivity of the process. Generally, several process steps, such as basic flotation, secondary flotation, and purification flotation are necessary to produce the final solid concentrate. From the viewpoint of apparatuses, one differentiates between the mechanical and the pneumatic procedures.

With the mechanical flotation apparatuses, the necessary energy for mixing and dispersing is supplied, according to Figure 5.16, by means of rotor/stator systems to the aerated suspension.

The air is sucked into the liquid through the hollow shaft of the mixer itself or is supplied from outside by means of a nozzle, then dispersed in the rotor/stator system into small bubbles, and mixed with the suspension under highly turbulent flow conditions. After turbulent dispersion and mixing, the froth must have the possibility to rise up and to pick up particles under calm conditions.

With the pneumatic flotation apparatuses, no rotor/stator system exists. The required air is supplied here in every case from the outside under pressure and

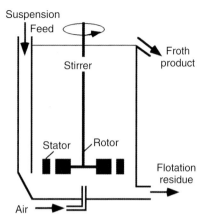

Figure 5.16 Mechanical flotation.

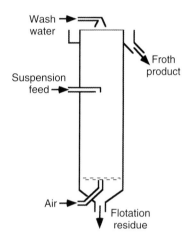

Figure 5.17 Pneumatic flotation.

dispersed into bubbles by means of a suitable aeration system. In Figure 5.17, such a flotation column is shown.

The height of flotation columns can vary from 5 to 15 m with a diameter of up to 4 m. Here, gas and suspension are fed in countercurrent direction. This leads to an improved selectivity compared with the mechanical flotation systems, which show a cross-flow arrangement.

With the modern process of dissolved air flotation (DAF) [16], the pressure dependence of the gas solubility in water is utilized for the bubble generation. The water is saturated under high pressure with air and afterward released to the atmospheric pressure. In comparison to mechanical or pneumatic flotation processes such as mentioned before, here, particularly small bubbles from approximately 50 to 80 μm diameter can be produced. The bubbles originate partly direct at the particle surfaces, where the nucleus formation work is lowest.

A rather special application in the area of water purification is represented by electroflotation [17, 18], with which the gas bubble formation takes place via electrolytic decomposition of the water. The oxygen bubbles, used for flotation, are reducing simultaneously organic impurities in the water by oxidation. The hydrogen bubbles float cationic materials very well. With the choice of a suitable electrode, material ions can become free, which are additionally supporting the flocculation of the particles to be separated.

5.5 Filter Aids – Body Feed Filtration

In several cases, it is advantageous to go the opposite way than discussed before and not to remove the finest particles from the slurry to be separated but to add coarser material as shown in Figure 5.18.

If the filter aid material is directly mixed with the slurry to be separated, the process is called body feed filtration. Although the focus often is more on the filterability of slurry, than on the cake deliquoring, better permeability of course also makes deliquoring easier. The function of the added coarse material is to form a kind of matrix, in which the fine particles are encased, as in a depth filter and drainage channels are implemented for easy displacement of the filtrate. As filter aid materials, various possibilities are given. One can distinguish between mineral and organic filter aids. Well-known examples for mineral filter aids are diatomaceous earth [19], as can be seen in Figure 5.19, perlite, or others.

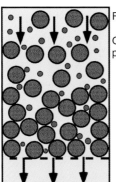

Figure 5.18 Cake filtration with added filter aids.

Figure 5.19 Diatomaceous earth.

Figure 5.20 Fibrilized cellulose fibers. Source: Courtesy of JRS-J. Rettenmaier & Söhne GmbH & Co. KG.

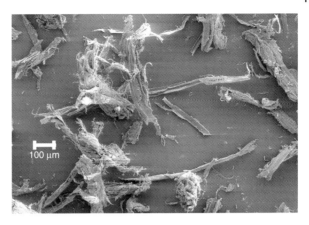

These classical materials are still often in use and exhibit some benefits, but also deficiencies in comparison to the fast upcoming organic filter aid materials are most often based on cellulose.

Mineral filter aids are forming more incompressible filter cakes which are available in reasonable costs and highly inert against elution of ions. Especially, diatomaceous earth has a very large specific surface, which means high dirt holding capacity. Deficiencies are a certain health risk, if fine particles are inhaled, relatively high weight, 100% ash content, and abrasive properties.

In comparison to mineral materials, cellulose-based organic materials have some interesting advantages. Figure 5.20 shows as an example fibrilized cellulose fibers.

Fibrilizing means that the fiber undergoes a pressing and shearing procedure, which leads to high specific surface. The fibers are fanned out by this procedure. To fan out means that the fiber undergoes a pressing and shearing procedure, which leads to high specific surface.

Cellulose is a product from renewable resources, exhibit little weight, shows no health risk, contains nearly now ash in case of combustion, but provides extra energy, is nonabrasive, and can be offered in different qualities as wood flower in the most simple and cheap case, as extract-free cellulose or ultrapure pharmaceutical quality. The waste disposal is variable from combustion, dumping, and composting to animal feeding. Although the weight-related price for the organic material itself might be often slightly higher than for mineral products, the overall costs for the process operation are very often significantly lower in comparison with usage of mineral products. Less mass is needed in comparison to minerals, less transport costs are generated for the waste, much lower disposal costs thereby are incurring for ash disposal after combustion, and less maintenance costs are developing for the machinery because of low abrasion. Depending on the fiber length, the filter cakes are more compressible than for mineral products, which mean that cellulose cakes often can be deliquored properly by squeezing and or desaturation. The compressibility depends on the fiber length. Very short fibers with thickness to length ratio of nearly 1 exhibit the character of particles, which are forming more incompressible cakes. The longer the fiber, the more compressible the cake becomes.

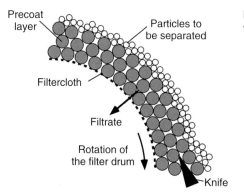

Figure 5.21 Precoat filtration on a vacuum drum filter.

In many applications, mineral-based filter aids such as diatomaceous earth are combined with cellulose to get the positive effects of both materials [20]. For example, a first layer of cellulose fibers can protect the filter medium against blockage by the following mineral particles. If a filter medium is damaged or filter media with greater meshes should be applied, cellulose fibers as first layer hinder particles from getting into the filtrate or even enable a body feed or precoat filtration, based on mineral filter aids.

Filter aid materials can not only improve the mechanical separation conditions for slurries but also show additional functionalities. Activated carbon, for example, or bentonite exhibits excellent adsorption properties for dissolved molecules or special resins can be used for ion exchange. Today, cellulose can also be functionalized to act additionally as adsorbent and is applied, for example, in combination with bentonite to refine edible oils.

Alternatively to the body feed filtration, filter aid materials can also be used as a precoat for the filtration of highly diluted slurries of extremely small particles, which in addition may contain slimy and/or sticky compounds. Standard cake filtration would fail in those cases because of high filtrate pollution, filter media blinding, and very thin filter cakes. Body feed filtration cannot hinder finest particles safely to get in contact with the filter medium or to get into the filtrate. In such cases, in a first step, a precoat layer (filter cake) of filter aid material is filtered to protect the filter media, and afterward the slurry to be separated is filtered. Depending on the process design, the separation can be more a depth filtration or a cake filtration. The principle of precoat filtration on a vacuum drum filter, as a typical cake filter, is shown schematically in Figure 5.21.

The particles to be separated are blinding the surface of the precoat layer and are forming a very thin cake. This thin cake, including the topmost particle layers of the precoat, is cut off by a sharp knife and discharged. Fresh and permeable precoat surface after the knife is immersing again into the filter trough for the next filtration cycle. To maintain this kind of operation, the knife is advancing, according to filters rotation speed, until a minimum thickness of the precoat layer has reached, and the precoat must be renewed. The thickness of such precoat layers is about 10 cm, and the advancement of the knife can be less than 100 μm per cycle for modern filters, which enables a continuous filtration for many hours.

5.6 Thermal Conditioning

As mentioned in Section 2.2, the liquid viscosity and the surface tension are depending on the temperature. Both parameters are reduced for greater temperatures, which increases the filtration velocity and reduces capillary forces. Except extreme cases, as very viscous oil or similar materials, it normally makes no sense to heat up slurries for better filtration conditions because this would need too much energy to be economical. On the other hand, it makes sense not to cool down slurry before filtration, if it is delivered at higher temperatures. In such cases, lower viscosity to increase the filtration velocity and to improve the cake deliquoring can be utilized without additional energy demand. More effective in comparison to the slurry heating as a pretreatment measure is to bring the cake in the deliquoring zone of the filter in contact with hot steam as a posttreatment measure. In this case, greater temperatures increase the cake deliquoring kinetics by reducing the liquid viscosity and the cake moisture content by reducing the capillary pressure. In addition, the generated condensate acts as an ideal washing liquid to purify the cake (cf. Sections 7.6 and 8.3.5).

5.7 Chemical Conditioning

Examples for chemical conditioning of slurries to promote agglomeration of the particles have been discussed already in Section 5.3.

One other chemical slurry treatment measure consist in the dosage of surfactants to decrease the surface tension of the liquid and thus to facilitate the cake desaturation [21]. However, it makes less sense to mix slurries with surfactants before cake formation as a pretreatment measure because a major part of the surfactants would get lost with the filtrate. Unfortunately, the surfactant molecules are not only staying at the boundary surface between liquid and gas but are also dissolving in the liquid and adsorbing at the particle surfaces. An adsorption equilibrium originates, which depends on the substances, the concentrations, and the temperature (Langmuir isotherm [22]). Surfactants would be eventually needed in the case of high capillary pressure, which means very small particles, but very small particles exhibit unfortunately a very high specific surface. As a consequence, the smaller the particles become, the more surfactant is needed and the more surfactant is not used for decreasing the liquid/gas surface tension but is adsorbed at the particles surfaces. This makes the process not really efficient. In addition, surfactants are costly, in many cases not biodegradable and often producing process disturbing foam. If surfactants should be applied anyhow, it would be more efficient to spray the surfactants as a posttreatment measure on the surface of the filter cake after emersion from the slurry and before desaturation. As a conclusion, it can be stated that surfactants are not a standard pretreatment or posttreatment measure to improve the deliquoring behavior of filter cakes but more an ultimate ratio, if no other possibility is given.

A third case for chemical treatment is given for liquid-filled biological cells, which should be separated as dry as possible. Unfortunately, the liquid inside

Figure 5.22 Vacuum precoat drum filter with wash pipes for yeast separation. Source: Courtesy of BOKELA GmbH.

of the cells cannot be displaced by mechanical means without cell destruction. If the cells should remain intact, they can be made smaller by osmotic effects. This can be realized by dissolving, for example, sodium chloride in the slurry to increase the ion concentration. Because of the osmotic pressure, liquid will permeate the cell membrane to dilute the surrounding liquid and the cell will shrink and increase the amount of dry substance per cell. A continuous operating vacuum precoat drum filter can be used to filter such slurry. Figure 5.22 gives an example for such a process for the separation of baker's yeast.

The capillary entry pressure of the separated yeast layer on the precoat surface is not overcome in any case by vacuum and so the slight increase of capillary pressure because of the shrunk particles does not matter. Just before cake discharge, the salt has to be washed out quickly to have it not in the final product. The wash bars and wash nozzles on the drum filter in Figure 5.22 can be seen clearly. During the short washing time through the thin yeast layer, the particles have not the time to expand again.

References

1 Stokes, G.G. (1845). On the theories of internal friction of fluids in motion. *Transactions of the Cambridge Philosophical Society* 8: 287–305.
2 Richardson, J. and Zaki, W. (1954). Sedimentation and fluidisation: Part I. *Transactions of the Institution of Chemical Engineers* 32 (1): 35–53.
3 Davies, R. (1968). Experimental study of the differential settling of particles in suspension at high concentrations. *Powder Technology* 2: 43–51.
4 Kleine, U. (1992). Der Einfluß der Flockenbildungsbeanspruchung auf die Festigkeit und das Sedimentationsverhalten von Flocken bei der Zentrifugalabscheidung, Dissertation Universität Karlsruhe (TH), Karlsruhe
5 Hunter, R.J. (1981). *Zeta Potential in Colloid Science: Principles and Applications*. London: Academic Press. ISBN: 0-12-361960-2.
6 Kirby, B.J. and Hesselbrink, E.F. Jr., (2004). Zeta potential of microfluidic substrates: 1. Theory, experimental techniques and effects on separations. *Electrophoresis* 25: 187–202. https://doi.org/10.1002/elps.200305754.

7 Hinze, F., Ripperger, S., and Stintz, M. (1999). Praxisrelevante Zetapotentialmessung mit unterschiedlichen Messtechniken. *Chemie Ingenieur Technik* 71 (4): 338–347.
8 Derjaguin, B.V. and Landau, L.D. (1993). Theory of the stability of strongly charged lyophobic sols and of the adhesion of strongly charged particles in solutions of electrolytes. *Progress in Surface Science* 43 (1–4): 30–59. https://doi.org/10.1016/0079-6816(93)90013.
9 Verwey, E.J.W. and Overbeek, J.T.G. (1948). *Theory of the Stability of Lyophobic Colloids; the Interaction of Sol Particles Having an Electric Double Layer*. New York: Elsevier.
10 Debye, P. and Hückel, E. (1923). The theory of electrolytes. I. Lowering of freezing point and related phenomena. *Physikalische Zeitschrift* 24: 185–206.
11 Pelssers, E.G.M., Cohen Stuart, M.A., and Fleer, G.J. (1989). Kinetic aspects of polymer bridging: equilibrium flocculation and nonequilibrium flocculation. *Colloids and Surfaces* 38: 15–25.
12 Gregory, J. (1988). Polymer adsorption and flocculation in sheared suspensions. *Colloids and Surfaces* 31: 231–253.
13 Fleer, G.J. and Scheutjens, J.M.H.M. (1993). Modeling polymer adsorption, steric stabilization and flocculation. In: *Coagulation and Flocculation* (ed. B. Dobiás), 209–263. NewYork: Marcel Dekker Inc.
14 Fuerstenau, M.C., Jameson, G., and Yon, R.-H. (2007). *Froth Flotation: A Century of Innovation*. Englewood, Colorado: Society for Mining, Metallurgy & Exploration (SME). ISBN: 978-0-87335-252-9.
15 Rao, S.A. (2004). *Surface Chemistry of Froth Flotation*. Heidelberg: Springer. ISBN: 978-0-306-48180-2.
16 Edzwald, J.K. (1995). Principles and applications of dissolved air flotation. *Water Science and Technology* 31 (3–4): 1–23. https://doi.org/10.1016/0273-1223(95)00200-7.
17 Thakur, S. and Chauhan, M.S. (2016). Treatment of wastewater by electro coagulation: a review. *International Journal of Engineering Science and Innovative Technology (IJESIT)* 5 (3): 104–110, ISSN: 2319–5967, DOI: https://doi.org/10.1007/s11356-013-2208-6.
18 Chen, X., Chen, G., and Yue, P.L. (2002). Novel electrode system for electroflotation of wastewater. *Environmental Science and Technology* 36 (4): 778–783.
19 Wang, L.K. (2006). Diatomaceous earth precoat filtration. In: *Advanced Physicochemical Treatment Processes, Handbook of Environmental Engineering*, vol. 4 (ed. L.K. Wang, Y.T. Hung and N.K. Shammas). Totowa: Humana Press https://doi.org/10.1007/978-1-59745-029-4_5. ISBN: 978-1-58829-361-9.
20 Braun, F., Hildebrand, N., and Wilkinson, S. (2011). Large-scale study on beer filtration with combined filter aid additions to cellulose fibres. *Journal of the Institute of Brewing* 117 (3): 314–328.
21 Stroh, G. (1994). Die Wirkung von Tensiden auf die Entfeuchtung von Filterkuchen durch einen aufgeprägten Gasdruck, Dissertation Universität Karlsruhe (TH), Karlsruhe
22 Langmuir, I. (1918). The adsorption of gases on plan surfaces of glass, mica and platinum. *Journal of the American Chemical Society* 40: 1361–1403.

6

Filter Cake Formation

6.1 Introduction

The entire cake filtration process can be subdivided into several process steps. The first step consists of the slurry preparation, which is often combined with a slurry pretreatment measure to improve the filterability (cf. Chapter 5). The subsequent step consists of the filter cake formation, and afterward, post-treatment steps such as cake washing and/or cake deliquoring, in the form of squeezing (compressible cake) or desaturation (incompressible cake), can be added. The basic principle of cake filtration is shown schematically in Figure 6.1.

Under the influence of a pressure difference on both sides of a porous filter medium, the slurry flows toward the filter medium, whereby the liquid permeates its structure as a filtrate and the particles are retained. The effective driving pressure difference can be realized as

- hydrostatic pressure of a liquid column;
- centrifugal pressure of a liquid column in a rotating system;
- gas pressure difference in the form of vacuum behind the filter medium and atmospheric pressure above the slurry or overpressure above the slurry and atmospheric pressure behind the filter medium; and
- mechanical pressure by a piston or diaphragm, which is pressed on the slurry.

Cake filtration can be carried out with a constant pressure difference Δp (e.g. vacuum drum filter) or a constant volume flow Q (e.g. chamber filter press).

Depending on the particle size and the zeta potential of the particles, more or less agglomeration takes place and more or less compressible filter cakes are formed. Compressibility of the cake is also given in cases where the particles themselves are deformable, as in the case of fibrous materials.

Normally, the pores of filter media for cake filtration are chosen to be relatively open in comparison to the particle size to avoid clogging. As a consequence, in the very first moment, particles can get into the filtrate; before sealing, particle bridges have formed across the pores, and the filter cake is built up on top of this bridge forming the first particle layer. This means that in the initial stage of cake formation, further filtration mechanisms, beside the cake filtration mechanism, are taking effect. If cake filtration is the task of operation, this initial stage should be short enough to be negligible. Unfortunately, this is not valid in all technical applications, and a closer analysis of this transitional state should be undertaken.

Wet Cake Filtration: Fundamentals, Equipment, and Strategies, First Edition. Harald Anlauf.
© 2019 Wiley-VCH Verlag GmbH & Co. KGaA. Published 2019 by Wiley-VCH Verlag GmbH & Co. KGaA.

6 Filter Cake Formation

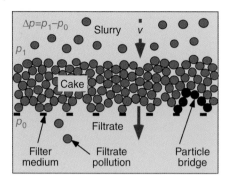

Figure 6.1 Basic principle of cake filtration.

The principal modeling of cake filtration processes is based on the law of Darcy (cf. Eq. (4.14)) for the permeation of porous layers and on the work of Ruth to apply this to a growing filter cake [1, 2].

6.2 Filtration Mechanisms During the Initial Phase of Cake Filtration

During the initial stage of cake filtration, the particles are retained on the surface or inside the structure of a porous filter medium corresponding to different deposition mechanisms, which are illustrated schematically in Figure 6.2.

Depending on the respective boundary conditions, these mechanisms appear in a complex superposition at the same time or consecutively, whereby one mechanism at a time is dominating.

Cake filtration implies the deposition of particles on top of other particles at the surface of a filter medium. This may be due to a particle bridge across a greater pore of the filter medium, a porous particle deposited in the upper layers of a filter medium, or the surface of the filter cake itself.

During the phase of bridge formation, filtrate pollution can occur because single particles, smaller than the pores of the filter medium, are still able to penetrate the filter medium and to get into the filtrate. This situation is nominated as intermediate filtration.

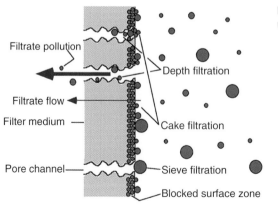

Figure 6.2 Filtration mechanisms.

If the particles are much smaller than the pores of a filter medium and the solid concentration of the slurry is quite very low, depth filtration takes place, and the particles are captured inside the filter mediums structure if it shows a certain three-dimensional structure.

Last but not least particles in low concentrated slurries can be larger than the pores of the filter medium. In this case, single particles approach the surface of the filter medium, and following the liquid flow, they block the single pores completely. This mechanism is called sieve or blocking filtration.

In cake filtration processes, only the cake filtration mechanism should be relevant and dominant, but this is, especially at the beginning, only a theoretical idealization. During the initial stage of a cake filtration process, the particle bridges across the filter medium pores have to be formed, and during this time, simultaneously intermediate, depth, and blocking filtration can take place. For cake filtration applications, this complicated situation can be held normally very short in the range of a few seconds down to a part of a second. The cake filtration mechanism can be assumed here, as valid with adequate accuracy, from the very beginning. If in practical applications turbid filtrate can be observed for longer times, the reason is often not an incomplete formation of particle bridges but particle deposits in the filtrate piping system, which are flushed out over longer times. However, in any case, if a certain turbidity of the filtrate cannot be accepted, the filtrate has to be recycled, until sufficient filtrate clarity is reached. In any case, a reduction of the bridge formation time is desirable and improves filtrate clarity. Rushton suggested some measures to achieve that [3].

Flocculation supports the spontaneous start of cake filtration and thus promotes filtrate clarity because the flocs are acting as preformed bridge elements (cf. Section 5.3).

The preconcentration of slurry to be filtered leads to faster bridge formation and improved filtrate clarity, as is demonstrated in Figure 6.3, for the filtration of a flotation coal slurry.

The solid mass, which is getting into the filtrate per square meter filter area and during one filtration cycle, is plotted against the pore diameter of several plain

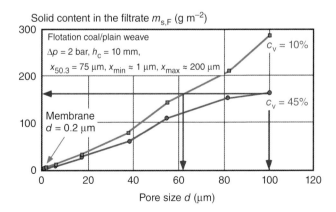

Figure 6.3 Filtrate pollution depending on filter medium pore size and slurry concentration.

weave fabrics with open square-shaped pores (cf. Section 9.2) and a microporous membrane with a pore diameter of 0.2 μm.

As expected, the filtrate pollution increases with increased pore diameter of the fabrics. If the solid content of the slurry is increasing from 10 to 45 vol%, a remarkable decrease of filtrate pollution can be observed. This is due to the fact that the particle bridges are formed much faster for greater solid concentrations. For a constant filtrate pollution of about 170 g m^{-2} for the diluted slurry, a filter medium pore size of about 62 μm is necessary, whereas for the concentrated slurry, 100 μm pores can be chosen. The obvious advantage is that the much more open fabric tends much less to clog and is much easier to regenerate.

If these findings are transferred to rotary filters, such as drum, disc, belt, or pan filters, the remarkable reduction of filtrate pollution for increasing slurry concentration becomes obvious. This is demonstrated in Figure 6.4 on the example of lab-scale experiments, which are transferred to the conditions of a rotary filter.

If a constant solid mass throughput, $Q_{m,s}$, has to be handled and the filter rotates with a constant speed, n (constant cake formation time t_1), the increase of slurry solid volume concentration, c_v, leads to increased cake height, h_c, and reduced necessary filter area, A. In Eq. (6.1), the cake height is given as a proportional to the square root of the concentration parameter κ (cf. Eq. (2.36)), which includes cake porosity ε and slurry concentration c_v

$$h_c \propto \sqrt{\kappa} = \sqrt{\frac{c_v}{1 - \varepsilon - c_v}} \tag{6.1}$$

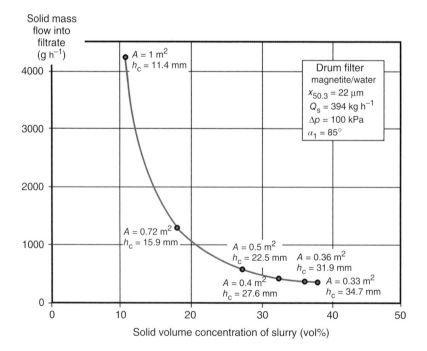

Figure 6.4 Filtrate pollution of a drum filter, depending on slurry concentration.

The solid mass throughput of a rotary filter, $Q_{m,s}$, is according to Eq. (6.2) proportional to the filter area A, the square root of the concentration parameter κ, and the square root of the rotation speed n (cf. Eq. (6.47))

$$Q_{m,s} \propto A \cdot \sqrt{\kappa} \cdot \sqrt{n} \qquad (6.2)$$

If the rotation speed is held constant, an increased concentration must be compensated by an area reduction to get constant throughput. If the filter area is held constant, an increase of slurry concentration must be compensated by a speed reduction.

In other words, because of an increase in the slurry concentration less number of particles "see" the filter medium and the bridge formation takes place faster.

The pore bridging not only exerts influence on the filtrate pollution but is also a decisive factor for the practical relevant filter medium flow resistance R_m. The diagram in Figure 6.5 explains this more in detail.

The filter medium resistance is plotted against the pore diameter of several twill weave fabrics (cf. Section 9.2) as well as of polypropylene and polyamide microporous membranes. For the filtration tests, particle-free water, flotation coal slurry, hematite iron ore slurry, and glass bead slurry had been used. For the permeation of the filter media with particle-free water, the expected result of great pore size influence on the filter medium flow resistance can be observed clearly. The filtration of slurries shows a completely different behavior. Now, the filter medium resistance increases for all media approximately to a similar value of resistance and remains more or less constant for all medium pore sizes. The reason for this phenomenon is that the practically relevant filter medium resistance for cake filtration is influenced not only by the pore geometry of the clean filter media but also by particles, which are intruded in the filter medium

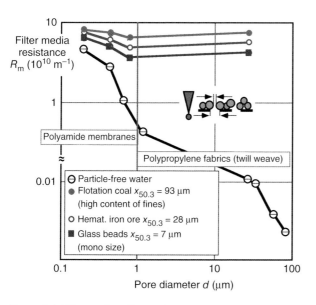

Figure 6.5 Filter medium flow resistance.

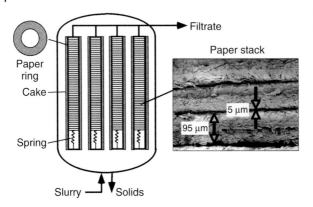

Figure 6.6 Paper stack candle filter.

structure, and, last but not least, sensitively by the first bridge-forming particle layer. These particle bridges are essential for cake formation, and thus, they belong to the filter medium. The small drawing in Figure 6.5 beside the exclamation mark makes clear that the resistance determining bottleneck is located in the particle bridges, and it is more or less irrelevant which kind of filter media are used below the bridges. The flotation coal shows the greatest variance of particle sizes and includes the smallest particles of all slurries. Therefore, the filter medium resistance exhibits the greatest value and even the membrane pores produce hardly greater resistances. The glass beads are nearly monosized and show by that reason the lowest overall resistance. The pores in the glass bead particle bridges are obviously greater than the membrane pores. Therefore, the filter medium resistance increases here clearly with decreasing membrane pore size.

As a rule to choose a well-suited filter cloth, the particles of the slurry should either be considerably greater or smaller than the filter medium pore size, but the average particle size should never be the same as the average pore size of the filter medium because this combination would lead to the maximal clogging probability.

Beside the explicit cake filtration apparatuses, for which the previous discussion is valid, the filter systems exist, which are realizing all the filtration mechanisms with similar duration in a consecutive and overlapping process, until at the end exclusively cake filtration takes place. One example for such a filtration behavior is given by a special candle filter, which is shown schematically in Figure 6.6 for direct filtration and cleaning of highly diluted grinding oils in the hard metal processing [4, 5].

Each candle consists of 9000 paper rings, which form a cylindrical edge filter element of about 1 m length, which is held together by an internal spring.

The front end of the paper rings (80 g m^{-2} standard copy paper) of about 95 μm thickness and the gaps of about 5 μm between them can be identified in the microscopic photo. The distance between the paper rings is produced by the roughness of the paper surface. Because of the low particle concentration in the usually filtered slurries, a complete filtration period takes about eight hours. At the beginning, particles of less than 5 μm are migrating into the gaps between the paper rings and are captured by fibers of the paper, which are not only responsible for the distance of the paper rings from each other but they are also forming a three-dimensional pore network in the gaps. Particles of more than 5 μm are blocking a part of the gap. The period of depth and blocking filtration takes many minutes to some hours, while the cake filtration mechanism becomes more and more dominant. The process is operated with constant volume flow, and after reaching a defined flow resistance or pressure loss, the filter has to be flushed back for regeneration. To describe the filtration process, it is necessary to recognize the different filtration mechanisms and to distinguish between them.

For this purpose, the approach of Hermans and Bredée [6] is very helpful. They supposed that each filtration mechanism influences the change of filter medium resistance R_m as a function of the generated filtrate volume V_L in a characteristic way. This could be formulated for constant pressure difference Δp (Eq. (6.3)) or constant volume flow $Q_{v,L} = dV_L/dt$ (Eq. (6.4)) by a general exponential equation

$$\Delta p = \text{const.} \Rightarrow \frac{dR_m}{dV_L} = \frac{d^2 t}{dV_L^2} = K \cdot \left(\frac{dt}{dV_L}\right)^q \tag{6.3}$$

$$Q_{v,L} = \frac{dV_L}{dt} \Rightarrow \frac{d(\Delta p)}{dV_L} = K \cdot (\Delta p)^q \tag{6.4}$$

Each filtration mechanism leads to a characteristic value of the exponent q. Luckert and Gösele have adapted the general equations to the single filtration mechanisms and expressed by linear functions [7, 8]. The experimentally measured data of filtrate volume and time have to be compared with these linear functions. The function, which fits at best, identifies the relevant filtration mechanism. Figure 6.7 assigns the respective value of q to the different filtration mechanisms.

Cake filtration is represented by $q = 0$. From this follows Eq. (6.5)

$$q = 0 \Rightarrow \frac{d^2 t}{dV_L^2} = K \tag{6.5}$$

Here, the filter medium resistance R_m remains constant. Very important for further discussion of cake filtration is the finding from Figure 6.5 that the filter medium resistance, which is relevant for practical use, is not the resistance that can be measured by permeating the medium with particle-free liquid and calculated on the basis of the Darcy equation (cf. Eq. (4.14)). As discussed previously, the interaction of the filter medium with the slurry to be filtered and

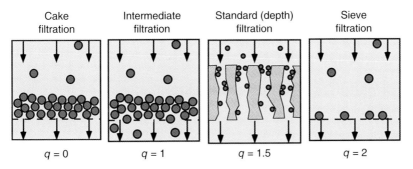

Figure 6.7 Characterization of different filtration mechanisms.

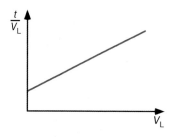

Figure 6.8 Linear representation of cake filtration for Δp = const.

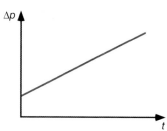

Figure 6.9 Linear representation of cake filtration for $Q_{v,L}$ = const.

the particle bridges across the pores of the filter medium have to be considered. This will be discussed later on in more detail. A linear function can be formulated from the Hermans and Bredée approach for Δp = constant (Figure 6.8) and $dV_L/dt = Q_{v,L}$ = constant (Figure 6.9) and the respective linear functions are formulated in Eqs. (6.6) and (6.7)

$$\Delta p = \text{const.} \Rightarrow \frac{t}{V_L} = K \cdot V_L + \frac{1}{Q_0} \tag{6.6}$$

$$Q_{v,L} = \frac{dV_L}{dt} = \text{const.} \Rightarrow \Delta p = \Delta p_0 \cdot K \cdot V_0^2 \cdot t + \Delta p_0 \tag{6.7}$$

To decide for a certain filtration process, whether the mechanisms of cake filtration are relevant, it is not necessary to know anything in detail about parameters such as K, Q_0, or Δp_0. Only if the experimental data are plotted according to Figure 6.8 or Figure 6.9 and a linear function results, cake filtration is identified as the dominant filtration mechanism. For the other filtration mechanisms, which should not be discussed here in more detail, the same method can be applied, and for each filtration mechanism, a special plot of the experimental filtration data leads to an appropriate linear function. If more detailed information about the influencing parameters of cake filtration, such as slurry concentration, liquid viscosity, cake permeability, or pressure difference, is required, an extended model of cake filtration is necessary to itemize global parameters, such as K or Q_0.

6.3 Formation of Incompressible Filter Cakes by Pressure Filtration

6.3.1 Principle Model of Time-Dependent Filter Cake Growth

In the following the previously mentioned global parameters, K and Q_0 should be itemized to enable the full description of the filter cake growth in dependence of

Figure 6.10 Principal situation during cake formation.

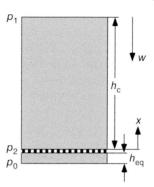

all practically relevant parameters. For this purpose, some boundary conditions have to be defined, which can be realized not only in theory but also in practice. The slurry must be homogeneously mixed to exclude particle segregation, the liquid must have Newtonian flow behavior, the liquid flow must behave laminar, and the resulting filter cake structure must be incompressible. If one or more conditions in this list are violated, the modeling is principally still possible, but of course more complex. The formation of compressible filter cakes as a practically very relevant example for the deviation from the ideal boundary conditions will be examined in Section 6.4.

Figure 6.10 explains the principal situation.

The superficial flow velocity w of the liquid through a filter cake of height h_c and the pressure loss $\Delta p'$ can be expressed according to Darcy by Eq. (6.8)

$$w = \frac{dV_L}{A} = \frac{dV_L}{A \cdot dt} = \frac{p_c \cdot \Delta p'}{\eta_L \cdot h_c} \tag{6.8}$$

The flow through the filter cake leads to the pressure loss $\Delta p'$ in Eq. (6.9)

$$\Delta p' = p_1 - p_2 \tag{6.9}$$

The flow through the filter medium leads to the pressure loss $\Delta p''$ in Eq. (6.10)

$$\Delta p'' = p_2 - p_0 \tag{6.10}$$

The specific cake permeability p_c, which is related to the cake height, provides a hint, whether the filtration is in a meaningful range between c. 10^{-11} and 10^{-16} m². 10^{-11} m² characterizes very easy to filter slurries, such as coarse crystals, and 10^{-16} m² describes extremely hard to filter slurries, such as slimy microorganisms. If the specific permeability is located outside of this range, cake filtration represents normally not more the adequate separation technique. The equivalent cake height h_{eq} can be used optionally instead of the filter medium resistance R_m. It gives a better imagination, how relevant the medium resistance is in comparison to the cake resistance. The equivalent cake height can be calculated from the product of absolute resistance of the filter medium and specific permeability of the filter cake as is expressed in Eq. (6.11)

$$h_{eq} = R_m \cdot p_c \tag{6.11}$$

Normally, h_{eq} should result in a few millimeters or less. Equation (6.8) cannot be integrated in the basic form because the filter cake height is growing during cake formation and is not constant. Therefore, at first, a correlation between filtrate volume V_L and cake height h_c must be found out. This correlation is given by the concentration parameter κ in Eq. (2.39). This parameter can be derived from a mass balance around the total solid/liquid system, which is formulated in Eq. (6.12)

$$A \cdot h_c \cdot (1 - \varepsilon) \cdot \rho_s = (V_L + A \cdot h_c \cdot \varepsilon) \cdot Y \cdot \rho_L \tag{6.12}$$

On the left side of Eq. (6.12), the solid mass m_s of the filter cake can be found and on the right side the liquid mass m_L, which consists of the filtrate volume V_L and the liquid volume, which is enclosed in the voids of the filter cake. The parameter Y represents, according to Eq. (6.13), the ratio between solid and liquid mass to make the equation consistent

$$Y = \frac{m_s}{m_L} \tag{6.13}$$

Finally, the correlation between cake thickness and filtrate volume can be formulated with κ as summarizing parameter in Eq. (6.14)

$$h_c = \frac{Y \cdot \rho_L}{(1 - \varepsilon) \cdot \rho_s - Y \cdot \varepsilon \cdot \rho_L} \cdot \frac{V_L}{A} = \kappa \cdot \frac{V_L}{A} \tag{6.14}$$

Now, the differential equation to describe the growing of the cake can be formulated in Eq. (6.15) and principally be integrated after separation of the variables

$$\Delta p' = \frac{\eta_L \cdot \kappa \cdot V_L}{p_c \cdot A^2} \cdot \frac{dV_L}{dt} \tag{6.15}$$

However, the filtration takes place not only through the growing filter cake but also through the porous filter medium with the pressure loss $\Delta p''$ (cf. Figure 6.10). Also, the flow through the filter medium is formulated in Eq. (6.16) on the basis of the Darcy equation

$$\Delta p'' = \frac{\eta_L \cdot R_m}{A} \cdot \frac{dV_L}{dt} \tag{6.16}$$

To get the total pressure loss Δp of the permeated system, both pressure losses have to be added, according to Eq. (6.17)

$$\Delta p = \Delta p' + \Delta p'' \tag{6.17}$$

Now, the final differential equation to describe the filter cake formation can be formulated in Eq. (6.18)

$$dV_L = \frac{A \cdot \Delta p}{\eta_L \cdot \left[\dfrac{\kappa \cdot V_L}{p_c \cdot A} + R_m\right]} \cdot dt \tag{6.18}$$

Before integrating Eq. (6.18), the mode of filtration must still be fixed, according to Figure 6.11.

The filtrate flow is plotted here against the pressure difference. As mentioned in Section 6.1, cake filtration can be carried out with constant pressure difference Δp

Figure 6.11 Mode of filtration.

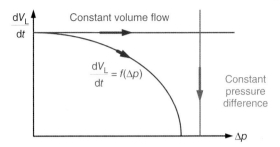

or constant volume flow dV_L/dt. The process in principle is the same. Although the pressure difference is held constant, the cake height increases and thus the flow resistance increases too. As a consequence, the filtrate flow must decrease. If the filtrate flow is held constant, the cake increases and the flow resistance increases too. To maintain the constant flow rate, the pressure difference must increase. If the characteristic of the pump does not provide constant conditions in a certain range of flow rate or pressure difference, the process follows the pump capacity curve $dV_L/dt = f(\Delta p)$.

Equation (6.18) should now first be integrated for constant pressure difference. Equation (6.19) shows the result

$$V_L = \frac{A}{\kappa} \cdot \left[\sqrt{(R_m \cdot p_c)^2 + \frac{2 \cdot \kappa \cdot \Delta p \cdot p_c \cdot t_1}{\eta_L}} - R_m \cdot p_c\right] \qquad (6.19)$$

From this result of integration, the typical declining course of filtrate flow with time can be recognized. If not the filtrate volume but the cake height is interesting, Eq. (6.19) can be transformed with the help of Eq. (6.14), which correlates filtrate volume and cake height in Eq. (6.20)

$$h_c = \sqrt{(R_m \cdot p_c)^2 + \frac{2 \cdot \kappa \cdot \Delta p \cdot p_c \cdot t_1}{\eta_L}} - R_m \cdot p_c \qquad (6.20)$$

If this equation is solved for the cake forming time t_1, two terms are formed, as can be seen in Eq. (6.21)

$$t_1 = \frac{h_c^2 \cdot \eta_L}{2 \cdot \kappa \cdot p_c \cdot \Delta p} + \frac{h_c \cdot R_m \cdot \eta_L}{\kappa \cdot \Delta p} \qquad (6.21)$$

The first term contains the specific cake permeability p_c and the squared cake height h_c. The second term contains the filter medium resistance R_m and the linear cake height. As a consequence, the first term increases much more rapidly with time than the second one, and if the filter medium is selected well, the second term can be neglected more or less from the beginning. In that case, the filter cake formation can be described according to Eqs. (6.22) and (6.23) by a very simple equation, which nevertheless is valid for practical use, if the specific cake permeability is known

$$h_c = \sqrt{\frac{2 \cdot \kappa \cdot \Delta p \cdot p_c \cdot t_1}{\eta_L}} \qquad (6.22)$$

$$t_1 = \frac{h_c^2 \cdot \eta_L}{2 \cdot \kappa \cdot p_c \cdot \Delta p} \tag{6.23}$$

If in the reverse case the flow resistance of the filter medium is dominating the cake filtration process, the entire filter cake resistance can be neglected. There is one example in the filtration practice, where the filter medium resistance dominates the cake formation process and the filter cake resistance can be neglected widely. This is realized in the form of a vacuum disc filter, which is using filter media of sintered ceramics with pore sizes of about 1 μm (cf. Section 9.1) [9]. These filter plates must exhibit from stability reasons a certain thickness of a few mm, but this generates a comparatively high flow resistance. The cake formation equation can be written in the modified form of Eq. (6.24) for this special case

$$t_1 = \frac{h_c \cdot R_m \cdot \eta_L}{\kappa \cdot \Delta p} \tag{6.24}$$

Now, the correlation between cake formation time and cake thickness is not more quadratic, but only directly proportional.

If the volume flow $Q_{v,L} = dV_L/dt$ = constant, Eq. (6.18) likewise can be integrated. In Eq. (6.25), the result is expressed

$$\Delta p = \frac{Q_{v,L} \cdot \eta_L}{A} \cdot \left[\frac{\kappa \cdot Q_{v,L} \cdot t_1}{p_c \cdot A} + R_m \right] \tag{6.25}$$

Now, the increase of the pressure difference with increasing cake formation time can be calculated. This increase is limited by the capacity of the pump.

6.3.2 Experimental Determination of Process Characterizing Parameters

To calculate the filter cake formation on a realistic basis, filter medium resistance and specific filter cake resistance r_c or specific filter cake permeability p_c, respectively, must be determined by a filter experiment as material characterizing parameters. The filter media, intended for the technical application, and the original slurry must be used here. The right execution of the experimental determination of R_m and r_c, as well as the suggested laboratory equipment and the procedure of data evaluation, has been described detailed in [10, 11] and the Verein Deutscher Ingenieure (VDI) guideline 2762 Part 2 [12]. In Figure 6.12, an example of the pressure filter cell as central part of the lab filtration unit is shown schematically.

The filter cell exhibits a filter area of 20 cm², which is great enough to avoid wall effects for particles in the range of c. 1–100 μm (cf. [13], p. 369). The double wall of the cell optionally enables to adjust and maintain the temperature during filtration. The lid is made of glass and equipped with a light source, to observe the filtration process visually. This is especially important to register exactly the end of the cake formation process and also subsequent phenomena, such as shrinkage cracks during desaturation. In the case of low filtration velocity and not exactly horizontal and even cake surface, the cake does not emerge at once and the question is when the cake formation time is finished. As a rule, one can define that

Figure 6.12 Lab-scale pressure filter cell.

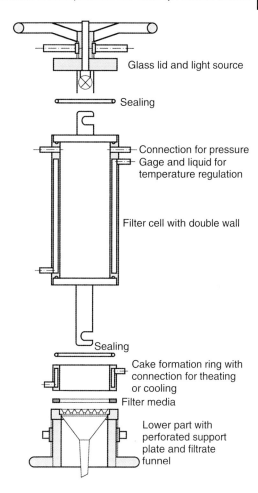

the cake formation time is completed when about one-half of the dry filter cake surface can be seen. At this moment, a button is pressed, and the data logging system takes this time as the cake formation time. The filter cake is formed in a separate cake formation ring to enable easy take-out after filtration for further analysis. The filter medium is placed on a perforated support plate. The boreholes in the plate are countersunk to guarantee unhindered flow of liquid through the filter media. If a damageable microporous membrane is used as a filter medium, a supporting fabric has to be placed between the membrane and the support plate. Beneath the support plate, a removable plastic funnel is mounted in the lower part of the cell. The idea behind this design is the possibility to remove the remaining liquid drops after cake deliquoring from the backside of the supporting plate, to avoid cake rewetting after aeration of the filter cell. An alternative would be a fixed funnel with a small angle and thus a small volume and a filtrate outlet pipe of small diameter, which can be closed by a valve. This enables the filling of the filtrate system with liquid before the experiment. It hinders filtrate from getting on the balance beneath the filter cell before the pressure is applied. The filtrate

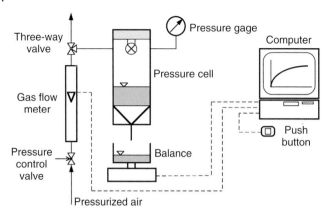

Figure 6.13 Lab-scale pressure filtration setup.

valve then has to be opened together with the pressurized gas inlet valve. This additional volume is displaced from the funnel, if a gas flow through the cake takes place during desaturation and has to be considered for the recalculation of the cake moisture.

The complete filtration setup is shown in Figure 6.13.

The pressurized gas flows through a pressure control valve, a gas flow meter, and a three-way valve into the filter cell. The maximal pressure depends on the construction but is limited normally to 0.5–1 MPa. The filter cell is designed exclusively for overpressure filtration, but vacuum filtration can also be described by this design normally without any problem. Decisive from the physical point of view is the pressure difference between both sides of the filter media and not the pressure level [14]. It is important that the final pressure in the filter cell is reached after starting the experiment as fast as possible. This should not take more than 0.5 seconds to fulfill the condition of constant pressure during the complete filtration. The gas flow meter allows measuring the gas throughput during cake desaturation.

The three-way valve enables the pressure release at the end of the experiment. The filtrate is collected on a balance beneath the filtrate outlet. It is essential to register the appearing filtrate as fast as possible to record the filtration procedure with maximal precision. The information about filtrate production and gas flow with time is registered by electronically data logging. As mentioned before, the identification of cake formation end and start of deliquoring is made visually through the glass lid and by pressing a button, when half of the cake surface can be seen. Normally, the cake is not exactly horizontal and the cake surface appears not completely at the same moment. Especially for difficult to filter slurries, the cake formation needs long time and the appearance of the cake surface also takes some time.

For the filter experiment, a slurry sample is prepared and filled through the open lid into the pressure filter cell. The lid has to be closed as quickly as possible and the preselected gas pressure has to be applied through the three-way valve. The opening and closing of the valve starts and stops the data logging program. After aeration of the cell at the end of the experiment, the filter cake is removed and can be analyzed relating to height, porosity, and residual moisture content. For the determination of filter medium resistance and specific cake resistance, the measured filtrate flow during the cake formation has to be evaluated. For this purpose, Eq. (6.19) has to be redrafted into a linear function. For this purpose, Eq. (6.19) has to be solved first for the cake formation time t_1 according to Eq. (6.26)

$$t_1 = \left[\frac{r_c \cdot \kappa \cdot \eta_L}{2 \cdot A^2 \cdot \Delta p}\right] \cdot V_L^2 + \frac{R_m \cdot \eta_L}{A \cdot \Delta p} \cdot V_L \tag{6.26}$$

In the next step, Eq. (6.26) has to be divided by V_L, as shown in Eq. (6.27)

$$\frac{t_1}{V_L} = \left[\frac{r_c \cdot \kappa \cdot \eta_L}{2 \cdot A^2 \cdot \Delta p}\right] \cdot V_L + \frac{R_m \cdot \eta_L}{A \cdot \Delta p} = b \cdot V_L + a \tag{6.27}$$

If the filtration data are plotted according to Eq. (6.27), as can be seen in Figure 6.14, a linear function results.

The inclination b contains the specific cake resistance r_c (Eq. (6.28)), and from the axial section a, the filter medium resistance R_m (Eq. (6.29)) can be calculated

$$r_c = b \cdot \frac{2 \cdot A^2 \cdot \Delta p}{\kappa \cdot \eta_L} \tag{6.28}$$

$$R_m = a \cdot \frac{A \cdot \Delta p}{\eta_L} \tag{6.29}$$

Going back to Eq. (6.6), from the model of the Hermans and Bredée approach, the parameters K and Q_0 can now be itemized, as shown in Eq. (6.30)

$$\frac{t_1}{V_L} = K \cdot V_L + \frac{1}{Q_0} = \left[\frac{r_c \cdot \kappa \cdot \eta_L}{2 \cdot A^2 \cdot \Delta p}\right] \cdot V_L + \frac{R_m \cdot \eta_L}{A \cdot \Delta p} \tag{6.30}$$

If deviations from the linear course of the function are observed, the boundary conditions for cake filtration, which have been formulated at the beginning of this chapter, must have been violated. One example for such deviations is superimposed particle sedimentation. Gravity cannot be switched off, and if no stirrer homogenizes the slurry permanently, particles are settling down toward the filter media as long as slurry is present in the filter cell. The more concentrated the slurry and the smaller the particles are, the less particle segregation and sedimentation will be observed (cf. Section 5.2). If in the pressure filter cell the filtration

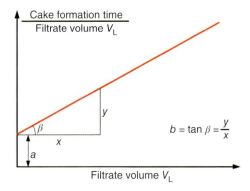

Figure 6.14 Evaluation of R_m and r_c from filtration data.

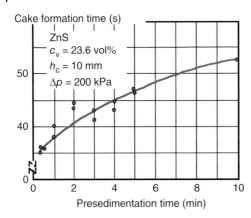

Figure 6.15 Influence of presedimentation on cake formation time.

is not started immediately after the slurry is fed in, a sediment will form on top of the filter medium, like a precoat, and will be identified during the data evaluation as increased filter medium resistance. If the sedimentation velocity is relatively high in comparison to the filtration velocity, a zone of clear liquid will form above the slurry. As a consequence of a delayed start of the filtration and thus increased filter medium resistance (filter media + particle bridges + sediment), the total cake formation time, until the cake surface appears from the liquid, increases with increasing sedimentation time. This can be observed exemplarily in Figure 6.15 for experiments with zinc sulfide slurry.

A start of filtration as quick as possible after filling the slurry into the filter cell minimizes the presedimentation effect, but a simultaneous filtration and sedimentation cannot be avoided at all. A further consequence of superimposed sedimentation consists in the fact that the $t_1/V_L = f(V_L)$ function no longer be a straight line during the complete cake formation time, which is defined from the start of filtration, until the dry surface of the cake appears from the liquid. This can be seen qualitatively in Figure 6.16.

Unfortunately, the transition from slurry filtration to clear liquid permeation cannot be identified very clearly in the $t_1/V_L = f(V_L)$ representation of the filtration data because each additionally originating filtrate mass is added up on

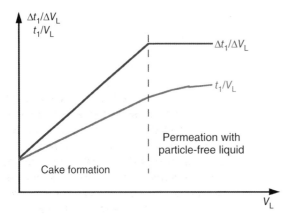

Figure 6.16 Influence of superimposed sedimentation on the linear data representation.

6.3 Formation of Incompressible Filter Cakes by Pressure Filtration

the balance and therefore the relative increase of mass is small. This transition point can be identified much more clearly by plotting $\Delta t_1 / \Delta V_L = f(V_L)$. To calculate the filter medium resistance and the specific cake resistance for this case, Eq. (6.26) has to be differentiated, as shown in Eq. (6.31)

$$\frac{dt_1}{dV_L} = \left[\frac{r_c \cdot \kappa \cdot \eta_L}{A^2 \cdot \Delta p} \right] \cdot V_L + \frac{R_m \cdot \eta_L}{A \cdot \Delta p} = \frac{\Delta t_1}{\Delta V_L} \tag{6.31}$$

The dt_1/dV_L method identifies very clearly the transition point between cake filtration and cake permeation, but the measured filtrate data are scattering more, than for the t_1/V_L representation especially in the case of difficulty to filter slurries with dropwise filtrate production. On the other hand, in such cases of difficult to filter slurries, the settling velocity is also poor, and the $t_1/V_L = f(V_L)$ representation of the experimental data can be used.

The true filter medium resistance and specific cake resistance for fast settling products can be determined by first forming the filter cake and then permeating the complete cake with clear liquid at constant pressure. Doing this for two cakes of different height and evaluating the filtration data according to the Darcy equation (cf. Eq. (6.8)), the resistances can be calculated from the two straight lines, shown in Figure 6.17.

The corresponding calculations can be seen in Eq. (6.32) for the specific cake resistance and in Eq. (6.33) for the filter medium resistance

$$r_c = \frac{A \cdot \Delta p \cdot (Q_{v,L2} - Q_{v,L1})}{Q_{v,L2} \cdot Q_{v,L1} \cdot \eta_L \cdot (h_{c1} - h_{c2})} \tag{6.32}$$

$$R_m = \frac{A \cdot \Delta p}{Q_{v,L1} \cdot \eta_L} - h_{c1} \cdot r_c \tag{6.33}$$

Obviously, things become still more complicated if not only sedimentation but also segregation takes place. In such a case, the local filter cake resistance rises up during filtration because the cake consists more and more of finer particles [15].

Summarizing the consequences of deviations from the formation of homogeneous filter cakes from homogeneous slurries, one can state that all these deviations have negative influence on the process and should be avoided as much as possible. In practice, well-suited measures to promote homogeneous cake

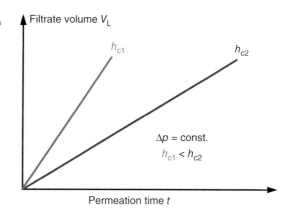

Figure 6.17 Determination of R_m and r_c by permeation experiments.

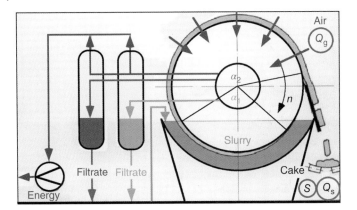

Figure 6.18 Vacuum drum filter.

formation conditions are prethickening, agglomeration, eventually fractionation, and stirring the slurry during filtration. In stirred nutsche filters, in drum and disc filters, the slurry can be homogenized during cake filtration. In candle and leaf filters, the slurry can be held homogeneous by pumping the slurry upward in a closed loop alongside the filter elements. No chance of slurry homogenization is given for nutsche filters without stirrer, horizontal belt, and pan filters.

Not in all cases, the time-dependent filtrate flow, to determine the filter medium and cake flow resistances, can be measured directly, as in the pressure filter cell. For example, in the case of a rotary filter apparatus, such as the vacuum drum filter, sketched in Figure 6.18, the complete filtrate is collected in central filter pipes, and the filtrate flow cannot be related directly to the growth of the filter cake.

In such a case, the measurement of different cake heights for different rotation speeds of the filter solves the problem. Equation (6.27) has to be modified using Eq. (6.14) and results in Eq. (6.34)

$$\frac{t_1}{h_c} = \frac{r_c \cdot \eta_L}{2 \cdot \kappa \cdot \Delta p} \cdot h_c + \frac{R_m \cdot \eta_L}{\kappa \cdot \Delta p} \tag{6.34}$$

The filter cake formation time t_1 can be calculated, according to Eq. (6.35), from the rotation speed of the filter n and the cake formation angle α_1.

$$t_1 = \frac{\alpha_1}{360} \cdot \frac{1}{n} \tag{6.35}$$

As demonstrated in Figure 6.19, again a linear function originates, if t_1/h_c is plotted as a function of h_c.

6.3.3 Throughput of Discontinuous Cake Filters

With the knowledge of cake formation theory and measured material characterizing parameters, a scale-up of cake filter apparatuses can be conducted and the expected slurry, filtrate, or solid throughput can be calculated.

Figure 6.19 Evaluation of R_m and r_c from rotary filter results.

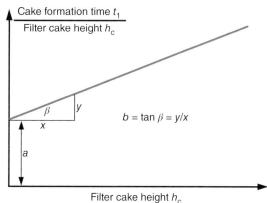

The average specific filtrate throughput $q_{v,L,av}$ in Eq. (6.36) can be calculated from the filtrate volume $V_L(t_{tot})$, the total batch time t_{tot}, and the filter area A

$$q_{v,L,av} = \frac{V_L(t_{tot})}{A \cdot t_{tot}} \tag{6.36}$$

The total batch time t_{tot} in Eq. (6.37) consists of the sum of the times for all process steps, such as slurry feed t_f, cake formation t_1, cake washing t_w, cake deliquoring t_2, cake removal t_{re}, and apparatus cleaning t_{cl}

$$t_{tot} = t_1 + (t_f + t_w + t_2 + t_{re} + t_{cl}) = t_1 + t_d \tag{6.37}$$

Cake is produced only during the cake formation time. The sum of times for all other process steps, where no further cake is formed, is called down time t_d. The filtrate production during cake deliquoring and eventually cake washing is not considered for this examination and neglected. Figure 6.20 illustrates one entire batch cycle.

An advantage of this configuration is given by the possibility to choose the time of each process independently from each other. This enables maximal flexibility to adjust the process parameters to the needs of the particular product.

Otherwise, the throughput is limited due to the down time. Such down times are missing in the case of continuous operating filter apparatuses, as schematically shown in Figure 6.21.

Here, the same process steps exist, as for the batch process, but now all process steps are linked together by the same transport velocity of the product through the process room. The time for each process step results from the transport velocity and the length of the process zone. This configuration leads to comparatively high throughputs because filter cake is produced permanently. The deficiency here consists of the restricted flexibility in comparison to batch processes.

One question, which has to be answered for the discontinuous process, is how to find the conditions for maximal filtrate production, which means maximal throughput for a slurry with given properties.

First, the case of constant filtration pressure difference Δp should be examined. Neglecting the filter medium resistance, the produced average filtrate volume flow $q_{v,L,av}$ during a complete batch can be formulated by combining Eqs. (6.13),

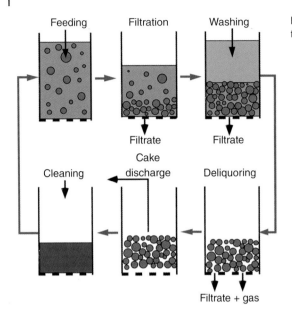

Figure 6.20 Discontinuous cake filtration process.

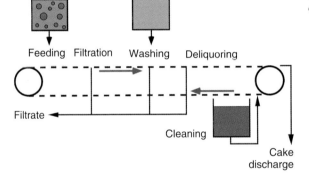

Figure 6.21 Continuous cake filtration process.

(6.21), and (6.36) in Eq. (6.38)

$$q_{v,L,av} = A \cdot \sqrt{\frac{2 \cdot p_c \cdot \Delta p}{\kappa \cdot \eta_L}} \cdot \frac{\sqrt{t_1}}{t_1 + t_d} \tag{6.38}$$

To find a maximum of $q_{v,L,av}$, this function has to be differentiated with respect to t_1 and set equal to zero, as shown in Eq. (6.39)

$$\frac{d}{d_{t_1}} \cdot \left[\frac{\sqrt{t_1}}{t_1 + t_d} \right] = 0 \Rightarrow \frac{1}{2 \cdot \sqrt{t_1}} \cdot (t_1 + t_d) - \sqrt{t_1} = 0 \tag{6.39}$$

The result of this calculation is given in Eq. (6.40)

$$t_1 + t_d = 2 \cdot t_1 \Rightarrow t_1 = t_d \tag{6.40}$$

The maximal filtrate throughput is expected for the case where the cake forming time t_1 equals exactly the down time t_d. This leads to the calculation of the

6.3 Formation of Incompressible Filter Cakes by Pressure Filtration

Figure 6.22 Filtrate throughput of discontinuous cake filters (Δp = const.)

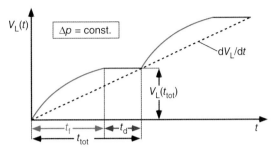

maximal filtrate throughput in Eq. (6.41)

$$q_{v,L,max} = A \cdot \sqrt{\frac{2 \cdot p_c \cdot \Delta p}{\kappa \cdot \eta_L}} \cdot \frac{1}{\sqrt{2 \cdot t_1}} \qquad (6.41)$$

From the theoretical point of view, the maximal filter throughput can be increased by reduction of the cake forming time, if the down time can be reduced adequately. This might be eventually possible by faster cake discharge or shorter apparatus cleaning time. The discussed optimization can be illustrated in graphical form, as shown in Figure 6.22.

During the cake formation time, the typical declining trend for the filtrate production over time can be observed for constant pressure difference. During the down time, no further filtrate volume from cake formation accumulates. If the start and the end point of the batch is interlinked, a straight line is the result and the slope of this straight line represents the average filtrate flow $q_{v,L,av}$. Optimization of this situation means a maximum slope of the straight line or $q_{v,L,av}$, respectively. How to get the maximum slope is demonstrated in Figure 6.23.

It can be clearly seen that the maximum slope of the straight line results in the case of $t_1 = t_d$. Further increase of the slope could be realized by reduction of the down time or by an increase of cake formation velocity. The latter would be a result of increased cake permeability due to agglomeration or prethickening.

Beside constant pressure filtration, constant volume flow filtration has to be discussed with respect to maximum throughput. As can be seen from Figure 6.24, the filtrate throughput $q_{v,L}$ for constant feed volume flow and batchwise operation shows no optimum.

The maximal cake forming time depends, according to Eq. (6.42), on the maximal pressure difference Δp_{max}

$$t_{1,max} = \frac{p_c^2 \cdot A^2}{\kappa \cdot \eta_L \cdot V_L^2} \cdot \Delta p_{max} \qquad (6.42)$$

Figure 6.23 Maximal filtrate throughput of discontinuous cake filters (Δp = const.)

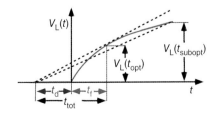

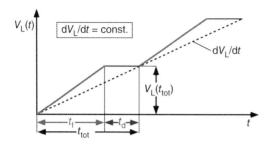

Figure 6.24 Filtrate throughput of discontinuous cake filters ($q_{v,L}$ = const.)

With Eq. (6.36), the filtrate throughput for the batch process can be calculated in Eq. (6.43)

$$q_{v,L} = \frac{V_L}{A \cdot t_{tot}} = \sqrt{\frac{\Delta p_{max} \cdot p_c}{\eta_L \cdot \kappa}} \cdot \frac{1}{\sqrt{t_{1,max}} \cdot (1 + t_d/t_{1,max})} \quad (6.43)$$

The shorter the down time t_d becomes, the more the throughput is approaching its maximal value, which is given by Δp_{max} and from this following $t_{1,max}$.

The ideal conditions to get maximal throughput, which have been discussed here, are not possible to fulfill in any case of industrial practice. There may be a deviation from ideal constant pressure or constant slurry feed volume flow. The filtration time to form a cake, which can be safely discharged, might be too long in comparison to the down time. As an example, Stickland demonstrated optimization calculations for filter presses [16].

6.3.4 Throughput of Continuous Vacuum and Pressure Filters

As explained on the example of a continuously operating vacuum belt filter, all process steps now are linked together by the joint transport velocity. According to Figure 6.25, filtrate chambers underneath the filter medium are separating the different process zones of a belt filter for cake formation, washing, and deliquoring.

A similar situation can be found for rotary drum, disc, or pan filters, where the filter area is subdivided into several filter cells. This is demonstrated in Figure 6.26 on the example of a vacuum drum filter.

Rectangular filter cells are individually connected via filtrate pipes with the control valve, which is located at the interface between the rotating and the static part of the filter. A filter cloth, on which the filter cake is formed, covers the filter

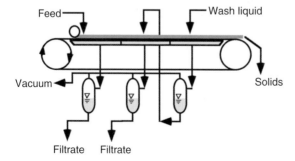

Figure 6.25 Vacuum belt filter.

Figure 6.26 Vacuum drum filter.

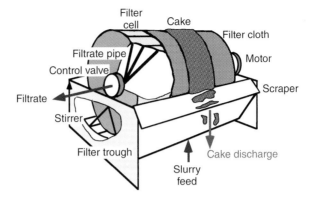

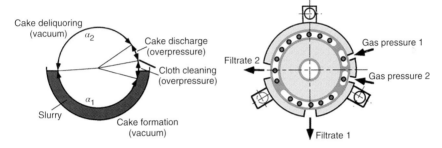

Figure 6.27 Control valve of rotary cake filters.

cells. At the opposite side of the drum, the motor of the filter is mounted. The drum is partly immersed into the filter trough, which contains the slurry to be filtered. The slurry should be fed at the deepest point of the filter trough to generate an upward-directed flow, which counteracts particle segregation. In addition, a stirrer mixes the slurry permanently during filtration. If the filter cake is well desaturated and behaves brittle, a scraper, combined with a gas blowback, can be used for the solid discharge. The control valve separates the different process zones as is illustrated in Figure 6.27.

Here, an axial control valve is shown, which needs a control disc with fixed process zones. To change the process zones, the control disc has to be changed. The control valve can also be realized in a radial version, where the process zones can be adjusted by plugs, which can be relocated from outside. Nevertheless, the axial version is more common because it is easier to seal and it is normally not necessary to adjust the process zones frequently.

To calculate the throughput of rotary drum or disc filters, the results from the lab-scale pressure filter cell must be transferred to the rotating apparatus. According to Eq. (6.44), the cake formation time t_1 is correlated with the rotation speed of the filter area and the cake formation angle α_1

$$t_1 = \frac{\alpha_1}{360°} \cdot \frac{1}{n} \tag{6.44}$$

Neglecting the filter medium resistance, this equation can be inserted in the cake formation Eqs. (6.21) and (6.45) results

$$h_c = \sqrt{\frac{2 \cdot p_c}{\eta_L}} \cdot \sqrt{\kappa \cdot \Delta p \cdot \frac{1}{n}} \cdot \sqrt{\frac{\alpha_1}{360°}} \tag{6.45}$$

For better understanding, the parameters are grouped into product, process, and design parameters. As expected, the cake height becomes smaller, if the rotational speed of the filter increases because the residence time of the filter area in the slurry is shortened. The maximal rotational speed is set by the possibility of a complete cake discharge. For discharge by air blowback and scraper in the case of drum or disc filters, this limit is reached for a few millimeters of cake thickness. If the cake is too thin, it is not more discharged but is transported behind the scraper back into the filter trough. A certain safety distance between filter area surface and scraper is necessary, especially for large filter units, to exclude any contact between scraper and filter cloth. Normally, the maximal speed of rotary filters is considerably less than 10 rpm.

The question now is whether a high or a low rotational speed of the filter is beneficial for high throughput. To answer this question, the specific solid mass throughput of the rotary filter must be calculated in form of Eq. (6.46):

$$q_{m,s} = \frac{Q_{m,s}}{A} = n \cdot h_c \cdot \rho_s \cdot (1 - \varepsilon) \tag{6.46}$$

The specific solid mass throughput is given in kg m^{-2} h^{-1}. One has to consider that the rotational speed is additionally inserted in the cake height. Equation (6.47) gives the final result

$$q_{m,s} = \rho_s \cdot (1 - \varepsilon) \cdot \sqrt{\frac{2 \cdot p_c}{\eta_L}} \cdot \sqrt{\kappa \cdot \Delta p \cdot n} \cdot \sqrt{\frac{\alpha_1}{360°}} \tag{6.47}$$

Now, it becomes clear that the filter throughput increases with increasing rotational speed, although the cake height decreases. The maximal throughput is limited by the minimal cake height, which can be still safely discharged. For drum and disc filters and good desaturated brittle filter cakes, which are discharged by air blowback and scraper, this limit is given at about $h_{c,min} = 5$ mm or less. The cake must have a certain weight (thickness) to be separated from the filter media and must not be transported behind the scraper (safety distance to the rotating filter) back into the slurry.

For an example, a vacuum drum filter is operated with a certain pressure difference Δp_{vac}, and the specific solid throughput $q_{s,vac}$ is maximized for maximal rotation speed and minimal possible cake height. The pressure difference for vacuum filters is limited to the vapor pressure of the liquid of less than 100 kPa, and therefore, the variation of pressure differences is limited. If greater pressure differences of up to about 1 MPa are desired, disc or drum filters can be installed completely in a pressure vessel, as schematically shown for a hyperbaric disc filter in Figure 6.28.

In contrast to vacuum filters, hyperbaric filters need a special gastight sluice to transfer the filter cake from the pressurized vessel to the atmosphere. In the

Figure 6.28 Hyperbaric disc filter.

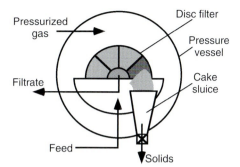

Figure 6.29 HiBar drum filter. Source: Courtesy of BOKELA GmbH.

case of great filters, the pressure vessel exhibits a manhole to carry out maintenance work without opening the big bumped boiler head. In Figure 6.29, such a completely opened pressure vessel can be seen, in which a drum filter is installed.

If the filter is smaller, it can be installed in a pressure vessel, which is separated horizontally and can be opened like an oyster, as shown in Figure 6.30.

This is especially advantageous if a quick access to the filter is desired and the pressure vessel must be relieved from an unhealthy atmosphere. In addition, the downtime before new start up is reduced and thus the effectiveness of the process is increased.

If in the case of hyperbaric filters for constant rotation speed the pressure difference Δp_{pr} is increased, the solid mass throughput is increased and simultaneously the cake thickness rises. Now, the rotation speed can be increased, until the minimal possible cake height is reached again. This additionally increases the specific solid mass throughput $q_{m,s,pr}$. Running the filter for constant minimal cake height means that the specific solid mass throughput becomes directly proportional to the pressure increase, as derived in Eq. (6.48)–(6.50)

$$\frac{h_{c,pr}}{h_{c,vac}} \propto \frac{\sqrt{\Delta p_{pr} \cdot n_{vac}}}{\sqrt{\Delta p_{vac} \cdot n_{pr}}} \qquad (6.48)$$

Figure 6.30 HiBar oyster filter. Source: Courtesy of BOKELA GmbH.

$$h_{c,pr} = h_{c,vac} \Rightarrow \Delta p_{pr} \cdot n_{vac} = \Delta p_{vac} \cdot n_{pr} \tag{6.49}$$

$$\frac{q_{m,s,pr}}{q_{m,s,vac}} \propto \sqrt{\frac{\Delta p_{pr} \cdot n_{pr}}{\Delta p_{vac} \cdot n_{vac}}} = \sqrt{\frac{\Delta p_{pr} \cdot n_{pr}}{\Delta p_{vac}} \cdot \frac{\Delta p_{pr}}{\Delta p_{vac} \cdot n_{pr}}} = \frac{\Delta p_{pr}}{\Delta p_{vac}} \tag{6.50}$$

Coming back to the belt filter at the beginning of this chapter and also in the case of pan filters, the length of the cake formation zone is not fixed, as for drum or disc filters, but depends on the operation conditions. To calculate the length of the cake formation zone L_1 on a belt filter with continuously moving belt, in a first step, again the cake formation time t_1 to get a certain cake height h_c is measured in the laboratory filter cell for a certain pressure difference and slurry concentration. These data have to be transferred to the belt filter according to Figure 6.31.

To get the desired cake height, the slurry height on the belt must be adjusted to the same height for given slurry feed volume flow rate $Q_{v,sL}$. From this information, the belt velocity can be calculated according to Eq. (6.51)

$$v_{belt} = \frac{Q_{v,sL}}{B \cdot h_c} \tag{6.51}$$

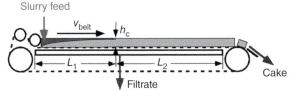

Figure 6.31 Cake formation on a continuously operating belt filter.

If the belt velocity is increased or the slurry feed flow rate is decreased, the cake height decreases and vice versa. The necessary length L_1 for the cake formation zone can be calculated from the belt velocity v_{belt} and the cake formation time t_1, which was measured in the lab filter cell, as formulated in Eq. (6.52)

$$L_1 = v_{belt} \cdot t_1 \tag{6.52}$$

Last but not least, the solid mass throughput can be determined from the solid content in the filter cake, as expressed in Eq. (6.53) for a belt filter of belt width B

$$Q_{m,s} = v_{belt} \cdot B \cdot h_c \cdot \rho_s \cdot (1 - \varepsilon) \tag{6.53}$$

6.3.5 Aspects of Filter Design and Operation Regarding Cake Formation and Throughput

Precondition for an optimal filter performance with respect to all process steps is in every case that the cake on the filter area of any filter apparatus is as homogeneous as possible in cake height and structure. Apparatus design and operation play a remarkable role regarding this task and should be discussed for some examples in more detail.

The optimal operation of discontinuous and also continuous filters requires an adapted cake height, which is not too small or too large for proper operation. In any case, the filter cake should be removed completely from the filter media because otherwise the remaining parts of the filter cake are acting during the next filtration cycle as preresistance for the formation of the new cake, and the entire cake production is reduced. Figure 6.32 shows, as an example, a pressure candle filter, which can be operated with dry or wet cake discharge.

Figure 6.33 allows, in addition, a view from the top into the filter vessel of a modern pressure candle filter.

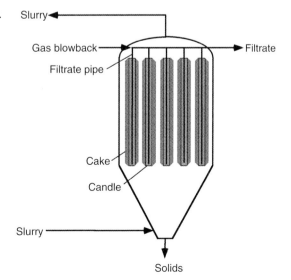

Figure 6.32 Pressure candle filter.

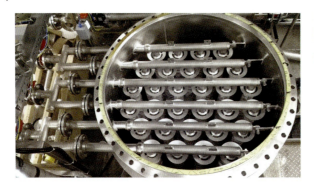

Figure 6.33 FUNDABAC® pressure candle filter. Source: Courtesy of DrM Dr. Müller AG.

The arrangement of the filter candles can be identified very nicely. Common filtrate pipes connect a certain number of filter candles each in the form of registers.

In the case of wet cake discharge, as slurry preconcentration measure, the cake is removed by back-pulse with liquid into the completely slurry-filled vessel. The cake settles down and is discharged as a highly concentrated sludge from the bottom of the vessel. In the case of dry cake discharge, the desaturated cake is removed by gas back-pulse into the gas-filled vessel and likewise discharged from the bottom of the vessel. Dry or wet cake discharges are influencing the detachment mechanisms of the cake from the filter media.

In the case of dry cake discharge, the cake is accelerated together with the filter media by a gas back-pulse of a steep pressure increase velocity. The filter media must exhibit certain elasticity, but must not be elongated by plastic deformation. When the limit of filter medium expansion is reached, the filter media stops, but the cake moves on by its inertia. The inertia force must be greater than the adhesion force between cake and filter media, which is formed by van der Waals and particularly capillary forces. The capillary forces are increasing with increasing cake moisture. This is the reason why extensively desaturated and brittle filter cakes are easier to detach than moist and sticky filter cakes. The smaller the particles become, the more contact points between the filter medium surface and cake exist and the greater the entire adhesion force becomes. To generate enough inertia force of the cake to overcome the adhesion force between the filter media and cake, the acceleration and the mass of the cake must exceed a critical value. If the filter cake mass and thus the filter cake thickness falls below a critical value, a gas blowback is not more possible. In such cases, a back-washing with liquid into the gas-filled vessel followed by discharge of the concentrated slurry becomes necessary (cf. Figure 6.6). Back-washing with liquid needs only low pressure because capillary forces are no longer present due to the fully saturated situation between the filter media and cake. Only van der Waals forces are acting, which are much smaller in liquid environment than in gaseous atmosphere, because the Hamaker constant decreases. Furthermore, particles are not ideally smooth but exhibit a certain roughness, which extends the distance between the particle and filter media and thus decreases the van der Waals forces again.

In the case of wet cake discharge, the detachment mechanism is completely different from the dry cake discharge. Because of the tremendously greater viscosity of liquid in comparison to gas, now inertia forces are playing no more a decisive role. Capillary forces are not present and a comparatively very low back-washing pressure meets the condition for cake detachment. The adhesion forces between filter media and cake must be smaller than the cohesion forces in the cake itself. A bulging of the filter media only costs back-washing liquid with no positive effect. The filter media should behave more stiff than elastic. As a further condition for a complete cake detachment, the cake must have a certain and material-dependent minimum thickness. Otherwise, the cake cracks first at the weakest location and the back-washing liquid mainly flows through this leakage, which impedes the detachment of the leftover parts of the filter cake.

A further problem of cake discharge originates if the filter cakes become too thick because in this case, the danger of growing together between the candles is given, or the total weight can lead to a deformation or destruction of the apparatus. To get the optimal performance, Eq. (6.41) for throughput optimization delivers the optimal cake height and thus the necessary distance of candles from each other to guarantee unhindered discharge. For the filters, as shown in Figure 6.33, a special sensor system has been developed for a reliable detection of an overload of the filter candles because it is not possible to look into the filter vessel during the filters operation. If the filter cake becomes too thick, a sensor detects the deformation of the register and stops at a previously defined critical value the filtration process.

To be able to get optimal filter results, a further necessary condition is a homogeneous filter cake formation. If the slurry in the filter vessel is not homogenized adequately, the performance of the filter will not be optimal. In the case of filters with vertical filter elements in the vessel, like candle or leaf filters, no stirrer can be installed, and thus, an unsatisfactory slurry homogenization, due to superimposed sedimentation of the particles, can occur with the consequence of a particle size and slurry concentration gradient across the vessel height. The coarser particles and the greater concentration will be found near the bottom of the vessel, and thus, the cake formation on the lower parts of the filter elements will be faster than in the upper regions. In such a case, the filtration is limited very early because of exceeded maximal cake height in the lower parts of the candles. As a countermeasure, the slurry should be fed from the bottom and guided in a circulation loop of a little faster flow velocity than the particle settling velocity.

In stirred pressure nutsche filters, as can be seen in Figure 6.34, the slurry homogenization can be promoted by careful stirring during cake formation and adequate removal of the stirrer from the growing cake surface.

Slurry prethickening, agglomeration, or fractionation (cf. Sections 5.2–5.4) also counteracts particle segregation.

Similar to the discussion for the batchwise operating apparatuses, also for continuously operating vacuum and pressure filters, proper process conditions and technically mature filter design are necessary to realize an optimal filter performance. Homogeneous slurry and an even filter cake have to be strived for in any case.

Horizontal filters, such as belt or pan filters, have no possibility to homogenize the slurry during filtration. Adequate slurry pretreatment can prevent the

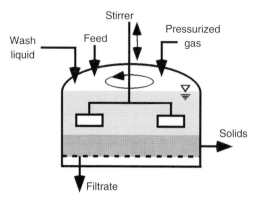

Figure 6.34 Stirred pressure nutsche filter.

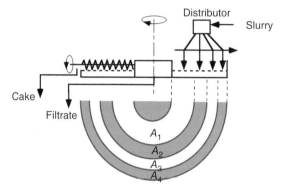

Figure 6.35 Vacuum pan filter slurry feed.

particles from segregation, reduce the settling velocity, and increase the cake formation velocity and thus the throughput.

Especially in the case of pan filters, which are applied for relatively coarse particles of some hundred micrometer size and thus relatively high settling rates, a careful and homogeneous distribution of the slurry on the filter area is essential to get an even cake surface. The fed particles are settling more or less straight to the filter media without deviations in horizontal direction. For that reason, a special slurry feed dosing device, as depicted schematically in Figure 6.35, is necessary to guarantee the same quantity of slurry for each filter area element $A_1 = A_2 = A_3 = A_4$.

The same considerations are valid for belt filters of great width, although the particle size here is normally smaller than in the case of pan filters. As Figure 6.36 illustrates, slurry feed distributors can support the formation of an even filter cake.

In comparison to belt and pan filters, drum and disc filters exhibit better possibilities for slurry homogenization. Here, the filter disc or drum rotates through a slurry filled filter trough, which should be fed from the deepest point of the trough to impede sedimentation by the upstream. In addition, different stirring devices can homogenize the slurry. Nevertheless, these filters are limited toward coarse particles by two reasons. First, the capability of the stirring is limited, and toward greater particles, the danger increases such that a growing sediment is

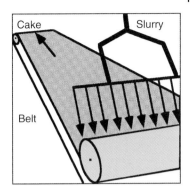

Figure 6.36 Vacuum belt filter slurry feed.

formed in the trough, and the rotating filter disc or drum gets stuck after some time in the sediment. Second, consequence of too coarse and thus very easy to filter particles are very thick and heavy cakes, which may not be held attached at the filter surface after emersion from the slurry and tend to fall back into the filter trough.

In the following, the disc filter should be discussed in more detail to demonstrate the various aspects, which are influencing the filter cake formation. The slurry should be fed in any case from the bottom upward (cf. Figure 6.26) because this measure counteracts particle settling. In the slurry trough of drum filters, normally an oscillating stirrer is integrated (cf. Figure 6.26). For disc filters with several discs, the situation is a bit more complicated. A propeller mixer with axial transport direction between each of the two discs can be found frequently in practice, but this solution generates relatively high costs and effort for maintenance. A more elegant solution, which can be seen in Figure 6.37 for modern disc filters, consists in the so-called joint single trough.

Each disc is rotating in an individual narrow trough and acts as a stirrer itself. For optimal slurry distribution, the slurry is fed individually into the lower part of each small trough and directly on the outer edge of the rotating disc. To guarantee the same slurry level and thus the same cake formation conditions for the entire filter, all single troughs are connected in the upper region. In comparison to the other types of continuously operating filters, the disc filter exhibits a special geometric problem, which hinders the formation of uniform cake height on the filter cells. This problem was examined by Schweigler [17] and should be explained in Figure 6.38.

Looking to the locations A and E, it becomes clear that the residence time of point E in the slurry is much shorter than that of point A. As a consequence, the cake thickness at point A is the greatest and that of point E is the smallest on the sector. A cake height distribution results across the sector area. After leaving the slurry, point E exhibits the longest and point A the shortest deliquoring time. Obviously, cake moisture and a gas flow distribution result across the sector. The lowest overall cake moisture and gas consumption are achieved for a perfect even and homogeneous cake. The question is how to get such a homogeneous cake on the filter sectors of a disc filter. The first measure is to reduce the difference between the cake forming angles $\alpha_{1,E}$ and $\alpha_{1,A}$. The smaller the angle of the sector and thus greater the number of cells, the smaller the differences between points

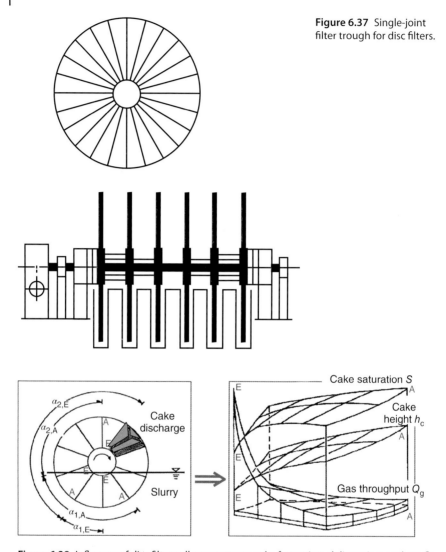

Figure 6.37 Single-joint filter trough for disc filters.

Figure 6.38 Influence of disc filter cell geometry on cake formation, deliquoring, and gas flow.

E and A become. Modern high-efficiency disc filters exhibit up to 30 sectors per disc. Figure 6.39 shows, as an example for such modern high efficiency disc filters, a unit of three discs for the separation of aluminum hydrates.

The difference between E and A becomes even smaller, if the immersion depth of the disc in the slurry is increased. As can be seen in Figure 6.40, if the immersion depth becomes 50%, the advancing and the trailing edge of the sector are getting out of the slurry at the same moment.

One has to consider in that case that the rotating shaft of the filter has to be sealed against the slurry. A further measure to homogenize the cake thickness on the filter sector consists in an increase of slurry concentration because the cake formation rate is increased and the relative difference between the cake heights

Figure 6.39 "Boozer" vacuum disc filter. Source: Courtesy of BOKELA GmbH.

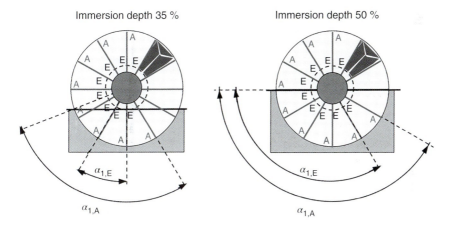

Figure 6.40 Immersion depth variations of the filter disc in the filter trough.

at A and E become smaller because of the fact that the cake height is a function of the root of cake formation time.

An additional negative effect for an equal cake height across the sector is given by hydrostatic cake formation. This is valid especially for large filters of disc diameters of up to 6 m. Before vacuum is applied, the aerated filter cell must be immersed completely in the slurry to avoid leak air in the vacuum system. Because of the hydrostatic pressure of the slurry, which is maximal at the deepest point of the sector, liquid is pressed into the filter cell and some cake is formed until the cell volume is filled up with liquid. As a consequence, an additional cake is built at the outer radius of the cell at the radius of point A. To minimize this effect, the filter cell should be designed conical, as shown in Figure 6.41.

The cell volume is minimal where the outer hydrostatic pressure is maximal. The conical widening of the cell toward the outlet respects the increase of the filtrate volume flow from the outer to the inner radius of the filter cell.

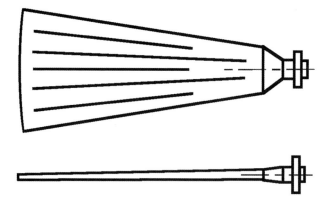

Figure 6.41 Conical disc filter cell.

Last but not least, if the homogenization of the slurry is not perfect, a particle size and concentration gradient forms in the filter trough. The cake formation at the outer radius of the cell is again favored. As a consequence, one has to strive for maximal slurry homogenization in the filter trough.

To get optimal cake formation results, it is at the final end of the filter cycle necessary to remove the filter cake as complete as possible from the filter medium. If parts of the cake would remain on the filter media, the cake formation at these locations would be reduced remarkably during the next cycle because of the high flow resistance of filter cloth plus remaining cake. At those locations would form comparatively moist "islands" because of aggravated cake formation and deliquoring conditions and thus a decrease of the average throughput and an increase of the average cake moisture content would be observed. On the example of the disc filter cake discharge can be discussed very well the necessity of an favorable interaction of filter media, filter design, and filter operation for getting maximal throughput and lowest cake moisture because these parameters are interacting very sensitively [18, 19]. Figure 6.42 shows a perfect cake discharge (a) and an insufficient discharge (b) from disc filter sectors.

The cake discharge is detached exclusively by air blow back. This denomination could cause the misunderstanding that the gas flow is responsible for the cake detachment. In reality, a fast acceleration of the system cloth/cake and its abrupt deceleration results in the detachment of cake from the cloth. The mass force of the cake must be greater than the adhesion forces between the particles and the threads of the filter cloth. The filtrate piping system and the cell volume, which have to be loaded abruptly with pressurized air, must be as small as possible, and pressure losses between the storage vessel for pressurized air and the filter should be avoided. According to Figure 6.43, the vessel should be installed as near as possible to the filter, and the pipes should be straight and of sufficient diameter.

In Figure 6.43, the bigger vessel in the foreground represents the filtrate receiver and the smaller vessel behind the vessel for the pressurized air. The time to open the full cross section between filtrate pipe and process zone in the control head should be minimal, to increase the pressure increase velocity. As

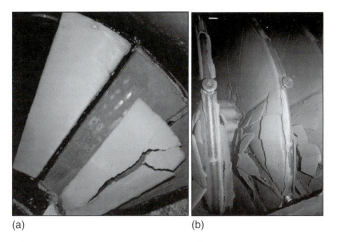

(a) (b)

Figure 6.42 Perfect and imperfect cake discharge from disc filter sectors.

Figure 6.43 Disc filter, blow back vessel, and filtrate receiver. Source: Courtesy of BOKELA GmbH.

Figure 6.44 illustrates, the overlapping of two circles needs more time than the overlapping of a trapezoidal geometry.

In Figure 6.45, a real vacuum disc filter barrel with external trapezoidal filtrate pipes without mounted filter sectors can be seen.

If the filter disc is rotating with comparatively low rotational speed of a few rpm, the pressure increase velocity would be insufficient also in the case of optimal pipe cross section. In such cases, a fast opening or snap blow valve should be used. This valve opens abruptly, when the cross sections are nearly completely overlapping and pressurized air is shot into the filter sector.

Beside the filter design and operation, the filter medium is influencing sensitively the cake detachment. The filter medium must be mounted tight and with perfect fit on the filter sector to avoid the bulging of the filter medium during blowback. Beside inefficient acceleration of the cake, in this case, the danger of cloth destruction due to contact with the scraper exists. The filter cloth should have a high elasticity in circumferential direction and must not wear out by

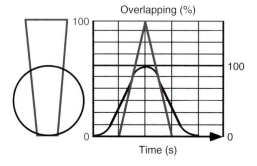

Figure 6.44 Time for overlapping of different pipe opening geometries.

Figure 6.45 Vacuum disc filter barrel with external trapezoidal filtrate pipes. Source: Courtesy of BOKELA GmbH.

plastic deformation. If the chosen filter medium provides these properties, it should be checked whether warp and weft threads are arranged in the correct direction on the filter sector. Only then, the optimal function can be ensured. Sometimes, the pieces of cloth, which are forming the filter bag, are not cut out of a large piece of cloth in warp and weft direction to avoid scrap.

Last but not least, the filter fabric should have a smooth surface to facilitate the cake detachment. In the case of polymeric fabrics, a smoothened surface by calendering would be preferable (cf. Section 9.2).

The adhesion forces between filter media and cake are not only determined by the structure of the filter medium surface but sensitively by the residual moisture of the filter cake. The capillary forces are depending on the liquid saturation of the porous system (cf. Section 8.5.1). For high saturation, which means poor deliquoring, the cake behaves sticky and pasty. Here, the cake discharge from disc filters comes to a limit and will fail. The better the deliquoring and the dryer the cake, the easier the cake discharge becomes.

At the end of this discussion, regarding the influences on the filter throughput, one last aspect should be mentioned especially for vacuum filters. It should be validated that the hydraulic design from the filter cell to the vacuum pump is optimized. Otherwise, the manometer at the pump shows perfect and maximal possible vacuum, but because of several pressure losses on the way to the filter

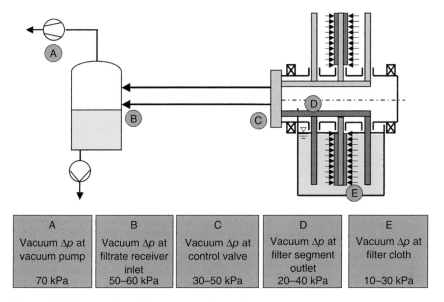

Figure 6.46 Pressure losses from pump to filter cell.

cell, only a very small pressure difference remains finally for the filtration. This should be explained in more detail in Figure 6.46.

As can be seen from practical experiences, the real pressure difference, which is available for the cake filtration, can be much smaller than it seems to be, if the pressure is registered directly at the compressor. By this reason, the arrangement of the plant should be similar to that in Figure 6.42. Short and straight pipes of sufficient diameter should be chosen and the filtrate receiver should be placed as close to the filter as possible. The pressure directly behind the filter cloth can be measured by a hollow needle, which is punctured from the outside through cake and cloth into the filter cell.

6.4 Formation of Compressible Filter Cakes by Pressure Filtration

6.4.1 Fundamental Considerations Regarding Compressible Cake Filtration

If a filter cake structure reacts on a pressure load in the way that the change of structure or the reduction of void volume, respectively, cannot be neglected anymore, the filter cake is characterized, according to the sketch in Figure 6.47, as compressible instead of incompressible.

The physical reasons for compressible cake structures are manifold. It can be generally stated that compressible filter cakes must not consist of compressible particles, but compressibility normally is a consequence of particle rearrangement depending on the particle shape, on the state of agglomeration, and the mechanical load on the structure (cf. Section 3.3). In reality, often several of these

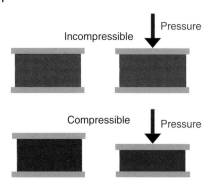

Figure 6.47 Incompressible and compressible filter cake structure.

mechanisms are superposed. Only in exceptional cases, fine particles in a filter cake are arranged in the form of the densest packing structure, and thus certain particle mobility remains. Under the influence of a mechanical load, the particles can be relocated into a more dense packing structure. Small particles in stabilized slurry are forming relatively dense cake structures, which nevertheless can be rearranged in a certain way under the influence of greater filtration pressures. If the particles are agglomerated, the agglomerate structure can be deformed by increased pressure. In the case of not spherical particles, a reorientation can condense a packing structure, and last but not least, the particles themselves can be of compressible nature. From these different consolidation mechanisms, different material laws can be derived, which are formulated in empirical equations, and characteristic material depending parameters must be determined by experiments.

There do exist today two principal approaches to describe compressible cake filtration. The traditional theory was developed over many years by the comprehensive and fundamental work of Grace, Tiller, and Shirato [20–23]. The other approach is based on the so-called compressional rheology, originally developed by Buscall and White [24] and further elaborated by Scales, de Kretser, and Stickland [25–27]. The compressional rheology uses two material functions to describe the compressible cake filtration The compressive yield stress $P_y(c_v)$ describes the resistance against the compression of the cake structure, and the hindered settling function $R(c_v)$ describes the drag between the solid and liquid phase, which is proportional to the well-known specific filter cake resistance in the Darcy equation. Theoretical considerations and extensive experimental evaluation have proved that both principal approaches show good agreement to each other for slightly and highly compressible materials [28]. For this comparison pressure filter cell, compression permeability cell and piston-driven filtration rig had been used for the experiments and analyzed regarding their similarities and differences.

In the following, the conventional approach will be described in more detail, which provides a good principal understanding of the process. If the permeation of a compressible filter cake during formation should be described exactly, the continuity condition demands to consider that not only the liquid is flowing toward the filter medium but also the particles are moving in this direction because of the cake compression [29]. This leads at the end to a system of

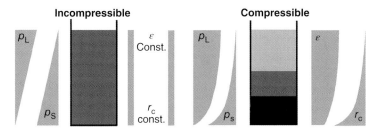

Figure 6.48 Incompressible and compressible filter cake structure.

linked differential equations, which can only be solved numerically. Alternatively, the problem can be solved analytically by the simplification of neglecting the solid flow in the filter cake [30–32]. The integration of the simplified equation for known material laws allows to calculate average values of porosity and flow resistance, which then can be implemented into the basic equation of filter cake formation (cf. Eq. (6.18)). Although the use of local data leads to more exact results [33–35], the simplified integral approach is often used with adequate accuracy for scale-up calculations.

Before going into more details, the principal phenomena of compressible cake formation should be described roughly, to get a first understanding of the process [36]. If a filter cake is formed, the liquid pressure is decreasing in flow direction, whereas the pressure on the solid structure is increasing correspondingly. As shown in Figure 6.48, the pressure of the liquid is decreasing linearly in the case of homogeneous incompressible structures and nonlinearly in the case of compressible structures.

At the end of cake formation, porosity and specific cake resistance are constant for incompressible structures, whereas compressible filter cakes show a gradient of porosity and specific cake resistance across the cake height. The porosity is minimal at the filter medium and maximal at the cake surface. The pressure difference, responsible for the compression, consists in the pressure difference across the filter cake $\Delta p'$ (cf. Eq. (6.9)). At the very first beginning of the filtration process, no cake is present, and the total pressure difference Δp is applied only to the filter medium and thus is equal to $\Delta p''$ (cf. Eq. (6.10)). The higher the pressure loss of the filter media, the less pressure is available for the compression of cake structure. For normally well-selected filter medium, $\Delta p'$ increases in a very short transition time in comparison to $\Delta p''$ until $\Delta p''$ can be neglected. This change of pressure losses during the initial phase of filter cake formation is formulated in Eqs. (6.54)–(6.56).

$$t_1 = 0 \Rightarrow \Delta p' = 0, \Delta p'' = \Delta p \tag{6.54}$$

$$t_1 > 0 \Rightarrow \Delta p' \uparrow, \Delta p'' = \text{const.} \tag{6.55}$$

$$t_1 \gg 0 \Rightarrow \Delta p' = \Delta p \tag{6.56}$$

Because of the fact that the cake compression reacts on $\Delta p'$, the first cake layers are compressed by an increasing pressure difference $\Delta p'$. When the pressure loss

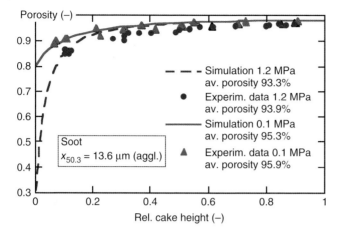

Figure 6.49 Local porosity as a function of the relative cake height [37].

of the filter medium is negligible in comparison to the pressure loss of the cake, $\Delta p'$ remains constant. The particle layers across the cake are getting increasing load toward the filter medium because the flowing liquid is transferring pressure and shear forces to particle structure, which are increasing toward the filter medium.

Figure 6.49 shows exemplarily the measured and calculated local porosity of highly compressible filter cakes of soot for two different pressure differences of 1.2 and 0.1 MPa.

As can be seen very clearly, the first cake layers near the filter medium are compressed much more for $\Delta p = 1.2$ MPa than for $\Delta p = 0.1$ MPa. The porosity for the layer directly at the filter medium was measured by an independent experiment in a compressibility/permeability (CP)-cell (cf. Figure 6.54), where filter cakes had been pressed to the equilibrium state at 0.1 and 1.2 MPa. In contrast to that, a difference between the average porosities of the cakes is apparently present, but comparatively small. The bottleneck for the filtrate flow in every case is located at the highly compacted particle layers next to the filter media. This is normally known as skin effect and is held responsible for difficult filtration conditions in the case of compressible filter cakes. If, on the other hand, the cake would not be compressible, the complete cake would have the small porosity of the skin layer and the filtration conditions would become even worth. In the case of cake compressibility, the porosity of the cake becomes greater toward the cake surface. This, at the end, supports the cake formation conditions.

Because of the fact that for compressible filter cakes the local cake properties are not constant during cake formation, the description of the filtration process can be done considering local cake properties or average values, such as average porosity or average permeability of the entire filter cake. The global description considers the entire filter cake with an average porosity and an average specific permeability. No detailed information about the inner structure of a filter cake is given by this approach, but the handling is easier and only relatively simple measurements are needed to get the necessary data. In contrast to that, the local

Figure 6.50 Local solid pressure.

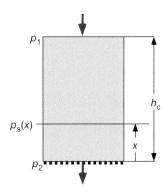

description gives detailed information about the local changes of cake properties inside the filter cake during the filtration process, but the measurements are more complex. For the description on the basis of local data, the filter cake can be divided into layers parallel to the filter media, for which beside a permeation law, mass and force balances have to be formulated. For that matter, each single layer has to be as thin as its structure can be considered as approximately homogeneous.

The cake is considered here as consisting of rigid particles. A Newtonian liquid is transferring for a laminar flow regime pressure and shear stresses to the particles, which are passing them over to the neighboring particles via contact points. A force balance around the filter cake between the cake surface and a location x results in a force $F_s(x,t)$, which is absorbed for a certain time t at the location x by the cake structure. According to the sketch in Figure 6.50, $F_s(x,t)$ can be calculated by the local pressure of the liquid and the cross-sectional area A of the filter cake.

Dividing this force by the area, according to Eq. (6.57), a pressure on the solid structure p_s results

$$\frac{F_s(x,t)}{A} = p_s(x,t) = p_1 - p(x,t) = \Delta p' - (p(x,t) - p_2) \tag{6.57}$$

For constant pressure difference $\Delta p'$, the solid pressure is a function of location x and time t. The solid pressure increases in the direction of liquid flow by the same value, as the hydraulic pressure decreases and grows at a constant location with increasing cake height.

The assumption can be made that the local values of porosity ε and thus the local specific flow resistance r_c are exclusively dependent on the local pressure on the solid structure p_s. This causes consequences for the average porosity and thus the average flow resistance of the filter cake during the formation, when the cake height is increasing. It is assumed that for a constant time, the solid pressure should be only a function of pressure difference at the cake $\Delta p'$ and the location inside the cake x/h_c. The porosity ε should be a function of the solid pressure and thus a function of $\Delta p'$ and the location inside the cake x/h_c. This is expressed in Eq. (6.58)

$$\varepsilon = f(p_s) \Rightarrow \varepsilon = f\left(\Delta p', \frac{x}{h_c}\right) \tag{6.58}$$

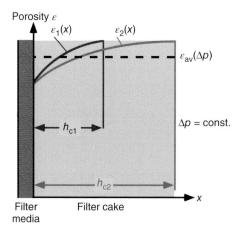

Figure 6.51 Average cake porosity independent of cake height.

Now, the integral porosity ε_{av} for the entire filter cake can be calculated in Eq. (6.59)

$$\varepsilon_{av} = \int_0^1 \varepsilon\left(\Delta p', \frac{x}{h_c}\right) \cdot d\left(\frac{x}{h_c}\right) \Rightarrow \varepsilon_{av} = f(\Delta p') \qquad (6.59)$$

As a result and as shown in Figure 6.51, the average porosity ε_{av} is not more a function of the filter cake height but only a function of the applied pressure difference $\Delta p'$.

The result of Eq. (6.59) has been validated many times in practice and simplifies the description and analysis of the compressible filter cake formation remarkably because from a constant average porosity follows a constant average cake resistance. If the filtrate flow during filter cake formation is measured as a function of time, the representation of the data in the form of $t_1/V_L = f(V_L)$ should result according to Figure 6.52 equivalent to incompressible filter cakes in a linear function.

Two deviations in comparison to Figure 6.14 for incompressible filter cakes can be observed here. To identify compressible cake structures, the pressure difference Δp is now included additionally to the ordinate of the diagram. Equation (6.59) describes the linear function for compressible filter cakes, from which $r_{c,av}$

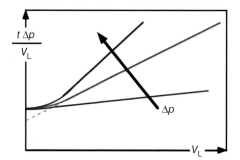

Figure 6.52 Determination of average specific cake and medium resistance for compressible cakes.

and R_m can be determined

$$\frac{t_1 \cdot \Delta p}{V_L} = \frac{\kappa_{av} \cdot \eta_L \cdot r_{c,av}}{2 \cdot A^2} \cdot V_L + \frac{R_m \cdot \eta_L}{A} \tag{6.60}$$

In case of incompressible filter cakes, all linear functions would result in one single master curve for all pressure differences. In the case of changing cake structure by compression for each pressure difference, an individual function originates. The second difference to Figure 6.14 is related to the start of the cake formation. As discussed before, the pressure difference applied to the cake $\Delta p'$ varies for compressible cakes from $\Delta p' = 0$ to $\Delta p' = \Delta p$. Therefore, at the very first beginning of filtration, the function is horizontal and approaches after a certain time the final inclined straight line. In practice, for very fine-grained particles and filter media of comparable small resistance, the resolution of this phase is in most cases nearly not possible and the extrapolation of the linear function back into the ordinate is precise enough. Nevertheless, this initial phase of the filtration should be measured as accurate as possible because the simple extrapolation of the straight line back to the ordinate could cause some miscalculation of the filter medium resistance. The filter medium resistance would be underestimated in such cases.

In comparison to incompressible filter cakes for compressible filter cakes, the additional information about the compressibility of the cakes must be measured for different pressure differences. According to Figure 6.53, Alles could demonstrate that the number of necessary experiments can be reduced remarkably by increasing the filtration pressure step by step during one single filtration experiment.

For each pressure, the expected straight line of different slope originates. This evaluates the compressible character of the filter cake. The curves are starting horizontally and not exactly, but nearly from the origin. Here, the filter medium resistance is obviously very small in comparison to the cake resistance.

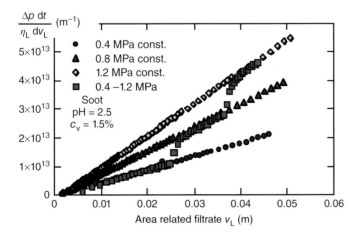

Figure 6.53 Flow resistance measurement of compressible filter cakes (cf. [36], p. 120).

In addition, it can be observed that starting with the lowest pressure and increasing the pressure during the filtration experiment to the next level, the same curve progression originates, as having started the experiment with the higher pressure from the beginning. This enables to measure the average cake flow resistance for different pressures in one single experiment.

To describe the dependency of the average porosity and average flow resistance of compressible filter cakes from the pressure difference, several empirical approaches are available. Often power functions are used for this purpose. The following equations are using reference values for the average porosity $\varepsilon_{av,0}$, cake flow resistance $\alpha_{c,av,0}$, and pressure p_0 [38]. The experimental results can be represented as can be seen in Eqs. (6.61) and (6.62) by power functions with characteristic exponents G and N, which can be denominated as global compressibility numbers

$$(1 - \varepsilon_{av}) = (1 - \varepsilon_{av,0}) \cdot \left[1 + \frac{\Delta p}{\Delta p_0}\right]^G \tag{6.61}$$

$$\alpha_{c,av} = \alpha_{c,av,0} \cdot \left[1 + \frac{\Delta p}{\Delta p_0}\right]^N \tag{6.62}$$

They can be used to rate the compressible material behavior. For incompressible cakes, N approximates 0 and for highly compressible cakes, N approximates 1. The same kind of correlation can be used to describe filter cakes, which have been pressed to the equilibrium state.

For compressible filter cakes, it is often usual to express the specific flow resistance of the cake not by the cake height-related parameter $r_{c,av}$ (m^{-2}) but by the cake mass-related parameter $\alpha_{c,av}$ (m kg^{-1}). Both parameters can be converted into each other, as Eq. (6.63) makes clear

$$\alpha_{c,av} = \frac{r_{c,av}}{(1 - \varepsilon_{av}) \cdot \rho_s} \tag{6.63}$$

The reason for that is not of principle nature, but compressible cakes originate normally from difficult to filter slurries and exhibit after a reasonable filtration time only small heights and not exactly even surfaces. In such cases, it is simpler to weigh the mass with adequate precision than to measure the height. A second reason is that a compressible filter cake sometimes relax a little bit after pressure release, and if the cake height is not measured in situ, but measured after the filtration experiment, a wrong cake height is registered and used for further calculations.

6.4.2 Experimental Determination of Process Characterizing Parameters

To analyze the filter cake structure during and after filtration, different techniques are available. In principle, the formation of compressible filter cakes is possible in the same pressure filter cell, as for the incompressible filter cakes (cf. Figure 6.12). In this case, normally, the filtration pressure difference is held constant. Especially in the case of highly compressible filter cakes, normally press filters are

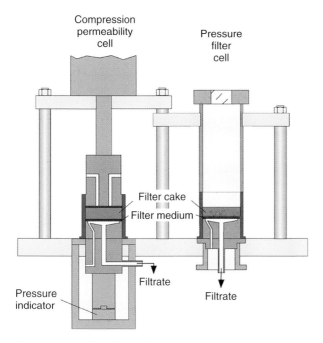

Figure 6.54 Tandem filter cell.

applied, which are often fed with a volumetric pump and constant volume flow. For this type of filtration or for the application of any pressure function during cake formation, a so-called CP-cell is the best option. In this case, a movable piston applies the pressure. Such a CP-cell is schematically shown in Figure 6.54 in combination with a pressure filter cell in the form of a tandem arrangement [39].

The filter cake either can be formed first in the pressure filter cell of about 50 cm^2 filter area and afterward transferred for consolidation into the CP-cell or is formed directly in the CP-cell. The advantage of the tandem arrangement is on the one hand the possibility to filter relatively large volumes of diluted slurry in the pressure filter cell and on the other hand to check whether shrinkage cracks would occur in the case of a subsequent gas pressure deliquoring. After formation, the filter cake can be put into the CP-cell and squeezed by a movable piston under defined conditions with pressures of up to 1.6 MPa. In addition, the cake can be permeated by liquid. Not only the bottom of the cell is covered by a filter medium but also the piston contains at its end a liquid distributor plate, which is covered by a filter medium, and liquid can be pressed through the cake. The bottom of the cell with the filtrate drainage channel is connected via a floating piston with a second pressure indicator to be able to measure the real pressure difference, which acts on the cake.

In the ideal case, the solid pressure is identical to the applied pressure, but one has to consider the interaction between cake and wall of the pressure cell. Typically, the quotient of cake thickness h_c and diameter of the pressure cell D is used here to estimate the magnitude of this influence. Measurements have demonstrated that even for $h_c/D \approx 0.1$, the applied pressure is not transferred ideally

[40, 41]. As a solution of this problem, the measured data for porosity and flow resistance can be allocated to an average solid pressure or effective pressure, which can be calculated for dominant wall friction as logarithmic mean of the pressures on the surface and bottom side of the cake. If wall friction is negligible, the arithmetic mean is preferred. For the generation of scale-up data for press filters, the quotient h_c/D of the lab-scale experiment should be adapted to the technical conditions [42].

The filter cake height and thus the average porosity of the cake in the CP-cell can be measured by measuring the movement of the piston. Extremely sensitive works are applied in the method of laser triangulation. With this method, not only the consolidation of the cake but also the elastic recovery of the cake can be registered in the CP-cell as well, as in the pressure filter cell.

Local porosities of the cake can be measured in the simplest case, and if the filter cake is stable enough, by cutting the cake after cake formation into small slices. For each slice, the porosity can be determined and thus the porosity function across the cake height.

If the local data of porosity and pressure should also be measured during the cake formation, an extended version of the CP-cell is needed. Local porosity can be measured in situ by stepwise X-ray transmission and local pressure by water-filled capillaries and pressure transducers in different heights of the cake. In Figure 6.55, such an arrangement is sketched after a proposal of Sedin et al. [43].

The local porosity in cake layers of about 1 mm thickness is measurable by an X-ray source, which radiographs the cake and is registered by a detector. This arrangement is mounted on a movable rack and enables in that way to measure the local porosity distribution of the entire cake in a nondestructive way. The local pressure in the cake can be measured by several liquid-filled capillaries, which are mounted in the bottom and are passing through the forming filter cake. Each capillary is connected with a pressure transducer. However, the capillaries cause a slight defect in the cake structure, but the shown vertical arrangement has been

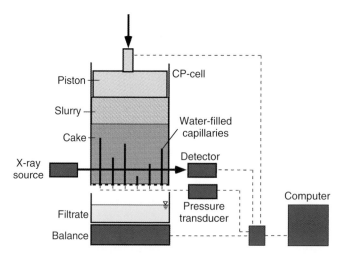

Figure 6.55 CP-cell with integrated local porosity and pressure measurement.

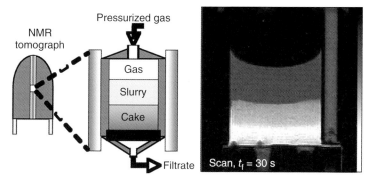

Figure 6.56 Magnetic resonance imaging (MRI) of a growing filter cake.

proven as a better solution than a horizontal arrangement, as it was installed in the previously described tandem filter cell (cf. [36], p. 63).

The most informative rather complex method to analyze a filter cake structure is given by the X-ray-based microtomography. This technology enables to create cross sections of a filter cake and recreate, with the support of a special software, a virtual three-dimensional model, without destroying the original object.

An alternative technique for in situ measurements is realized in the form of nuclear magnetic resonance tomography (NMR). As shown in Figure 6.56, a special filter cell, developed by Erk, has to be installed completely in a superconducting magnet, and the water content in each volume element of the cake can be analyzed and visualized by magnetic resonance imaging (MRI) [44, 45].

This technique is able to follow the local changes of porosity across the growing cake height during the filter cake formation or the forming of sediment in the case of particle separation by sedimentation instead of filtration.

6.4.3 Optimization of Compressible Cake Filtration

For compressible filter cakes, the question can be asked, whether an optimal pressure exists to get maximal filtrate flow because two tendencies are contradicting each other. On the one hand, an increased pressure increases the filtration velocity. On the other hand, an increased pressure consolidates the cake structure, increases its flow resistance, and thus decreases the filtration velocity.

The area-related filtrate volume flow $q_{v,L}$ in Eq. (6.64) for compressible filter cakes can be determined by the Darcy equation in combination with Eq. (6.62)

$$q_{v,L} = \frac{\Delta p}{\eta_L \cdot m_s \cdot \alpha_{c,av} \cdot \left[1 + \frac{\Delta p}{\Delta p_0}\right]^N} \qquad (6.64)$$

The compressibility is characterized by the integral compressibility factor N. For $N = 0$, the filter cake behaves ideally incompressible and $q_{v,L}$ is directly proportional to Δp. For $N > 0$, the cake becomes more and more compressible and $q_{v,L}$ increases with declining slope, if the pressure difference Δp is increasing as can be observed in Figure 6.57.

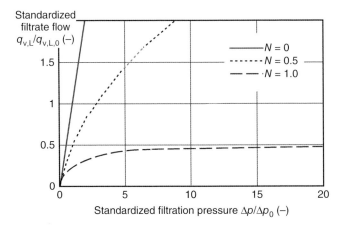

Figure 6.57 Pressure-dependent filtrate flow for filter cakes with variable compressibility (cf. [36], p. 99).

In the case of $N = 1$ and greater pressure differences, a further increased filtration pressure will be compensated nearly completely by the increased cake flow resistance. The most intensive consolidation takes place near the filter medium (cf. Figure 6.50). As a consequence, the pressure loss is located mainly near the filter medium, and a loose structure is forming above this highly consolidated particle layer, which is sometimes called skin layer and dreaded in practice [46].

For theoretically supposed values of $N > 1$ at high pressure differences, again a decrease of filtrate flow could be expected, but in extensive experimental studies, even for extremely compressible products, such as soot, china clay, or cellulose, N never exceeded 1. As a result, it can be stated that normally an optimal pressure does not exist, and even for very compressible materials, a very slight increase of filtrate flow with increasing pressure can be observed. However, in practice, sometimes, an optimal pressure can be identified. Probably, secondary effects are responsible for this observation. One reason for such an effect might be, as illustrated in Figure 6.58, a destruction of particle bridges at the filter medium and a following pore blinding in the case of exceeding a critical pressure.

Here, a smooth pressure increase at the beginning of the filter cycle may lead to a stable bridge formation and in combination with an afterward applied high pressure, a minimum of total cake formation time can be achieved.

Continuously operating press filters, such as belt or screw presses, unfortunately cannot be operated with constant maximal pressure because they need principally a very careful pressure increase. If the pressure is increased here too fast, the sludge is in danger to be pressed through the screen or backward.

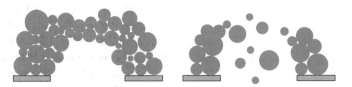

Figure 6.58 Particle bridge destruction.

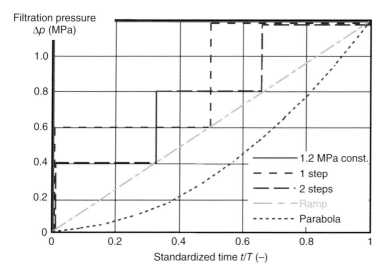

Figure 6.59 Pressure increase functions for filter cake formation (cf. [36], p. 117).

If a start of filtration at lower pressure would lead to a stable situation at the filter medium, Alles had checked whether this strategy could also circumvent the formation of the highly compressed first cake layer to minimize the cake formation time. According to Figure 6.59, different pressure increase functions for the filter cake formation phase had been tested in lab-scale experiments to check this hypothesis.

In comparison to the application of the full pressure of 1.2 MPa from the start to the end of the cake formation, a one-step pressure increase, a two-step pressure increase, a ramp, and a parabola had been tested. The pressure at the end in every case amounted to 1.2 MPa.

Figure 6.60 shows the results for different materials from which titanium dioxide and product "I" represent less compressible, the soot, and material "S" highly compressible materials.

It becomes very clear that in all cases, a constant maximal pressure from the beginning leads to the shortest cake formation time. Although for lower pressure, less dense cake layers are formed; simultaneously, the total time for permeation increases, and the complete cake is consolidated in every case after increasing the pressure to the highest level to the same final porosity. The corresponding porosity data are given in Figure 6.61 for the different pressure increase functions.

The porosity is correlated here directly with the residual cake moisture content because all pores remained completely filled with liquid. For all pressure increase functions, the maximal pressure amounted to 1.2 MPa, and thus, the porosity in every case was identical as expected.

The general conclusion from the discussed results to get maximal throughput is the recommendation to hold the filtration pressure constant at the maximal reasonable level during the complete cake formation process. Exceptions can be given if the filter cloth is blinded for exceeding a critical pressure or the consolidation mechanism of the filter cake changes drastically to extremely high flow resistance.

6 Filter Cake Formation

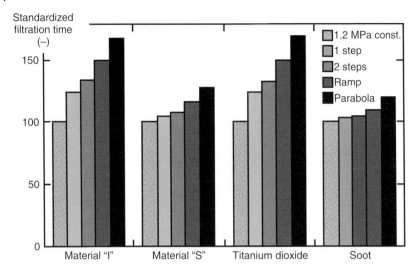

Figure 6.60 Cake formation time for different pressure increase functions. Source: Data from [36], p. 118.

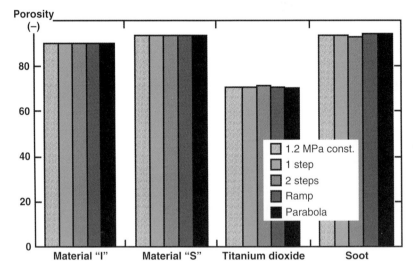

Figure 6.61 Cake porosity for different pressure increase functions. Source: Data from [36], p. 119.

6.4.4 Aspects of Filter Design and Operation Regarding Cake Formation and Throughput

A good example to demonstrate the historical progress of filter performance is given by the development process of filter presses depicted schematically and roughly in Figure 6.62.

These filter apparatuses are operating normally in the range of pressure differences between about 0.6 and 1.6 MPa and in some cases up to about 2.5 MPa.

6.4 Formation of Compressible Filter Cakes by Pressure Filtration

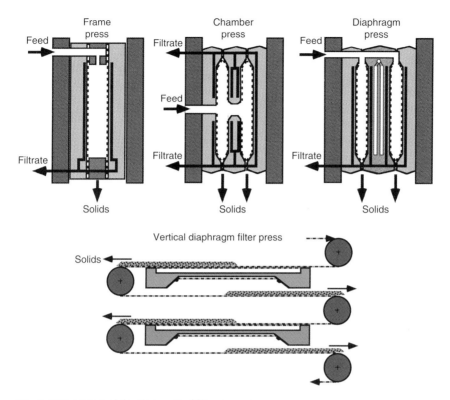

Figure 6.62 Historical development of filter presses.

Although all types are still practically used, a clear progress in filter performance can be observed from the horizontal frame, chamber, and diaphragm press to the vertical diaphragm press. However, not in all applications, the filtration conditions require the most efficient filter design and thus all versions can be the right choice for a practical filtration task.

The first aspect to discuss here should be the difference in down time for the filter cake discharge because down time means less throughput. For the frame press, the longest time is needed because the cakes have to be broken out of the frames, in which they are sticking after filtration. The design of the filter plates for chamber filter presses, where no more frames are required, is reducing the down time. The filter cakes should drop by themselves due to gravity. Unfortunately, it cannot be excluded that single cakes are remaining nevertheless in the opened chamber by sticking at the filter media or canting themselves between the filter plates. In such cases, they must be removed manually. The same is valid for the horizontal diaphragm press. In any case, the filter cake must have a certain thickness to offer enough weight to be detached from the filter cloth and to be able to fall down. By this reason, the filter cakes must have a thickness of about 3–6 cm. Because of the fact that filter presses are applied for difficult to filter slurries, the cake formation often needs several hours to fill the chambers completely with cake. To increase the throughput and to approach a maximum of possible throughput for constant rate filtration (cf. Section 6.3.3), one would

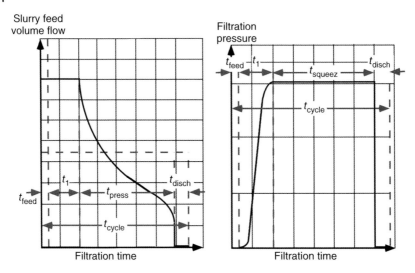

Figure 6.63 Filter press cycle for constant rate filtration.

be interested to realize cake formation times as long as possible and down times as short as possible. The ideal course of constant rate filtration for filter presses is given qualitatively in Figure 6.63.

During constant rate feeding of the chambers, the pressure increases during cake formation. In the ideal case, the consolidation of the cake starts when the maximum pressure is reached and additional slurry is pressed in the chambers at maximal pressure. Obviously, the feed flow decreases drastically during this procedure. After undergoing a critical value of feed flow, the consolidation phase is stopped, and in the case of no subsequent cake washing, the cake is discharged. In many cases of difficult to filter slurries and due to the fact that some cm of cake thickness are needed for a proper cake discharge, the maximum of pressure is reached before the chambers are filled completely with cake and the feed flow decreases (constant pressure filtration). As a general consequence, thinner cakes and shorter down time (consolidation) would be advantageous.

This is principally possible by using the vertical version of the diaphragm press, as shown in Figure 6.64, because here the mechanism of cake discharge is different from the horizontal presses.

After a quick opening of the chambers by gravity, the filter cloth, which consists of one endless piece for the entire press, is moved one chamber length forward. The consolidated and stiff filter cake is transported horizontally out of the chamber and the filter cloth is sharply bent around a roller and thus detached from the cake. For safety, an additional scraper at the discharge roller can be used to ensure a complete discharge. Now much thinner cakes can be handled in shorter times and the throughput rises.

A second advantageous aspect of the horizontal chamber consists in the homogeneous composition of the filter cakes. The better the homogeneity of a filter cake in general, the better the filter performance will become. Filter presses are built up to about $20\,m^2$ plate area and more than $1000\,m^2$ filter

Figure 6.64 Tower filter press PF 180. Source: Courtesy of OUTOTEC Oyj.

area per unit. These circumstances lead to the challenge to distribute the slurry homogeneously in all chambers. If the filtration velocity is poor, particles have time to settle because the transport velocity through the feed pipe toward the single chambers is low. If slurries contain a relatively broad particle size distribution and the slurry concentration is relatively low, the danger of particle segregation exists from the first to the last chamber and from the top to the bottom in each chamber. Normally, filter presses are fed with flocculated slurries of higher concentrations, which reduce such negative segregation effects, but in the case of many plates per unit also flocs are settling. Then, it might be a problem especially for large presses with a high number of filter plates to transport the same quantity of solids to the chambers with the maximal distance to the feed entry point. The production of homogeneous filter cakes in such unfavorable cases is easier to realize in the horizontal chambers of a vertical filter press with slurry layers of relative small height. Further advantages of the vertical diaphragm press regarding the formation of homogeneous filter cakes, which are important for washing and deliquoring, will be discussed in the respective chapters.

In the case of extremely hard to separate slurries, the batch time of filter presses, as discussed before, would be too long for an economical operation and greater pressure differences would be advantageous. This is no longer possible to realize in filter presses of the previously described types, which consist of a stack of flat filter plates from the viewpoint of mechanical stability. Pressure differences up to more than 10 MPa can be realized in so-called tube presses, which exhibit a circular cross section. The principle of a diaphragm tube press is depicted in Figure 6.65.

Relative thin cakes can be formed and squeezed with extremely high pressure in comparatively short time. For cake discharge, the diaphragm is sucked back, the inner part of the filter moves downward, and a circular gap between housing and filter element appears, through which the cake is able to drop downward.

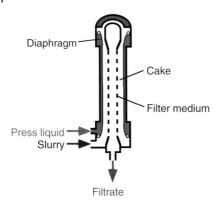

Figure 6.65 Tube diaphragm press.

The cake detachment is supported by a gas blowback. Unfortunately, here, the filter area per unit is relatively small, but greater throughput can be realized by a parallel arrangement of tube presses.

The filtration kinetics in filter presses with filter plates are promoted by covering both sides of the filter chamber with filter medium. This doubles the active filter area per chamber or halves the time for filling the entire chamber with cake. This principal strategy to increase the efficiency of a filter apparatus by increasing the filter area on the spent footprint of the apparatus had been observed already in the Chapter 5 during the discussion of continuous vacuum filters and especially disc filters. All of those continuous operating vacuum filters are doing the same from the physical point of view, but in the case of disc filters, the special arrangement of the filter area in the form of several vertical discs with filter media on both sides on one common shaft realizes a maximum of filter area per footprint area.

The transfer of the idea to install more filter area per apparatus volume or footprint, respectively, is also possible in tubular piston presses by implementing special drainage filter media. Two types of such filter apparatuses are known, which are realizing large filter areas in the process room and very thin filter cakes. The first type is shown in Figure 6.66 [47].

Figure 6.66 Bucher Press HPX5005 with tubular filter elements. Source: Courtesy of BUCHER UNIPECTIN AG.

6.4 Formation of Compressible Filter Cakes by Pressure Filtration

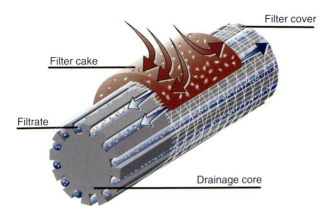

Figure 6.67 Bucher Press drainage element. Source: Courtesy of BUCHER UNIPECTIN AG.

The filter media in these piston filter presses consist of a high number of parallel-arranged flexible tubular filter elements, which are connected with the piston on one side and with the wall of the pressure vessel on the opposite side. As Figure 6.67 illustrates, these drainage elements consist of a supporting structure of polyurethane, which is covered by a hose of woven polypropylene fabric.

In a first step, slurry is filled between the strained filter elements. Afterward, the piston is moved forward, the flexible filter elements are distorted, and the filter cake is formed. Short filtration distances and great filter area per volume lead to high filtration kinetics of this thin layer filtration. Filtration and squeezing is carried out with pressure up to about 0.6 MPa. In the next step, the piston is moved back and the press slowly rotates. The solids are rearranged and liquid inclusions are liberated. The process can be repeated several times and finally opening the vessel and stretching the filter elements discharge the deliquored cake.

The second and principally similar system, in the form of the WRING press, uses flat rectangular filter sheets instead of tubular filter elements, as can be seen in Figure 6.68 [48].

Again, a concentrated sludge is fed in a first process step between the stretched drainage media. Then, the piston is moved forward and filtration as well as squeezing takes place in one cycle with pressure up to 10 MPa. After squeezing (Figure 6.68a), the piston is moved back, the filter media are stretched, and the vessel opens for cake discharge (Figure 6.68b). The cake is broken into small flaky pieces and detached from the filter media by shaking them mechanically (Figure 6.68c).

Press filtration cannot be realized only by batchwise operating apparatuses but also continuously by screw or double wire presses. Because of the limited residence time in the process room and pressures up to only about 400 kPa, these presses are very good suited for relatively easy to filter slurries, which form highly compressible filter cakes. Many applications can be found for slurries, which contain fibrous particles such as cellulose or highly flocculated systems. This type of apparatus is characterized not only by squeezing but also by shearing of the cake, which promotes consolidation.

(a)

(b)

(c)

Figure 6.68 WRING press. Source: Courtesy of BOKELA GmbH.

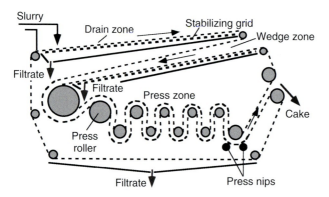

Figure 6.69 Belt press.

The first type of continuously operating press filters consists of the belt press. The principle here is the feeding of slurry between two filter media, which are moving in the same direction and are bent around several rollers. The sludge is squeezed and sheared between the clothes. In Figure 6.69, a sketch of a belt press is shown.

The sludge is fed on a horizontal strain zone, where the sludge is concentrated by gravity near the gel point. The hydrostatic pressure is due to a sludge layer thickness of about 3–10 cm limited to about 0.3–1 kPa. To support the drainage of liquid, the sludge has to be strongly flocculated or a body feed with fibrous particles, such as cellulose, is realized. In addition, baffles can be mounted to open drainage channels and turning the sludge. At the end of the drainage zone, the sludge consistency must allow further transport. The sludge is now transported to the wedge zone, where it is squeezed carefully between two filter media. The filter media are not sealed to both sides and the sludge must not leave the belts to the sides or pressed back but should be transported forward. Stabilizing grids often support the two belts. The throughput of the apparatus is mainly determined in the wedge zone, which has to be designed carefully regarding opening angle of the belts, length, and transport velocity of the sludge. At the end of the wedge zone, the filtration pressure has increased to about 20 kPa. The following squeezing takes place by the movement of the belts around a certain number of rollers. At the end of this zone, additional pressure can be applied by press nips from outside. Finally, the cake is discharged. The squeezing and shearing process will be discussed more in detail in Section 8.4.2.

The second type of continuous operating press filters is represented by the screw press, which is illustrated in Figure 6.70.

In a screw press of classic design, the sludge is filtered and squeezed in the spiral channel of a transport screw, which is narrowing in transport direction and thus reduces the volume between the flights. The screw rotates slowly in a screen basket.

Reversing the principle and doubling the filter area leads to an upgrade of screw presses [49]. As shown in a principle sketch in Figure 6.71, here the screw is fixed,

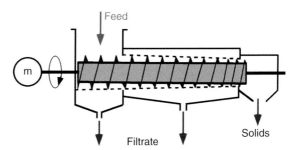

Figure 6.70 Screw press.

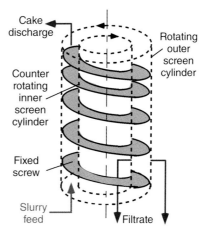

Figure 6.71 Screw press with a fixed screw and two independently rotating screens.

whereby an inner and an outer screen cylinder are independently operated and counter-rotating with variable rotation speed.

This concept follows on the one hand again the idea to increase the available filter area in the process room. On the other hand, the sludge is squeezed and revolved very carefully by narrowing the screw channel between the two screens. Doubling of the filter area and the revolving of the sludge result in comparatively low residual cake moisture and high separation kinetics.

Belt and screw press are operating continuously but are unfortunately only suited for relatively easy to filter suspensions. As discussed for discontinuously operating filter presses, the solid throughput is comparatively small for difficult to filter suspensions because of the necessarily relative thick filter cakes. The ideal case for maximal throughput would be a continuously operating filter for very difficult to filter suspensions. Because of the limited residence time of the product in the process zones of continuously operating filters, the resulting filter cakes would be very thin, which means in the range of 1 mm or less. One imaginable solution of such a task, and an alternative to a discontinuously operating filter press, could be a continuously operating drum filter in a pressurized vessel (hyperbaric filter), which is equipped with microporous hydrophilic membrane filter media and a roller discharge for the very thin and sticky filter cake, as illustrated schematically in Figure 6.72 [50].

A comparative calculation of the solid mass throughput of both apparatuses should demonstrate the expected potential of such a solution. The following boundary conditions were set. Both filters should operate at the same filtration pressure difference and identical slurry. For the filter press, a chamber depth of 50 mm is defined here in line with standard device dimensions, which means a cake thickness of 25 mm each, which are formed on both sides in each chamber. This cake thickness of 25 mm, required for a complete chamber filling, defines the cake formation time t_1. The consolidation time t_2 is supposed to be equal to the cake formation time t_1, just as in the case of the drum filter. This increases the comparability of both filter systems. The feed time t_f for filling the chambers with slurry and opening as well as closing of the chambers for cake removal should need 30 minutes. This allows the determination of the total down time t_d in Eq. (6.65)

$$t_d = t_2 + t_f \tag{6.65}$$

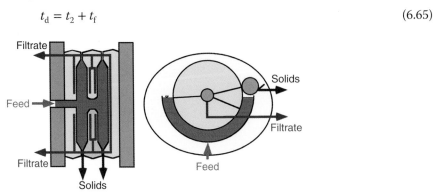

Figure 6.72 Filter press and rotary pressure filter.

6.4 Formation of Compressible Filter Cakes by Pressure Filtration

Now, the specific solid mass throughput $q_{m,s,FP}$ of the filter press can be calculated in Eq. (6.66)

$$q_{m,s,FP} = \frac{Q_{m,s,FP}}{A \cdot (t_1 + t_d)} = \frac{A \cdot h_{c,FP} \cdot (1 - \varepsilon_{av}) \cdot \rho_s}{A \cdot (t_{1,FP} + t_d)} = \frac{h_{c,FP} \cdot (1 - \varepsilon_{av}) \cdot \rho_s}{t_{1,FP} + t_d} \quad (6.66)$$

For the drum filter, the following boundary conditions were set. For the maximal possible rotary speed of $n = 12$ min^{-1}, a cake height of $h_c = 1$ mm should be formed. Cake formation angle α_1 and cake deliquoring angle α_2 should be 120° each, and thus, cake formation time $t_{1,PF}$ and cake deliquoring time $t_{2,PF}$ become equal. Now, the specific solid mass throughput for the pressurized drum filter $q_{m,s,PF}$ can be calculated in Eq. (6.67)

$$q_{m,s,PF} = \rho_s \cdot h_{c,PF} \cdot (1 - \varepsilon_{av}) \cdot \frac{\alpha_1}{360°} \cdot \frac{1}{t_{1,PF}} \quad (6.67)$$

The comparison of both systems is given according to Eq. (6.68) by the relation of both specific solid mass throughputs

$$\frac{q_{m,s,PF}}{q_{m,s,FP}} = \frac{\rho_s \cdot (1 - \varepsilon_{av}) \cdot h_{c,PF} \cdot \dfrac{\alpha_1}{360°} \cdot \dfrac{1}{t_{1,PF}}}{\rho_s \cdot (1 - \varepsilon_{av}) \cdot h_{c,FP} \cdot \dfrac{1}{t_{1,FP} + t_d}} \quad (6.68)$$

For a comparative calculation, the density of the solids is set to $\rho_s = 3000$ kg m^{-3}, the density of the liquid to $\rho_L = 1000$ kg m^{-3}, the dynamic liquid viscosity to $\eta_L = 10^{-3}$ Pa s, the solid volume concentration of the slurry to $c_v = 0.1$, and the average cake porosity $\varepsilon_{av} = 0.7$. The specific filter cake resistance r_c was varied. The results of these calculations are depicted in Figure 6.73.

Toward very high specific filter cake resistances, the throughput ratio once approaches the value of about 22. Toward lower specific filter cake resistances, the throughput ratio increases strongly. The marked limits for the maximal rotation speed of the drum filter indicate the lower limit of filter-specific cake resistances,

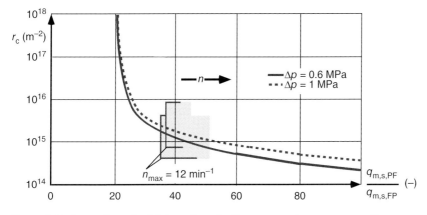

Figure 6.73 Comparison of solid throughput for filter press and rotary pressure filter.

which lead to a cake height of $h_c = 1$ mm. As a result of these examinations, it can be stated that the expected solid throughput of a pressure drum filter, for comparable conditions, would lead to 20–40 times greater throughput than a filter press. Problems, which can arise during deliquoring of compressible filter cakes with gas differential pressure, like formation of shrinkage cracks, will be discussed in more detail in Section 8.5.

6.5 Formation of Filter Cakes in Centrifuges

6.5.1 Fundamental Considerations Regarding Cake Filtration in Centrifuges

The formation of filter cakes in centrifuges is much more complex to describe than for vacuum or pressure filtration. The two main reasons are first the permanent and individual changing of the centrifugal pressure and second the superimposed and not negligible sedimentation due to the high mass forces. According to Figure 6.74, the hydrostatic pressure of a liquid column in the centrifugal field generates here the pressure difference Δp.

The centrifugal pressure can be calculated according to Eq. (6.69)

$$\Delta p = (\rho_L - \rho_g) \cdot a_{cent} \cdot h \approx \rho_L \cdot a_{cent} \cdot h = \rho_L \cdot a_{cent} \cdot (r_o - r_i) \tag{6.69}$$

The density of gas ρ_g can be neglected in comparison to the density of liquid ρ_L. The centrifugal acceleration a_{cent} in Eq. (6.70) not only depends on the angular velocity ω or the rotation speed n but also on the radius of the rotor r

$$a_{cent} = r \cdot \omega^2 = r \cdot 4 \cdot \pi^2 \cdot n^2 \tag{6.70}$$

To be able to calculate with a constant centrifugal acceleration for constant rotation speed, in many cases, the so-called long-arm approximation can be used. If the liquid height h in the centrifuge is small in comparison to its radius r_o, the

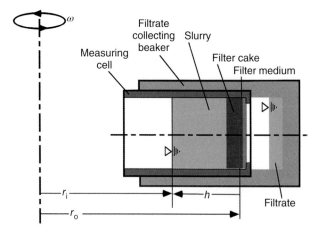

Figure 6.74 Principle of centrifugal filtration.

mean centrifugal acceleration can be calculated in Eq. (6.71) using the arithmetic average of the liquid height

$$a_{cent} = \frac{r_o + r_i}{2} \cdot \omega^2 = r_{av} \cdot \omega^2 = g \cdot C \qquad (6.71)$$

However, a certain error is still made, but one can calculate the deviation from the exact result and decide whether it is acceptable or not.

In the world of centrifugal applications, it is common to operate with the dimensionless centrifugal number C in Eq. (6.72), which is also known as the Froud number Fr

$$C = \frac{r_{av} \cdot \omega^2}{g} \qquad (6.72)$$

The C-number represents the multiple of g, which can be realized in a centrifuge, and thus, this number can compare the effectiveness of centrifuges properly.

First hints for the filterability of a slurry in filter centrifuges can be received from static filter experiments using a lab-scale pressure filter cell (cf. Section 6.3.2), but normally it makes more sense and generates more realistic data, if a lab-scale beaker centrifuge is used, as shown exemplarily in Figure 6.75.

There are several arguments for this practice-orientated procedure. The mass forces, acting on the particles, are by the factor C greater than in the earth field, which means that the sedimentation and segregation effects are of much more influence. The porosity of filter cakes in a centrifuge is often slightly lower in comparison to vacuum or pressure filtration due to superimposed vibrations of the centrifuge drive. A further argument to use a beaker centrifuge instead of a pressure filter cell is the principally different deliquoring behavior of the filter cake in a centrifugal field in comparison to that of vacuum or pressure filtration (cf. Section 8.3.6). To avoid noticeable start-up and shutdown times, it is essential

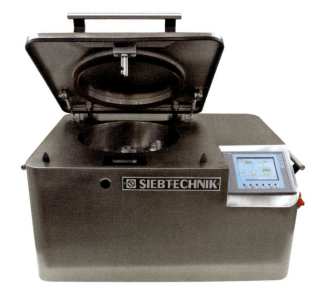

Figure 6.75 Laboratory beaker centrifuge CENTRILAB. Source: Courtesy of SIEBTECHNIK GmbH.

148 | *6 Filter Cake Formation*

for lab-scale beaker centrifuges to accelerate and decelerate the beakers as fast as possible, which means at least in one second. This is especially important for the design of continuously operating filter centrifuges with very short residence times of the product in the process room. To avoid spill over and churn up of the slurry in the beaker, a "soft-start-function" can be recommended, which needs a drive system with frequency converter. The process time should be adjusted in steps of less than one second and the maximum acceleration should cover most industrial applications up to C-numbers of about 3500–4000. If the filtrate could be collected in a special beaker, a load cell in the bottom of the beaker can register the time-dependent filtrate flow. The data transfer from the centrifuge to the data logging system either can made conventionally by a collector ring at the centrifuge shaft or wireless by telemetry.

A further possibility to analyze the filtration behavior especially of very small particles in the micrometer range is given by radiography of rotating transparent cuvettes, which are designed as a filter. This type of analytical photocentrifuge was developed on the basis of the so-called LUMiFuge of the company LUM GmbH, which was originally designed to analyze particle sedimentation. By analysis of the light transmission profile, specific cake permeability, and filter medium resistance, can be determined [51, 52]. Figure 6.76 shows the principle of this technique.

One advantage of this technology is the possibility to analyze several samples on the rotor at the same time. During the filtration process, the cuvette is radiographed by light and a radial CCD sensor detects the resulting light transmission profile. The light transmission profile is registered for several time steps. By analyzing these light transmission profiles, the time-dependent accumulation of filtrate is attained. To measure the filter medium resistance R_m, the filter medium is permeated with a certain amount of particle-free liquid at constant rotation speed. From the analysis of the light transmission profiles, the data are attained to calculate the filter medium resistance from the slope of the linear function Y_1

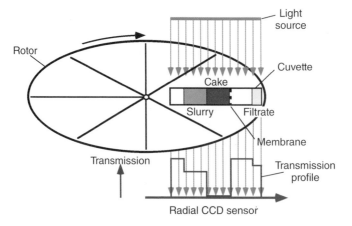

Figure 6.76 Principle of radiographic analysis of centrifugal filtration.

of the liquid height h_L above the filter media determined in Eq. (6.73)

$$Y_1 = \ln\left[\frac{2 \cdot r_m - h_L}{h_L} \cdot \frac{h_{L,0}}{2 \cdot r_m - h_{L,0}}\right] \tag{6.73}$$

According to Eq. (6.74), Y_1 depends on a constant factor k_1, which depends on the filter medium resistance R_m and on the centrifugation time t

$$Y_1 = k_1 \cdot t \tag{6.74}$$

The constant factor k_1 is given in Eq. (6.75)

$$k_1 = \frac{\rho_L \cdot \omega^2 \cdot r_m}{\eta_L \cdot R_m} \tag{6.75}$$

To measure the cake permeability P_c, a filter cake is permeated again with a certain amount of particle-free liquid at constant rotation speed. For the calculation of the cake permeability, it has to be considered that not only the filter cake but also the filter medium is permeated. From the experimental data, again a linear function can be formulated and the filter cake permeability can be calculated from the slope of the function Y_2, which is determined in Eq. (6.76)

$$Y_2 = \ln\left[\frac{r_m + r_c - h_L}{r_m - r_c + h_L} \cdot \frac{r_m - r_c + h_{L,0}}{r_m + r_c - h_{L,0}}\right] \tag{6.76}$$

According to Eq. (6.77), Y_2 depends on a constant factor k_2, in which the permeability P_c can be found, and on the centrifugation time t

$$Y_2 = k_2 \cdot t \tag{6.77}$$

The constant factor k_2 is given in Eq. (6.78)

$$k_2 = \frac{\rho_L \cdot \omega^2 \cdot r_m}{\eta_L \cdot \left[R_m + \dfrac{r_m - r_c}{P_c}\right]} \cdot t \tag{6.78}$$

The geometrical parameters are explained in Figure 6.77.

At the end, no guarantee is given to get absolutely comparable cake structures between beaker centrifuges and the real machine. In a beaker centrifuge, the slurry is not exposed to shearing as in production centrifuges, where the slurry is fed on a rotating liquid surface and accelerated abruptly. This can influence the

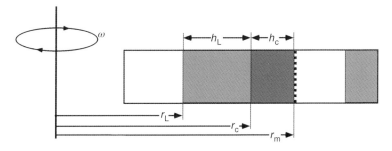

Figure 6.77 Geometrical parameters of the filtration cuvette.

cake structure and especially if the particles have been previously agglomerated, the flocs are stressed completely different in a lab-scale beaker or an industrial centrifuge basket.

However, in the lab-scale beaker centrifuge, the principle filtration behavior of slurry in the centrifugal field in many cases can be investigated under sufficient realistic conditions. In the next step, the results of the experiments have to be transferred to the different types of filter centrifuges. The challenge here is that the liquid level and thus the filtration pressure especially in batch filter centrifuges vary permanently. After closing the slurry feed valve, the liquid level and thus the centrifugal pressure decreases. The situation becomes much more complicated if one looks at the different feed procedures for batch centrifuges in Figure 6.78, which are used in practice.

Most frequently, the centrifuge basket is filled with solids by multiple refilling. The basket is first filled to a maximum level to generate maximal centrifugal pressure. When the liquid level has descended due to filtration, a second portion of slurry is filled into the basket. This procedure is repeated until enough cake is collected. Here, the liquid level rises up and down periodically.

A second procedure realizes continuous filling with first constant liquid and at the end decreasing level. To avoid an overflow of slurry, a peeling pipe permanently skims the upper layer of liquid. In this way, permanent maximal pressure is realized during the filling period, the slurry is preconcentrated by skimming, and eventually unwished small and light substances could be separated from the slurry. An example for such a procedure is the skimming of pore blocking proteins

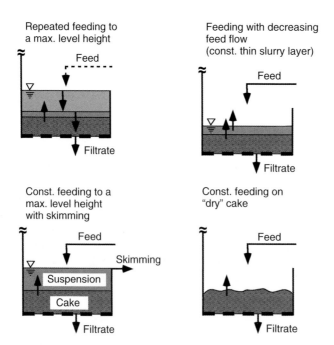

Figure 6.78 Feed procedures for batchwise operating filter centrifuges.

during filtration of starch slurry. Here, the centrifugal pressure remains constant during filling and decreases after closing the feed valve.

In the third case, slurry is fed in the way that only a small supernatant of slurry is present on the growing cake. Here, the fed slurry equals the discharged filtrate. The liquid height is rising permanently during the cake formation until the feed valve is closed and the liquid level decreases slightly at the end.

The fourth feed procedure is principally not preferred because it leads to very uneven cake surfaces, unbalances, and machine shutdown. In case of easy filterable products and relatively small feed volume flow, it can happen that after feeding at 6 o'clock, the dry cake surface appears at 10 o'clock and fresh slurry is given afterward on the dry cake surface again. In such a case, the particles have no chance to be distributed evenly on the whole filter area to form an even cake. Here, in principle, the centrifugal pressure increases during the cake formation.

The formation of incompressible filter cakes from a homogeneous slurry with negligible sedimentation in comparison to filtration as one extreme of cake formation in a centrifuge has been described principally in Section 6.3.1. For thin cakes ("long-arm approximation"), the filtration process can be handled as an even problem. For thick cakes, the change of permeated area with changing radius has to be considered. Pure cake filtration as dominating cake formation mechanism can be assigned to continuously operating filter centrifuges with relatively open metallic slotted screens as filter media and easy to filter slurries of higher solid concentrations.

The other extreme is much faster sedimentation than filtration and subsequent drainage of the clear supernatant through the already formed sediment (cake). Beside the pressure variation during the cake formation process, the superimposed sedimentation has to be considered here. If the "long-arm approximation" is postulated and the Stokes equation is valid, the sedimentation velocity of a particle in the centrifugal field $w_{St,cent}$ is by the factor C faster than in the earth field $w_{St,g}$, as formulated in Eq. (6.79)

$$w_{St,cent} = \frac{(\rho_s - \rho_L) \cdot g \cdot x^2}{18 \cdot \eta_L} \cdot C = w_{St,g} \cdot C \tag{6.79}$$

For $C = 1000$, a particle settles 1000 times faster in the centrifuge than in a static tank. This separation behavior can be assigned to batchwise operating filter centrifuges with relatively tight fabrics as filter media and more difficult to filter slurries. Particularly, distinct to observe is this behavior for peeler centrifuges with a remaining heel, which remains on top of the filter media because the peeler-knife must not touch the filter media during cake discharge. This heel acts as additional filter medium resistance, which decreases the filtration velocity additionally in comparison to the particle settling velocity.

Figure 6.79 illustrates the principle situation of such a supernatant filtration after initial fast sedimentation.

The complete sandwich of different porous solid layers, which are causing the absolute flow resistance R_{tot} for the clear supernatant, consists of the perforated metallic filter basket, a drainage grid, the filter media itself, the first bridge-forming particle layer, (the heel) and the sediment. To calculate the cake formation time t_1 until the dry sediment (cake) surface appears and thus the

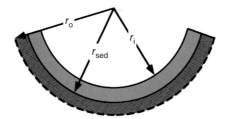

Figure 6.79 Sediment and clear supernatant in a centrifuge basket.

cake formation time has finished, sedimentation time t_{sed} and drainage time t_{dr} have to be added according to Eq. (6.80) and to be determined separately

$$t_1 = t_{sed} + t_{dr} \tag{6.80}$$

First, the time for the sediment formation has to be determined from sedimentation experiments. If the "long-arm approximation" is valid, experiments with a static glass cylinder in the lab can be transferred according to Eq. (6.79) to the centrifugal field.

Secondly, the time for the drainage of the clear supernatant from r_i at the beginning ($t = 0$) to the surface of the sediment r_{sed} must be calculated. For this purpose, the resistance of the porous system must be known. It can be estimated by a permeation experiment of a filter cake (including the heel) assuming that the filter medium resistance and the resistance of the perforated centrifuge basket are comparatively negligible. The exact calculation of the nonsteady flow of the supernatant through the sediment is relatively complex and leads, at the end, to the drainage time t_{dr} in Eq. (6.81) [53]

$$t_{dr} = \frac{R_{tot} \cdot \eta_L \cdot \ln(r_o/r_{sed})}{\rho_L \cdot \omega^2} \cdot \left[\frac{r_o^2 - r_{i(t=0)}^2}{r_o^2 - r_{sed}^2} \right] \tag{6.81}$$

6.5.2 Aspects of Centrifuge Design and Operation Regarding Cake Formation and Throughput

If the sedimentation in batch centrifuges is playing a considerable role in most cases, two consequences are following. As the first consequence, the filter cake is formed much faster by sedimentation than by filtration, and the remaining clear supernatant must drain through the cake of the final height. This elongates the total time, until the dry cake surface appears, as has been discussed previously in Section 6.3 for presedimentation in a pressure filter cell (cf. Figure 6.15). As a second consequence, the danger of particle segregation increases remarkably, if a relatively broad particle size distribution is present. As shown in Figure 6.80, a layer of the smallest particles and high flow resistance will form on top of the sediment after each portion of fed slurry.

This elongates the cake formation time again because the clear supernatant must now penetrate this fine particle layer of high flow resistance. In case of repeated feeding, as practiced very frequently in industrial applications, several such orifices are formed and the cake formation time is extended many times more than would be necessary in case of a homogeneous cake structure. Beside options, like agglomeration or prethickening, an effective operational measure

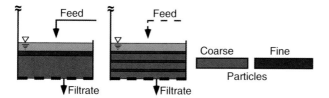

Figure 6.80 Particle segregation due to sedimentation in a filter centrifuge.

in case of expected segregation is a reduced feed flow and continuous feeding to the upper limit with only a relative small slurry layer above the growing cake. In that case, the particles have no more chance to segregate in the small slurry layer, and a homogeneous cake is formed in comparatively short time. In the case of varied feed volume flow, the feed pipe orifice in the centrifuge basket has to be adjusted in the way that it remains completely filled. In horizontal (horizontal rotation axis) peeler centrifuges, the end of the feed pipe is changing from a circular cross section to a horizontal slit. This leads to even distribution of the feed slurry jet in the centrifuge basket and the formation of a cylindrical cake of equal height. If the filling velocity is decreased to fill up the basket in one single stroke, the slit must be decreased too to remain completely filled. As a consequence, the slit must not be fixed but adjustable. One possibility for self-regulation in peeler centrifuges is shown in Figure 6.81.

Here, the feed pressure opens the orifice against a resilient spring force of a plate, which is fixed only at the circular feed pipe. The more slurry is fed, the more open the orifice becomes.

It is more difficult to form a cylindrical cake in a vertical (vertical rotation axis) centrifuge situation because centrifugal force and gravity are perpendicular to each other. Enough rotation speed and special feeding devices, as can be seen in Figure 6.82, can nevertheless solve the problem.

A static feed pipe with several outlets of different opening widths across its length or an inclined rotating feed plate is able to distribute the slurry homogeneously on the filter area.

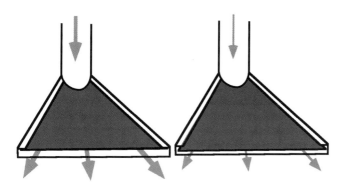

Figure 6.81 Self-regulating feeding device of a horizontal peeler centrifuge.

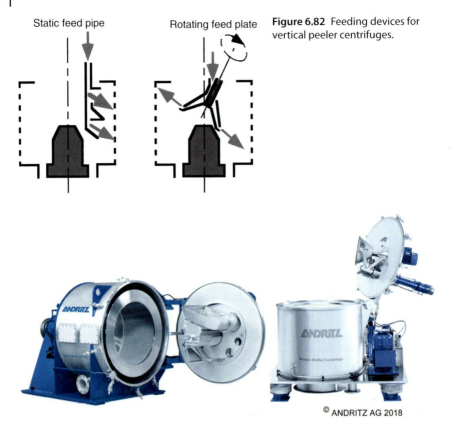

Figure 6.82 Feeding devices for vertical peeler centrifuges.

Figure 6.83 Horizontal and vertical peeler centrifuge. Source: Courtesy of ANDRITZ KMPT GmbH.

Beside these aspects to form a homogeneous cylindrical filter cake, a principal difference regarding throughput exists between horizontal and vertical peeler centrifuges as shown exemplarily in Figure 6.83.

The discussed feeding devices for the horizontal and vertical centrifuge are installed in the centrifuge door as well, as other devices, such as level controller, wash pipe, and scraper.

The highest throughput can be produced with horizontal peeler centrifuges in comparison to not only vertical peeler but also inverting filter or top discharge centrifuges. The reason is that the peeler centrifuge with horizontal rotation axis can be operated in the ideal case with maximal rotation speed during the whole batch time, whereas the other variants have to reduce the speed or even to shut down the machine completely for cake discharge. The speed reduction is time-consuming and thus reduces the throughput. Figure 6.84 illustrates the principle course of rotation speed during a centrifuge batch for horizontal and vertical peeler centrifuges.

The necessity of speed reduction is caused by the procedure of solid discharge. In horizontal peeler centrifuges, the peeler knife (cf. Figure 6.83) scrapes the cake out of the basket and the solid product is discharged through a chute, if

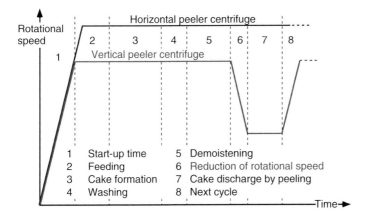

Figure 6.84 Speed profile for horizontal and vertical peeler centrifuges.

Figure 6.85 Cake discharge for vertical peeler centrifuges.

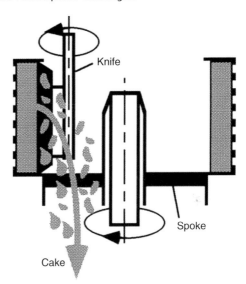

the flow behavior of the solids is sufficient. In the case of sticky and pasty products, a guided transport via screw is used. The scraper should move forward as fast as possible to minimize particle contact with the knife, which causes particle destruction and knife abrasion. This is not possible in every case. If the filter cake is quite hard, the limited mechanical stability of the discharge device requires a certain reduction of the rotation speed and thus additional time.

In the vertical centrifuge, the cake is discharged downward through a spoke wheel, as can be seen in Figure 6.85.

If the rotation speed during peeling would be too high, the solids would be centrifuged immediately back to the wall of the basket and the solids could not be discharged downward. For this reason, the rotation speed has to be reduced principally for vertical peeler centrifuges during solid discharge.

In the case of inverting filter centrifuges, as depicted in Figure 6.86, the cake is discharged by pushing an insertion out of the basket at reduced rotation speed.

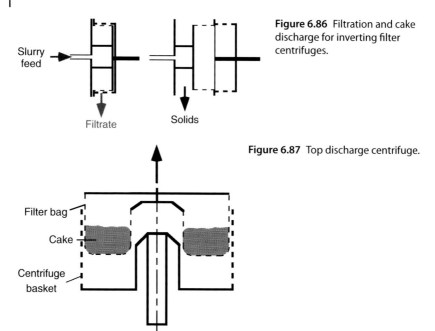

Figure 6.86 Filtration and cake discharge for inverting filter centrifuges.

Figure 6.87 Top discharge centrifuge.

The filter medium is fixed on one side at the insertion and on the other side at the basket. When the insertion is pushed out, the filter medium is inverted and the complete solids can be discharged without strong impact with the housing.

The top discharge centrifuges for mechanically sensitive products and small charges have to be shut down completely to open the lid and to lift out the cake-filled filter bag. This is illustrated in Figure 6.87.

Top discharge and inverting filter centrifuges are getting rid of the complete filter cake and thus are starting the next charge with minimal filter medium resistance. In peeler centrifuges, normally a small heel remains after peeling in the basket because the knife is not allowed to touch and destroy the filter cloth. Especially in the case of centrifuges for the pharmaceutical industry, if batch identification is an issue, the heel must be removed after every batch for complete gathering of all products of the batch. However, this can also be realized in peeler centrifuges by pneumatic means, as shown in Figure 6.88 for the horizontal version.

From the outside, nitrogen is blown through the basket. The filter cloth is deformed whereby the heel is fragmentized and blown into the solid discharge chute. The same principle is used for vertical peeler centrifuges. Alternatively, the heel can be separated from the cloth by pneumatic nozzles behind the scraper. The heel removal represents an extra process step, needs some time, and thus reduces the throughput.

If the heel is not removed after the batch, but remains in the basket, the next batch is confronted with the flow resistance not only of the filter cloth but also of the heel. This also elongates the time, until the cake surface appears from the liquid, and thus the heel should be as thin as possible in the range of a few millimeters. In addition, the flow resistance of the heel increases from batch to batch

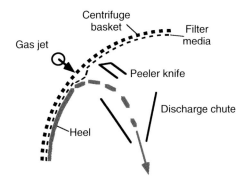

Figure 6.88 Pneumatic heel removal for horizontal peeler centrifuges.

and one has to remove or to regenerate it from time to time. For the removal, not only the pneumatic method can be used. If batch identification is not necessary, the heel can be rinsed off the basket by positioning the scraper at the heel surface and feeding clear liquid through the feed pipe. The heel is redispersed in that way and discharged via the solid chute. In extreme cases, the heel can be dissolved completely in a solvent.

A very efficient method to regenerate the heel consists in back-flushing, but for this procedure, a special design of the centrifuge basket is necessary. The filtrate is not flowing more through the filter medium and the basket directly into the housing of the centrifuge but is collected in an additional filtrate-collecting chamber behind the filter medium and skimmed from a ring channel at the backside of the basket by pipe, which is adjustable in height by swiveling. Figure 6.89 exemplifies the principle.

The outer wall of the open-ring channel must exhibit a greater height than the height of the heel. During filtration, the filtrate pipe is positioned near the outer radius of the ring channel and the fast rotating filtrate leaves the centrifuge through the fixed opening of the pipe under pressure.

After cake discharge, the filtrate pipe is turned to a position above the heel, and a part of the already separated filtrate is given back through a second pipe into the ring channel. The liquid level rises and penetrates the basket, the filter cloth, and the heel from the backside. Then, the back-flushing is stopped and fresh slurry

Figure 6.89 Back-flushing basket of a peeler centrifuge.

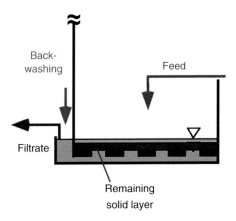

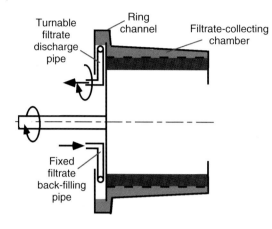

Figure 6.90 Basket design of a siphon peeler centrifuge.

is fed into the basket. The saturated and weakened heel becomes redispersed by mixing with the fresh slurry. After turning the filtrate pipe again into its deepest position, the filtration starts again. A new and permeable heel is formed. Beside the heel regeneration, particle washing and/or extraction can be improved by the possibility of holding a portion of liquid in the basket over some time without filtrate drainage (cf. Chapter 7).

A modification of the described filtrate-collecting system around the centrifuge basket in the form of the so-called siphon design leads to the possibility of adding an extra vacuum behind the filter medium to increase the filtration pressure and thus the throughput [54]. Figure 6.90 shows the design of a siphon basket.

The ring channel now exhibits a greater radius than the filtrate-collecting room. This enables to adjust the liquid level in the ring channel by a turnable filtrate discharge pipe to a greater radius, than explained for the back-flushing basket. In other words, a liquid column behind the filter medium originates, which is pulled by the centrifugal forces in radial direction and produces a negative pressure behind the filter medium. In Figure 6.91, the situation around the siphon cup is explicated in more detail.

The pressure behind the filter medium p_1 can be calculated by the Bernoulli equation, which is formulated in general terms in Eq. (6.82)

$$\frac{\rho_L}{2} \cdot v_L^2 + p + \rho_L \cdot g \cdot C \cdot h = \text{const.} \tag{6.82}$$

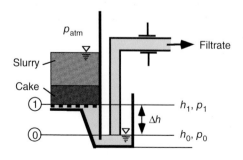

Figure 6.91 Generation of negative pressure in the Siphon basket.

The liquid flow velocity v_L is very small, and thus, the kinetic energy can be neglected. In Eq. (6.83), the Bernoulli equation is formulated between point "0" and point "1"

$$p_0 + \rho_L \cdot g \cdot C \cdot h = p_1 + \rho_L \cdot g \cdot C \cdot h_1 \qquad (6.83)$$

Finally, the pressure p_1, depending on the height of the liquid column behind the filter media $(h_1 - h_0)$, can be calculated in Eq. (6.84)

$$p_1 = p_0 - \rho_L \cdot g \cdot C \cdot (h_1 - h_0) \qquad (6.84)$$

If the liquid consists of water of 20 °C, the geodetic height is sea level, $\Delta h = 2$ cm, and $C = 500$, a hydrostatic pressure of 10 m water column results, which corresponds to 100 kPa. $\Delta h = 2$ cm and $C = 500$, are really not a challenge for peeler centrifuges, and thus, it makes no problem to reach the maximal possible pressure difference at the vapor pressure of water very easily. The maximal possible additional pressure for the given data of about 90 kPa remains constant during the cake formation and accelerates the process, whereby the centrifugal pressure decreases with decreasing liquid level in the basket. Unfortunately, the siphon effect is only temporary because the first emptied pore channel through the cake during deliquoring leads to a pressure equalization on both sides of the cake, and the vacuum pressure difference disappears. To realize the siphon effect, in a first step, the space behind the filter medium must be filled up with liquid to guarantee a hydraulic connection between cakes and the filtrate-collecting chamber. This procedure is combined with back-washing of the filter cloth and regeneration of the heel. It is equivalent to the before mentioned procedure in the back-flushing basket. Back-flushing and siphon basket are not applicable for high hygienic requirements and necessary batch identification because the filtrate-collecting room is not really accessible for cleaning and the heel can be regenerated, but not be discharged and is mixed with the following batch.

An alternative to peeler centrifuges, especially for hygienic applications, is given by the inverting filter centrifuge, as shown in Figure 6.92.

Because of a special cake discharge mechanism, the cake can be discharged completely and no heel remains in the process room. For cake discharge, the rotation speed is reduced, and a special drum insertion moves axially out of the drum and pushes the cake out. During this operation, the filter cloth, which is fixed at one side at the drum and at the other side at the insertion, is inverted.

A further feature of this centrifuge consists in the possibility to apply higher gas pressures in the drum because the process room can be designed as a pressure tight vessel. This allows, according to Figure 6.93, the application of a gas overpressure through the feed pipe of up to 600 kPa, which in contrast to the siphon peeler centrifuge can be held constant without temporal limitation.

The additional gas pressure is also beneficial for the deliquoring of the filter cake, which can be intensified by exchanging pressurized air by pressurized steam or by using hot pressurized air. This will be discussed in more detail in Section 8.3.6.3.

On the example of continuously operating filter centrifuges such as pusher, screen bowl, worm screen, vibrating, or sliding centrifuges, it can be demonstrated nicely how different product requirements regarding cake formation,

Figure 6.92 Inverting filter centrifuge. Source: Courtesy of HEINKEL Process Technology GmbH.

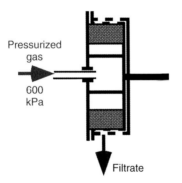

Figure 6.93 Inverting filter centrifuge with superimposed gas pressure.

washing, and deliquoring are considered by several machine design variations. Especially, centrifuges for easy or very easy to filter slurries exhibit only a relative short residence time in the range of seconds of the product in the process room. If the cake resistance is relatively high, machines for longer residence time are available. In any case, the residence time for continuously operating centrifuges is shorter than for batch centrifuges.

In the basket of the pusher centrifuge, the cake formation and the subsequent process steps, such as washing and/or deliquoring, must have been completed in about 20 seconds or shorter [55]. Figure 6.94 shows the principle of a one-stage pusher centrifuge.

Otherwise, the highly dreaded phenomenon of the so-called "flooding" can take place and the slurry is flowing more or less not separated into the solid outlet. However, flooding limits the throughput of a pusher centrifuge and can be influenced by process and product parameters as well as by centrifuge design and operation, as will be discussed later on.

Figure 6.94 Single-stage pusher centrifuge.

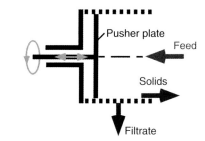

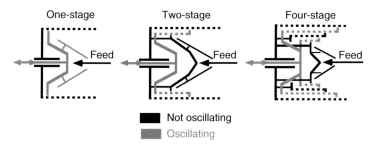

Figure 6.95 One, two, and three stage baskets of pusher centrifuges.

In principle, the slurry is fed into the rear part of the basket on the rotating screen. An axially oscillating pusher plate realizes the transport of the solids toward the open end of the basket. As demonstrated in Figure 6.95, pusher centrifuges are available not only in single stage but also in multistage design, whereby two stages represent the mostly applied standard design.

More than three stages are principally possible, but seldom due to the increasing complexity of the design and increasing costs. Here, as an extreme example, a four-stage design is shown. In the one-stage design, only the pusher plate is oscillating axially. During the forward stroke, the cake is transported toward the open end of the basket. During the backward stroke, again the space on the filter medium originates, which is filled up with fresh slurry. In the two-stage design, the inner screen is oscillating but not the pusher plate. The front edge of the inner screen now acts additionally as a pusher plate for the outer screen. In the four-stage design, the inner first basket and the third basket are oscillating, whereby the pusher plate and the second and the fourth basket are not oscillating.

As mentioned before, the two-stage design is the standard design for pusher centrifuges, although the design is more complicated and more costly than the one-stage version. Because of the rearrangement of the cake at the transition point from stage to stage, the danger of flooding is reduced. Furthermore, the force to push the cake along the screen is divided into two portions. This measure reduces the stress on the particles, which could be in danger to be disintegrated [56]. The second or last stage, respectively, can be designed cylindrical or slightly conical. If the basket has a conical shape, one component of the centrifugal force acts parallel to the screen and facilitates the solid transport. The described advantages of two and more stages are particularly remarkable from

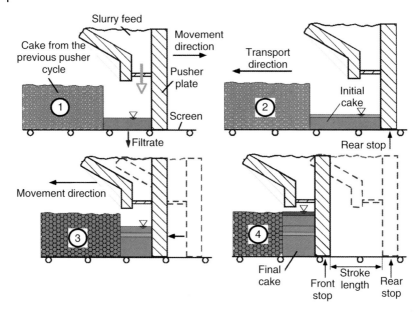

Figure 6.96 Cake formation in pusher centrifuges.

stage 1 to stage 2 and decrease for more stages, although the design becomes much more complicated and more costly. These are arguments for choosing the two-stage design in most cases as standard version.

The cake formation in pusher centrifuges is a very complex process because of the superposition of the cake formation and the axial stress. Figure 6.96 allows a closer look to the filter cake formation and transport in the first stage of a pusher centrifuge basket [57].

During the backward stroke between the previously formed filter cake and the pusher plate, a gap is opening and fresh filter area appears, on which slurry is fed. The slurry is normally fed permanently with constant flow rate. At the rear stop of the pusher plate initial, new cake and slurry are present both in the gap. Now the pusher plate again moves forward, deforms the still week cake, and further cake formation takes place, while slurry is fed continuously. At a certain point, the new cake is stable enough to withstand the pusher force without deformation, and the total cake is transported toward the solid outlet, until the front stop of the pusher plate has reached. This is the most critical point because the still flowing feed slurry is spreading on the cake surface and the danger of flooding is maximal. This is a limitation for the strategy of slurry feed with constant flow rate. There can be found several proposals in the patent literature to avoid the feeding at the front stop. One example is the pulsating feed. Here, slurry is only fed during the backward stroke [58]. It had been demonstrated experimentally that remarkably more slurry could be fed, and in the best case, a filter cake of sufficient thickness and stability could be formed, which was no longer deformed during the forward stroke and could be transported directly. The drainage capacity of the system in case of increased feed flow during backward stroke is remarkably greater than

Figure 6.97 Double-acting pusher centrifuge.

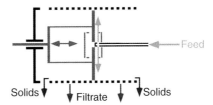

that of continuous feeding with reduced flow rate. For the perfect realization, a special measurement and control technology is required [59, 60].

This idea of pulsating feed can also be realized by a special design in the form of slurry feed in the middle of a long screen basket and solid discharge to the right and to the left end of the basket (cf. [51], p. 367–368). Figure 6.97 illustrates this principle of double-acting pusher centrifuge.

In the middle of a comparatively long and single stage basket, a pusher plate is oscillating, whereas the feed pipe is fixed. Depending on the position of the pusher plate, the slurry is fed via a feed distributor alternately on the front and on the back screen. By that, the centrifuge exhibits two feed zones, which are fed alternating. The feed is filled in every case into the new opened gap and thus the feed at the critical front stop of the pusher plate is avoided. Beside this effective protection against flooding, further advantages are given for this design. The investment and footprint area are reduced because of increase of the throughput. Secondly, the hydraulic system is more effectively utilized because the cake is pushed actively during forward and backward stroke of the pusher plate. Thirdly, if the screen is worn out, normally solids are getting behind the pusher plate and increase the filtrate pollution. Here, the solids remain on the screen and mixed again with fresh slurry.

Besides the discussed special operation mode or design of pusher centrifuges, the limit of throughput is increased for greater slurry concentrations, a smaller content of fine particles with diameters less than about 50 μm, and preferably open filter media, such as robust slotted screens (wedge wire screens), which are relatively stable against abrasion and exhibit a reduced danger to become clogged. On the one hand, the cake, which is pressed by centrifugal forces at the screen, can be transported along the screen bars with minimal friction. On the other hand, particles have only two contact points to the screen bars instead of several contact points in the pores of woven filter media. Figure 6.98 shows three types of screen bar profiles.

Figure 6.98 Screen bar profiles.

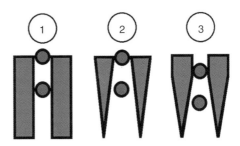

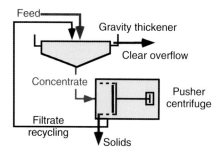

Figure 6.99 Cross-coupling of static thickener and pusher centrifuge.

The rectangular profile (1) is obviously disadvantageous because particles are able to stick across the whole channel length and thus such a screen tends to become clogged relatively easy. The second profile (2) exhibits because of the conical shape, only two contact lines at the screen surface for the particle contact. If a particle is once pressed through the opening at the screen surface, it is deliberated immediately because of the conical widening of the gap. This configuration seems to be ideal, but the deficiency is the permanent growing of the upper gap width because of consecutive abrasion, during operation. The turbidity of the filtrate would increase more and more and the screen would have to be changed after relatively short time. The third design (3) represents a compromise between (1) and (2). The short rectangular part of the gap slightly increases the danger of becoming clogged but guarantees constant gap width for comparatively long time.

To reduce the solid content in the filtrate and to decrease the cake formation time, the slurry solid concentration should be as high as possible. To meet this requirement, often a combination of prethickener and centrifuge in a cross-coupling is chosen. Figure 6.99 illustrates this principle on the example of a circular static sedimentation basin and a pusher centrifuge.

In this type of combination two separation apparatuses, of which each have a strong and a week side, are linked together crosswise. The strong sides are visible to the outside and the week sides are compensating each other internally. In the example on the one hand, the slurry is fed first into a sedimentation basin. Its strong side is a clear and nearly particle-free overflow. The week side is the sludge outlet, which discharges a concentrated, but still much liquid containing slurry. This concentrate is fed into the pusher centrifuge. The strong side of the pusher centrifuge is the very good deliquored solids, and the week side is a quite turbid filtrate because of the necessarily open screen. The turbid filtrate is given back to the sedimentation basin. As products clear liquid, highly deliquored solids are received.

Beside gravity thicker, other apparatuses such as hydrocyclones, bended screens, or back-flushing filters can be applied.

Prethickening of the slurry must not necessarily be carried out in a separate apparatus. If the feed slurry is not too diluted, the prethickening can be realized in the pusher centrifuge itself. In addition, fluctuating feed concentrations are compensated. The cake formation is improved and the slurry is accelerated more gently. In Figure 6.100, it can be seen that the slurry hits a screen at the inlet zone and a part of the liquid is discharged separately.

Figure 6.100 Slurry prethickening inside a pusher centrifuge.

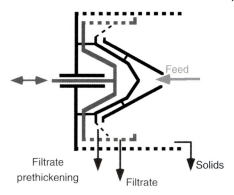

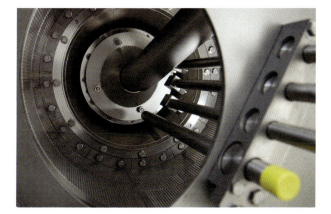

Figure 6.101 Pusher centrifuge with integrated prethickening. Source: Courtesy of FERRRUM Ltd. Centrifuge Technology.

A cake cannot form in this zone because of the impact of the fast flowing feed slurry, as in the case of a bent sieve. Figure 6.101 allows a view into the screen basket of a pusher centrifuge with integrated thickening cone and wash pipes to purify the filter cake.

The slurry can be preconcentrated inside of centrifuges not only by filtration, as discussed before, but also by sedimentation. For this purpose, a filter centrifuge has to be combined with a solid bowl centrifuge. One example for such centrifuges, especially for relatively small particles of the lower micrometer range, is represented by screen bowl decanters, which are from the basic principle sedimentation centrifuges but can be retrofit by a cylindrical screen basket. In Figure 6.102, the classic type of screen bowl decanters is depicted in the lower part and the principle of most modern screen bowl decanters is shown in the upper part.

In the classical design, the cylindrical screen basket is directly connected with the end of the conical part of the solid bowl. This represents unfortunately the smallest radius and thus the lowest centrifugal forces on the screen. In the modern design, which is also shown in Figure 6.103, the screen part shows a considerably greater radius and thus greater centrifugal forces.

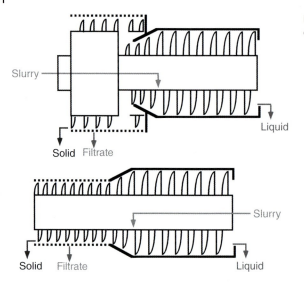

Figure 6.102 Screen bowl decanter.

Figure 6.103 Screen bowl decanter "Turbo Screen." Source: Courtesy of SIEBTECHNIK GmbH.

The particles are transported, as in every decanter centrifuge, by a transport screw, which is rotating with a certain differential rotation speed in comparison to the main rotation speed of the bowl. The liquid from the thickening part is discharged via a weir at the end of the cylindrical part of the solid bowl and filtrate through the screen at the opposite side of the centrifuge.

If the particles are settling relatively fast, the settling (decanter) zone can be reduced. Faster settling particles are also normally filtering easier and it is not necessary to increase the basket diameter at the transition point from the settling to the filtering part of the machine. The rotor of such a machine is short enough to design it in an overhung position. This makes the process room very good accessible. The machine can be denominated as screen decanter as well as worm screen centrifuge with an integrated prethickening zone.

Figure 6.104 Principle of a worm screen centrifuge.

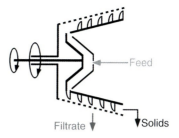

According to Figure 6.104, the classical worm screen centrifuge realizes, similar to the pusher centrifuge, a guided solid transport, but instead with a pusher plate with a transport screw, which rotates with a differential speed to the main rotation speed of the screen basket.

A view to a real machine is given in Figure 6.105.

In the standard design, the screen basket features a conical shape with an inclination of 20°. The solids are sliding more or less without pushing by the screw toward the solid outlet. The very good controlled process and the longer distance to the outlet due to the spiral channel remarkably reduce the danger of flooding. The torque is relatively low in comparison to more flat basket inclinations for the same discharge diameter because the main portion of the liquid is discharged at the small radius. As filter media, wedge wire screens, screen plate inserts, or laser screen plate inserts are used.

If advantageous in special applications, the sliding velocity of the solids can be reduced by adjusting the inclination of the flights or the inclination of the conical basket. Figure 6.106 shows three typical shapes of worm screen centrifuge baskets.

The cylindrical shape is advantageous for very slippery particles, such as ice crystals from the freeze concentration of fruit juices. These particles would slide

Figure 6.105 Worm screen centrifuge "Conturbex." Source: Courtesy of SIEBTECHNIK GmbH.

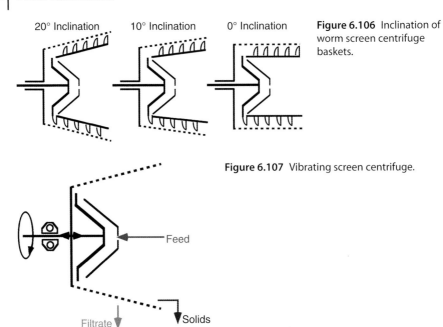

Figure 6.106 Inclination of worm screen centrifuge baskets.

Figure 6.107 Vibrating screen centrifuge.

too fast in the case of a conical basket. In the case of great inclination, the screw acts more as a dosing device and in the case of small inclination, more as a transport device.

The machine design becomes simpler if one can resign the devices for the guided solid transport. This is realized in screen centrifuges with a self-transport of the separated particles. The transition of guided to self-transport is represented by the vibrating screen centrifuge depicted in Figure 6.107.

Here, the transport is realized by the component of the centrifugal force parallel to the screen, but the cone angle is too small to overcome the static friction between cake and screen. To overcome this adhesion, the centrifuge basket is oscillating (vibrating) in axial direction with high frequency and small amplitude. The "stop and go" movement of the cake allows a very well controlled transport. The intensity of vibration can be adjusted to the needs of the product. These machines are well suited for relatively coarse particles of several hundred micrometers in size. The C-values are moderate in the range of 100–200 and the throughput is large. The largest units, for example, in the coal preparation are able to separate several hundred tons of solids per hour. For even coarser particles, centrifugal forces are often not more necessary and static vibrating screens can be applied, as realized, for example, in gravel plants.

The simplest variant of screen centrifuges is represented by the sliding centrifuge, as depicted in Figure 6.108 for vertical rotation axis on the left and horizontal rotation axis on the right side.

In these centrifuges, the cone angle is greater than the dynamic friction angle of the product. The critical point for the operation of such machines is the friction property change of the solids alongside the way through the cone. At the beginning of the cake formation, the friction between screen and particles is quite high.

Figure 6.108 Sliding centrifuges.

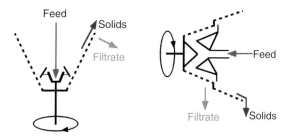

During cake deliquoring, the friction becomes lower and all the above-mentioned centrifugal forces become greater toward the greater radius of the basket. This leads to sensitive conditions for a controlled solid transport. By this reason, sliding centrifuges with vertical rotation axis are applied for slurries of high viscosity and/or fibrous small particles, as found in the sugar production. Such materials are not shooting across the filter area but more creeping.

Toward coarse and nearly monosized particles of several hundreds of micrometer size, horizontal sliding centrifuges, as can be seen on the right side of Figure 6.105, can be used, which need special measures to control the particle transport. The control of the solid transport is solved in this machine by separating the cone into several stages of increasing diameter and the product is hindered to leave the basket uncontrolled by deflector plates at each stage. With respect to cake filtration, this type of centrifuge represents a limit case because the particles are normally forming only a thin particle layer because of the spontaneous deliquoring and fast transport toward the outlet of the basket.

References

1 Ruth, B.F., Montillon, G.H., and Montonna, R.E. (1933). Studies in filtration: II. Axiom of constant-pressure filtration. *Industrial and Engineering Chemistry* 25 (2): 153–161.
2 Ruth, B.F. (1946). Correlating filtration theory with industrial practice. *Industrial and Engineering Chemistry* 38: 564–571.
3 Rushton, A. (1972). Size and concentration effects in filter cloth pore bridging. *Filtration & Separation* 9 (May/June): 274.
4 Anlauf, H. and Romani, X. (2010). Superposed filtration mechanism during clarification of very low concentrated suspension with a paperstack candle filter. *Chemical Engineering and Technology* 33 (8): 1334–1340.
5 Romani, X., Rosenthal, I., Anlauf, H. et al. (2011). Experimental and analytical modeling of the filtration mechanisms of a paper stack candle filter. *Chemical Engineering Research and Design* 89: 2776–2784.
6 Hermans, P.H. and Bredée, H.L. (1935). Zur Kenntnis der Filtrationsgesetze. *Recueil des Travaux Chimiques des Pays-Bas* 54 (9): 680–700. https://doi.org/10.1002/recl.19350540902.
7 Gösele, W., Leibnitz, R., Oechsle, E. et al. (2004). 8. Filterapparate. In: *Handbuch der Mechanischen Fest-Flüssig-Trennung* (ed. K. Luckert), 143–164. Essen: Vulkan, ISBN: 3-8027-2196-9.

8 Luckert, K. (1992). Modellierung der Fest-Flüssig-Filtration, Wissenschaftliche Zeitschrift der Technischen Universität "Otto von Guericke". *Magdeburg* 36 (5/6): 74–80.

9 Smith, J. (2015). Assessment of ceramic filtration for a metallurgical process. Dissertation. University of the Witwatersrand, Johannesburg.

10 Anlauf, H. (1994). Standardfiltertests zur Bestimmung des Kuchen- und Filtermediumwiderstandes bei der Feststoffabtrennung aus Suspensionen Part 1. *F&S Filtrieren und Separieren* 8 (2): 63–70.

11 Anlauf, H. (1994). Standardfiltertests zur Bestimmung des Kuchen- und Filtermedium-widerstandes bei der Feststoffabtrennung aus Suspensionen Part 2. *F&S Filtrieren und Separieren* 8 (3): 116–126.

12 VDI-Guideline 2762, Part 2 (2010). *Mechanical Solid-Liquid-Separation by Cake Filtration – Determination of Filter Cake Resistance*. Berlin: Beuth.

13 Rumpf, H. and Gupte, A.R. (1971). Einflüsse der Porosität und Korngrößenverteilung im Widerstandsgesetz der Porenströmung. *Chemie-Ing.-Techn.* 43 (6): 367–375.

14 Anlauf, H. (1986). A comparative consideration of pressure and pressure/vacuum filtration. *Aufbereitungs-Technik/Mineral Processing* 27 (3): 128–135.

15 Bothe, C., Esser, U., and Fechtel, T. (1997). Experimentelle und theoretische Untersuchungen zur Filtration mit überlagerter Sedimentation in Drucknutschen. *Chemie Ingenieur Technik* 69 (7): 903–912.

16 Stickland, A., Skinner, S., Cavalida, R.G., and Scales, P. (2017). Optimization of filter design and operation for wastewater treatment sludge. *Separation and Purification Technology* https://doi.org/10.1016/j.seppur.2017.01.070.

17 Schweigler, N. (1991). *Kuchenbildung und Entfeuchtung auf dem Scheibenfilter unter besonderer Berücksichtigung konstruktiver Aspekte, Fortschrittsberichte VDI*, Reihe 3, vol. 252. Düsseldorf: VDI, ISBN: 3-18-145203-3.

18 Kern, R. (1987). Die Optimierung des Filterkuchenabwurfs beim Scheibenfilter, eine Voraussetzung für dessen Weiterentwicklung. Dissertation. Universität Karlsruhe (TH), Karlsruhe.

19 Weigert, T. (2001). *Haftung von Filterkuchen bei der Fest/Flüssig-Filtration, Fortschrittsberichte VDI*, Reihe 3, vol. 680. Düsseldorf: VDI, ISBN: 3-18-368003-3.

20 Grace, H.P. (1953). Resistance and compressibility of filter cakes. *Chemical Engineering Progress* 49: 303–318.

21 Tiller, F.M. and Huang, C.J. (1961). Filtration theory. *Industrial and Engineering Chemistry* 53: 529–537.

22 Tiller, F.M. (1975). Compressible cake filtration. In: *The Scientific Basis of Filtration* (ed. K.J. Ives), 315–397. Leyden: Noordhoff.

23 Shirato, M., Sambuichi, M., Kato, H., and Aragaki, T. (1969). Internal flow mechanism in filter cakes. *American Institute of Chemical Engineering Journal* 15 (3): 405–409.

24 Buscall, R. and White, L.R. (1987). The consolidation of concentrated suspensions. Part 1. The theory of sedimentation. *Journal of the Chemical Society, Faraday Transactions* I (83): 873–891.

25 Stickland, A.D., de Kretser, R.G., and Scales, P.J. (2010). Nontraditional constant pressure filtration behaviour. *AIChE Journal* 56: 2622–2631.
26 Stickland, A.D. (2012). A compressional rheology model of fluctuating feed concentration during filtration of compressible suspensions. *Chemical Engineering Science* 75: 209–2019.
27 Stickland, A.D. (2015). Compressional rheology: a tool for understanding compressibility effects in sludge dewatering. *Water Research* 82: 37–46.
28 Höfgen, E., Kühne, S., Peuker, U., and Stickland, D. (2018). Conventional filtration theory and compressional rheology are interchangeable when using characterization techniques correctly. Preprints, FILTECH2018, L3, Cake Filtration – Modelling, Simulation, Characterization, Cologne, Germany.
29 Tiller, F.M. (1975). What the filter man should know about theory. *Filtration & Separation* 11 (7/8): 386–394.
30 Wells, S.A. (1990). Modelling and analysis of compressible cake filtration. Dissertation. Cornell University, Ithaca, NY.
31 Wakeman, R.J. and Tarleton, E.S. (1999). *Filtration – Equipment Selection Modelling and Process Simulation*. Oxford: Elsevier Advanced Technology.
32 Shirato, M. and Aragaki, T. (1972). Verification of internal flow mechanism in filter cakes. *Filtration & Separation* 9 (5/6): 290–297.
33 Durruty, J., Matson, T., and Theliander, H. (2015). Local and average filtration properties of kraft softwood. *Nordic Pulp & Paper Journal* 30 (1): 132–140.
34 Wetterling, J., Mattson, T., and Theliander, H. (2017). Modelling filtration processes from local filtration properties: the effect of surface properties on microcrystalline cellulose. *Chemical Engineering Science* 165: 14–24.
35 Stickland, A.D., de Kretser, R.G., Kilcullen, A.R. et al. (2008). Numerical modelling of flexible-membrane plate- and frame-filtration. *AIChE Journal* 54 (2): 464–474.
36 Alles, C.M. (2000). Prozessstrategien für die Filtration mit kompressiblen Kuchen. Dissertation. Universität Karlsruhe (TH), Karlsruhe.
37 Alles, C.M., Anlauf, H., and Stahl, W. (1999). Compressible cake filtration under variable pressure. In: *Advances in Filtration and Separation Technology, 13a* (ed. W. Leung and T. Ptak), 898–905. American Filtrations & Separations Society.
38 Tiller, F.M. and Hsyung, N.B. (1993). Unifying the theory of thickening, filtration and centrifugation. *Water Science and Technology* 28 (1): 1–9.
39 Alles, C. and Anlauf, H. (1998). Tandem-Filterzelle zur Charakterisierung kompressibler Kuchen. *F&S Filtrieren und Separieren* 12 (5): 220–222.
40 Grace, H.P. (1953). Resistance and compressibility of filter cakes Part I–III. *Chemical Engineering Progress* 49, pp. 303–318, 367–374, 427–436.
41 Tiller, F.M. (1972). The role of porosity in filtration VII: effect of side wall friction in compression-permeability cells. *American Institute of Chemical Engineering Journal* 18 (1): 13–20.
42 Riemenschneider, H., Wiedemann, T., and Jungermann, K. (1997). Process-specific design for press filters on the basis of laboratory tests taking into consideration wall friction effects. *Aufbereitungs-Technik (Mineral Processing)* 38 (11): 596–605.

43 Sedin, P., Johansson, C., and Theliander, H. (2003). On the measurement and evaluation of pressure and solidosity in filtration. *Chemical Engineering Research and Design* 81 (10): 1393–1405, ISSN 0263-8762, E-ISSN 1744-3563.

44 Erk, A. (2006). *Rheologische Eigenschaften feindisperser Suspensionen während ihrer Fest-Flüssig-Trennung in Filtern und Zentrifugen*. Aachen: Shaker, ISBN-13: 978-3-8322-5286-1.

45 Erk, A., Hardy, E., and Althaus, T. (2006). Filtration of colloidal suspensions – MRI investigation and numerical simulation. *Chemical Engineering and Technology* 29 (7): 828–831.

46 Tiller, F.M. (1973). The role of porosity in filtration IX. Skin effect with highly compressible materials. *American Institute of Chemical Engineering Journal* 18 (6): 1266–1269.

47 Anlauf, H. (2018). ACHEMA Berichte – Fest/Flüssig-Trennung. *Chemie Ingenieur Technik* 90 (12), in press. DOI: https://doi.org/10.1002/cite.201800115.

48 Langeloh, T., Kern, R., and Meck, F. (1992). Neueste Entwicklungen im Bereich der Schlammentwässerung. *F&S Filtrieren und Separieren* 6 (5): 271–276.

49 Anlauf, H. (2006). ACHEMA Berichte – Fest/Flüssig-Trennung. *Chemie Ingenieur Technik* 78 (10): 1492–1499.

50 Anlauf, H. (2010). Cake filtration beyond the possibilities of today's filter press – continuous ultra thin film filtration. In: *Global Guide of the Filtration and Separation Industry* (ed. H. Lyko and S. Ripperger), 142–148. Rödermark: VDL, ISBN: 978-3-00-029751-9.

51 Loginov, M., Lebovka, N., and Vorobiev, E. (2014). Multistage centrifugation method for determination of filtration and consolidation properties of mineral and biological suspensions using the analytical photocentrifuge. *Chemical Engineering Science* 107: 277–289. https://doi.org/10.1016/j.ces.2013.12.011.

52 Loginov, M., Samper, F., Gésan-Guiziou, G. et al. (2017). Centrifugal ultrafiltration for determination of filter cake properties of colloids. *Journal of Membrane Science* 536: 59–75. https://doi.org/10.1016/j.memsci.2017.04.064.

53 Stahl, W. (2004). Die Kuchenbildung und der Kuchenwiderstand. In: *Industrie-Zentrifugen Band II Maschinen- und Verfahrenstechnik* (ed. W. Stahl), 98–123. Männedorf: Dr-M. Press, ISBN: 3-9522794-0-4.

54 Wilkesmann, H. and Hultsch, G. (1979). Einfluss der Betriebsparameter auf die Wirkung eines Rotationssiphons in Schälzentrifugen. *Chemie Ingenieur Technik* 51 (2): 140–141.

55 Dubal, G. (2000). The pusher centrifuge – operation, applications and advantages. *Filtration & Separation* 37 (4): 24–27.

56 Benz, M. (2009). *Einfluss von Produkteigenschaften und Betriebseinstellungen auf das Verfahrensergebnis in Schubzentrifugen unter besonderer Berücksichtigung der Partikelzerstörung*. Aachen: Shaker, ISBN: 978-3-8322-8368-1.

57 Imhoff, O. (1979). Rechnerische und experimentelle Untersuchungen über die Kuchenbildung und die Filtrationsabläufe in einer einstufigen Schubzentrifuge. Dissertation. Universität Stuttgart, Stuttgart.

58 Benz, M. (2004). Schubzentrifugen. In: *Industrie-Zentrifugen Band II Maschinen- und Verfahrenstechnik* (ed. W. Stahl), 325–330. Männedorf: Dr-M. Press, ISBN: 3-9522794-0-4.

59 Stadager, C. and Stahl, W. (1994). Online Steuerung von Schubzentrifugen. *Chemie Ingenieur Technik* 64 (9): 769–888.

60 Stadager, C. and Stahl, W. (1994). Neue Messsysteme zur Online Steuerung von Schubzentrifugen. *Chemie Ingenieur Technik* 66 (4): 526–529.

7

Particle Washing

7.1 Introduction

The washing of filter cakes or more generally the washing of solid particles in solid/liquid separation processes is a measure to separate molecules from the solids, which are originally dissolved in the liquid of the slurry. This liquid is usually called the mother liquor. The principle of a washing process is depicted schematically in Figure 7.1.

The washing process can be realized as a slurry pretreatment measure before the cake filtration takes place or as a cake posttreatment measure after the filter cake formation.

Either the particles or the dissolved molecules in the liquid can play the role of the valuable product. In the first case, washing purifies the solid product, whereas in the second case, washing increases the yield of the dissolved product. In some cases, both aspects are of interest. An example may be the bauxite beneficiation, where the solid tailings (red mud) have to be deliberated from caustic soda, which is used to leach the aluminum out of the ore. Afterward, the mineral particles have to be decontaminated for safe waste disposal and the wash liquid contains not only rests of caustic soda but also aluminum as product in the dissolved form.

The dissolved molecules are adhering at the solid surfaces and are on the other hand present in the liquid in the form of residual cake moisture, which is discharged together with the solids after the separation process. Although a part of the mother liquor in the cake pores can be displaced eventually by desaturation, principally the rest of the mother liquor cannot be removed by mechanical means at all and remains in the cake as a remanent moisture content. A following thermal drying would not change the situation because only the pure liquid is evaporated, but the dissolved molecules will crystallize and remain in the solids.

As a wash liquid, a molecularly mixable liquid with the mother liquor is used. In case of aqueous slurry, normally pure water is used as the wash liquid. If crystals are separated from a saturated aqueous solution after a crystallization process and some washing is necessary, clean saturated solution can be used as the wash liquid or an organic liquid such as alcohol, which is molecularly mixable with water but in which the solids are much less soluble than in water. The lower surface tension of alcohol in comparison to water and a higher vapor pressure lead,

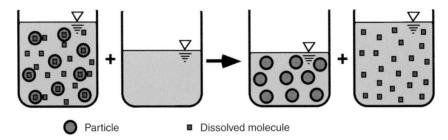

● Particle   ■ Dissolved molecule

Figure 7.1 Principle of particle washing.

at the end, to a lower cake moisture content, which can be finally evaporated by thermal means completely.

Washing means not extraction. In an extraction process, two molecularly not mixable liquids are present in the form of emulsion and the molecules to be extracted are transferred from the mother liquid of lower solubility to the emulsified liquid of higher solubility. A selective separation of molecules is possible in this way, which cannot be realized by washing.

Last but not least, washing in the above-mentioned way means not flushing. Similar to the displacement of mother liquor from the filter cake pores by gas (desaturation), the mother liquor can be displaced alternatively by a nonmolecularly mixable liquid (flushing) [1]. In both cases, capillarity plays an essential role because of the formation of a boundary layer between the two fluid phases.

7.2 Principles of Particle Washing

Particle washing can be principally distinguished between washing by dilution or permeation [2, 3]. According to Figure 7.2, the dilution washing process requires an alternating particle separation and redispersing with wash liquid.

This process can be realized on the basis of any solid/liquid separation process either with density separation or filtration and can be used for slurry pretreatment as well as for cake posttreatment.

In contrast to the dilution washing principle, the permeation washing, as is shown in Figure 7.3, can only be carried out in filtration processes.

Permeation washing is normally realized as the cake posttreatment measure, but in one special case, it is also realized as a slurry pretreatment process (cf. Section 7.6). In the posttreatment process, first a filter cake has to be formed, which is permeated in a second step by the wash liquid. Beside the direct application in the form of liquid, the wash liquid can be produced in some applications by exposing a cold filter cake to hot steam, which condensates at the cake surface or inside of already emptied cake pores. The condensate represents an ultrapure ideally suited wash liquid. Dilution and permeation washing can be combined with each other.

Beside different physical principles of washing, the wash liquid can be applied in cocurrent or countercurrent mode as well for dilution as for permeation washing processes. Figure 7.4 illustrates first the cocurrent mode.

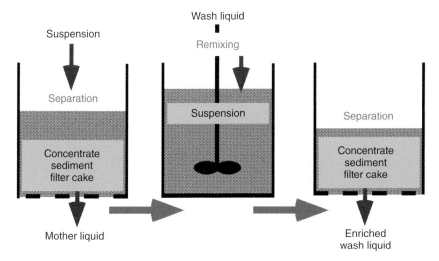

Figure 7.2 Principle of dilution washing.

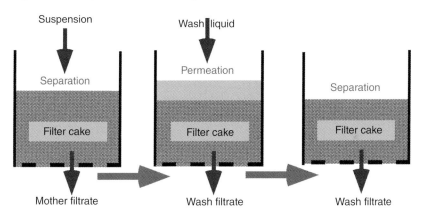

Figure 7.3 Principle of permeation washing.

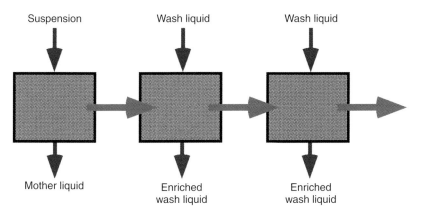

Figure 7.4 Principle of cocurrent washing.

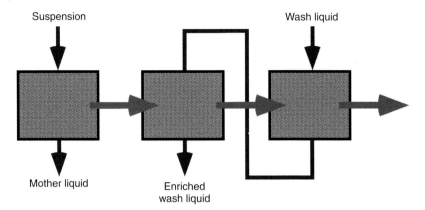

Figure 7.5 Principle of countercurrent washing.

Fresh wash liquid is applied consecutively to the filter cake. As a result, a relatively great amount of comparatively low concentrated wash filtrate is produced.

To save the wash filtrate and to increase its concentration of removed soluble substances, the more efficient countercurrent mode can be chosen as depicted in Figure 7.5.

Here, the fresh wash liquid is applied only once at the location, where the particle system is already nearly completely purified. This wash liquid is loaded only to a very little extent with soluble substances and can be reused again upstream, where the particle system is still more enriched with soluble substances. This procedure is repeated until the start of the wash zone has been reached.

The sharp separation of the different wash liquids is essential for a good wash performance. The countercurrent mode in principle can be applied for all processes of dilution and permeation washing, if several separation apparatuses are arranged in a series. In most cases, several apparatuses of same type are used for this purpose. In the case of permeation washing, only some apparatuses are well suited to realize countercurrent washing in a single-stage process. The examples are vacuum belt filters and to some extent modern vacuum and pressure drum filters, vacuum pan filters, or FEST rotary pressure filters. More or less, all other filter apparatuses are not able to integrate countercurrent washing in one single apparatus. For example, in a filter centrifuge, the wash liquid is distributed all along the circumference of 360° and there is no chance to separate different wash filtrates from each other. The same is valid for filter presses, where the parallel-arranged chambers could only be permeated in the cocurrent mode, for stirred nutsche filters, and all other types of discontinuously operating pressure filters, such as candle or leaf filters.

7.3 Limits of Particle Washing Processes

Particle washing processes are physically limited by two reasons. The first reason is certain adsorption equilibrium between molecules and particle surface, which corresponds to the respective adsorption isothermal curve. The physical

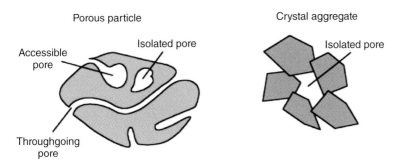

Figure 7.6 Particles and crystal aggregates with hydraulically isolated inner pores.

and chemical properties of the system are of significant influence on the specific wash performance. The second reason is given in the case where hydraulically isolated pores exist in the particle system. As shown in Figure 7.6, the isolated pores can exist in particles with some inner porosity or can develop if crystals are forming aggregates.

After reaching these limits, a further removal of molecules from the particles is only possible by shifting the adsorption equilibrium and/or by particle disintegration.

Beside physical limitations, technical limitations have to be considered. On continuously operating cake filters, such as disc, pan, drum, and belt filters, or continuous filter centrifuges, such as screen decanters, pusher, worm screen, sliding, and vibrating centrifuges, the wash time is principally limited by the transport velocity of the cake through the washing zone and its length. On the example of vacuum rotary filters, as depicted in Figure 7.7, it can be demonstrated that the limitations for cake washing are additionally given by the geometrical arrangement of the filter area.

Best possibilities are given for the belt filter. Because of the horizontal arrangement of the filter area, the wash liquid can be distributed easily, homogeneously, and with comparatively high amount such as a lake on the cake surface. Similar conditions are given for the horizontal pan filter. However, the homogeneous liquid distribution on the pan filter is more complicated, than for the belt filter, because at the inner radius, less wash liquid must be fed, than at the outer radius. An overdosage of wash liquid at the inner radius must be avoided; otherwise, the wash liquid would get into the deliquoring zone. On a drum filter, the wash liquid is applied via wash bars or pipes and spray nozzles on the cake surface, but the liquid flows downward to the filter trough and the danger of slurry dilution exists. The adjustment of the liquid dosage is more complicated than for belt filters. One has to avoid overdosage and thus slurry dilution on the one hand or underdosage and thus incomplete coverage of the cake surface with wash liquid on the other hand. Most difficulties with the application of wash liquid are given in the case of a disc filter, where the wash liquid has to be sprayed toward the vertical cake surface and the danger exists to rinse the filter cake back into the filter trough. In any case, the amount of wash liquid, which can be sucked through the cake on disc filters, is comparatively small.

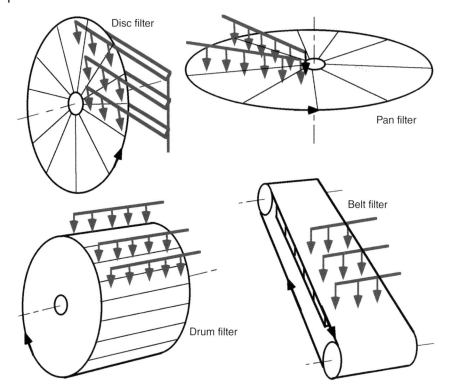

Figure 7.7 Cake washing on disc, pan, drum, and belt filters.

Last but not least, cake washing is limited by economical reasons. If the solids are the product, which have to be purified, a certain grade of purity is required and should be reached with a minimum amount of wash liquid. More purity than required unnecessarily increases the costs and influences other parameters, such as throughput or residual cake moisture content, negatively. In the case of the dissolved molecules as the valuable product, more washing increases, on the one hand, the yield principally. On the other hand, the efficiency of the wash process decreases drastically with time, and near the equilibrium, a lot of wash liquid is necessary to remove a very small amount of additional molecules, which, afterward, have to be isolated from this great volume of more diluted wash liquid. There exists an economical limit, after which more washing is more costly than the payback from the additionally removed product.

7.4 Characterization of Particle Washing Results

A cake washing process is characterized by the washing result and the effort to get this result. The washing result is described in Eq. (7.1) by the amount of remaining impurities in the particle system X

$$X = \frac{m_{ds}}{m_s} \tag{7.1}$$

X is defined as mass of dissolved substances m_{ds}, divided by the mass of solids m_s. To get a standardized value, the actual amount of dissolved substances in the solid product X can be related, according to Eq. (7.2), to the amount of dissolved substances at the start of washing X_0

$$X^* = \frac{X}{X_0} \tag{7.2}$$

To monitor the progress of impurity reduction during cake permeation, often the conductivity of the filtrate is registered because the conductivity is related to the amount of dissolved ions. This principally gives a good indication about the amount of removed impurities, but information about the real impurity of the cake is only possible if the initial impurity X_0 of the cake is known. Otherwise, the conductivity of the filtrate may indicate perfect purity, but in reality, hidden impurities, which cannot be reached by the wash liquid, are remaining in the cake. To get precise information about the impurities in the cake from the analysis of filtrate, an exact mass balance must be available. This is normally not the case. However, the most exact information of the cake purity can be achieved by the analysis of the cake itself.

The effort to get a certain washing result is given by the amount of spent wash liquid volume $V_{L,w}$ and characterized by the so-called wash ratio W. In practice, usually the mass of the spent wash liquid $m_{L,w}$ is related to the solid mass m_s, as indicated in Eq. (7.3)

$$W_m = \frac{m_{L,w}}{m_s} \tag{7.3}$$

The mass-related wash ratio W_m is easy to determine, but not as informative as the volume-related wash ratio W_V, expressed in Eq. (7.4)

$$W_V = \frac{V_{L,w}}{V_{void}} \tag{7.4}$$

Here, the wash liquid volume $V_{L,w}$ is related to the void volume of the filter cake V_{void}. In the ideal case, the void volume of the cake pores, which are initially completely filled with mother liquor, is displaced once with wash liquid and the wash ratio W_v becomes $W_v = 1$. This indicates the absolute minimum of necessary wash liquid volume to purify the filter cake completely. In reality, W_v becomes significantly larger because of not perfect washing conditions, but it gives an impression how far the result is from the ideal case and whether there may be a chance of further optimization of the process or not.

The results of either dilution or permeation washing regarding purity and effort can be represented graphically, as shown in Figure 7.8 [4].

The function No. 6 represents the ideal case of piston-like mother liquor displacement by the wash liquid for permeation washing. When the wash liquid replaces the void volume in the cake once ($W_v = 1$), the cake in this case is purified completely. In comparison, a typical experiment of permeation washing (circular dots) shows at the beginning a perfect agreement with the ideal curve. This segment of the curve represents ideal displacement of mother liquor. At a certain point, the curve deviates from the ideal curve toward greater wash ratios. This

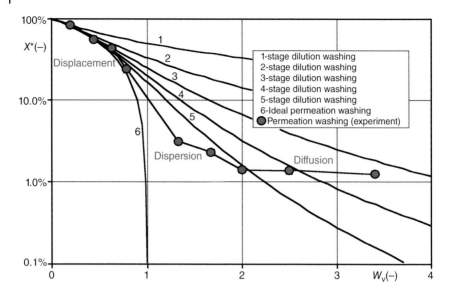

Figure 7.8 Washing results as a function of wash liquid consumption.

effect is called dispersion (cf. Section 7.6) and one reason for this is the fact that the flow of wash liquid is not ideally piston like, but a so-called fingering takes place. If for high wash ratios the curve shows nearly no more gradient toward lower impurity content in the cake, molecules from micropores or liquid inclusions can get into the wash liquid flow only by diffusion. This is a time-consuming process, and thus, much wash liquid is spent during permeation. At the end, in the equilibrium stage, a certain amount of immobile impurities remains in the cake.

In addition to the permeation washing, Figure 7.8 also includes theoretical calculative results for dilution washing. Obviously, the dilution washing becomes more effective if more wash stages are used for the same wash ratio. In general, permeation washing seems to be more effective than dilution washing, but as can be seen in Figure 7.8, from a certain wash ratio onward, the dilution washing can be more effective than the permeation washing. This is due to the disintegration of the filter cake during redispersing and therefore the perfect accessibility of all particle surfaces. Liquid inclusions are liberated and impurities, which had been trapped between the particle surfaces of laminar contact in the cake structure, can be removed.

7.5 Dilution Washing

As explained in Section 7.2, dilution washing is based on an alternating separation of the particles and remixing them with wash liquid. An advantage of this process consists in the optimal accessibility of the particle surfaces during the remixing with wash liquid. A disadvantage may be the often-needed high technical demand due to several separation and remixing stages.

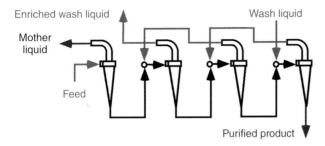

Figure 7.9 Particle washing in hydrocyclones.

An example for dilution washing as slurry pretreatment measure for relative coarse particles is a series of hydrocyclones in countercurrent mode, as shown in Figure 7.9.

In contrast to the function of hydrocyclones as a classifier (cf. Section 5.4), the cyclones here have to operate as separators. Having in mind that the lower particle size limit of separation for hydrocyclones is given at about 5 µm, only greater particles are allowed to be present in the slurry here. Beside the removal of dissolved molecules from the mother liquor, the slurry leaves the last cyclone with a remarkably increased solid concentration in comparison to the original slurry. Prepurification and prethickening are combined here. An example for practical application of such a process is the production of phosphoric acid, where gypsum crystals have to be separated from the acidic liquid. On the one hand, the gypsum can only be used in purified form, and on the other hand, the yield of phosphoric acid should be maximal.

If the particles become smaller than can be processed successfully by hydrocyclones, the analogous arrangement can be realized with other equipment, such as mixers and (centrifugal) settlers or cross-flow modules. The countercurrent flow arrangement of several cross-flow modules would be analogous to the serial hydrocyclone combination. A simple cocurrent solution for cross-flow filtration is shown in Figure 7.10 for a fed batch process, which is called diafiltration.

The slurry to be purified is pumped through the cross-flow module and recirculated to the feed tank, where the same quantity of wash liquid is added, as filtrate was separated in the cross-flow module. When the target purity is reached, the process is stopped and the purified slurry is discharged from the feed tank.

Figure 7.10 Diafiltration in fed batch mode.

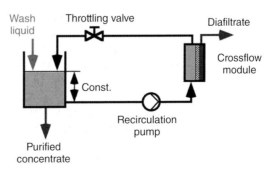

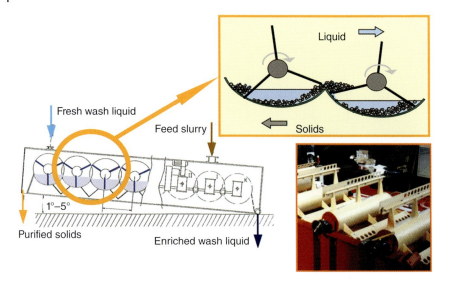

Figure 7.11 Paddle washer. Source: Courtesy of BOKELA GmbH.

A single apparatus for dilution washing as slurry pretreatment measure in countercurrent mode and simultaneous slurry concentration is represented by the so-called "paddle-washer," which is shown in Figure 7.11 [5].

This gravity-driven apparatus consists of several cups, which are arranged in a row of certain inclination. In each cup, a rotating stirrer mixes the slurry and transports a part of the particles uphill to the next cup. The fresh wash liquid is fed into the last cup, before the purified solids are leaving the apparatus and flows downhill in countercurrent mode. This apparatus is very robust, flexible, and adjustable to the needs of the product. The number of cups, the rotation speed of the stirrers, the inclination of the apparatus, and the wash liquid volume flow can be varied. Precondition for the operation of this apparatus is a sufficient settling velocity of the particles in the gravity field, which is given for particles greater than about 50 µm.

An example for cake posttreatment dilution washing process in cocurrent mode can be seen in Figure 7.12 on the example of vacuum drum filters.

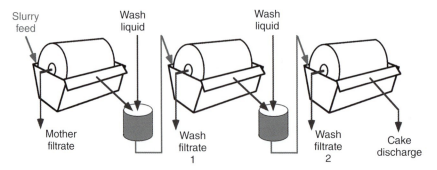

Figure 7.12 Two-stage cocurrent dilution washing by vacuum drum filters.

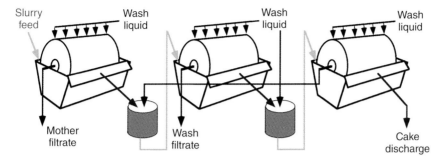

Figure 7.13 Optimized two-stage dilution washing on vacuum drum filters.

The wash result of this simple arrangement can be calculated quite easily, according to Eq. (7.5), on the basis of balances and the assumption of concentration equalization in each stage

$$X_2 = X_1 \cdot \frac{MC_2}{MC_1 + \frac{m_{L,w}}{m_s}} \qquad (7.5)$$

The amount of impurities in each next stage X_2 results from the amount of impurities in each previous stage, the cake moisture content MC of each next and previous stage, the added mass of wash liquid $m_{L,w}$, and the mass of the separated solids m_s.

To improve the process, shown in Figure 7.12, one can change from the cocurrent to the countercurrent mode. An additional intensification is possible by combining dilution with permeation washing on each filter. For this purpose, fresh wash liquid has to be sprayed on the filter cake via wash pipes, as demonstrated in Figure 7.13.

Dilution washing in one single cake filtration apparatus is also possible, but only in stirred pressure nutsche filters, which are extremely flexible batchwise operating cake filters and allow optionally dilution or permeation washing, as shown in Figure 7.14.

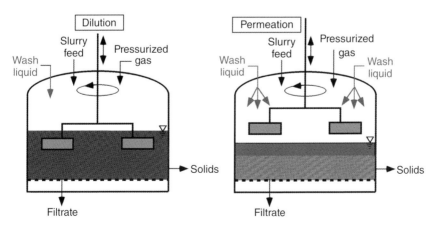

Figure 7.14 Stirred pressure nutsche filter.

During dilution washing, the stirrer helps to redisperse the cake in the wash liquid. During permeation washing, the stirrer must be moved upward out of the wash liquid. However, a countercurrent mode of operation is not possible for stirred nutsche filters.

7.6 Permeation Washing

Permeation washing can be realized, analogous to dilution washing, as slurry pretreatment or cake posttreatment measure. The permeation washing procedure can be subdivided into three ranges, as illustrated in Figure 7.15.

The mechanisms to get rid of soluble impurities are displacement and diffusion. The washing procedure starts at 100% impurities in the cake at the beginning and ends at the adsorption equilibrium of the respective boundary conditions. In the first range of displacement, only pure mother liquid appears as filtrate and the real washing results are following the function of ideal displacement. The ideal displacement is characterized by a piston-like permeation of the wash liquid through the filter cake. All impurities are then removed after one exchange of the mother liquid by the wash liquid in the pore system of the cake. Unfortunately, the real washing results are deviating soon from the ideal curve and more wash liquid is spent than necessary in the ideal case. This phenomenon is called dispersion.

The third range of diffusion is a time-controlled process and thus much wash liquid is spent to remove some more molecules of the impurities, before finally equilibrium of adsorption is reached.

To reduce dispersion effects and to make the washing process more effective, the reasons for dispersion must be understood in more detail. Dispersion is characterized by an undesired mixing of wash and mother liquid.

There are several reasons existing for the mechanical and diffusive dispersion in granular beds, which are based, according to the examples in Figure 7.16, on different transport velocities of molecules in the flow channels through porous beds [6].

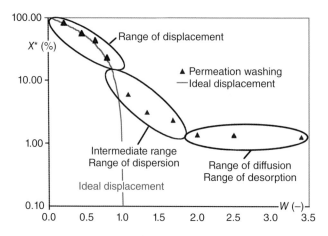

Figure 7.15 Permeation washing.

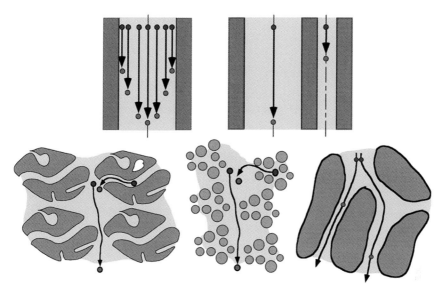

Figure 7.16 Dispersion in porous beds.

In a pore, which is permeated by a laminar flow, a parabolic flow profile is originating with different transport velocities of the molecules across the pore diameter. The flow in pores of different size will be different because of different flow resistance. If particles are porous themselves, molecules must get by slow diffusion from the micropores inside the particles into the fast main flow through the large pores between the particles. A similar situation exists in the case of agglomerates with small inner pores in comparison to the large pores between the agglomerates. Last but not least, the molecules are not taking the shortest way through the cake but have to cover longer distances due to tortuosity of the pores. Not included in this listing of dispersion effects is the possibility of diffusive back mixing. Because of the displacement of mother liquor by wash liquid, a concentration gradient of dissolved molecules originates. As a consequence, a diffusive transport takes place toward the lower concentration. Normally, this effect can be neglected in comparison to the convective flow of the liquid toward the filter medium.

For homogeneous filter cakes of uniform thickness, different models to describe the permeation washing can be found in the literature, which are valid for different special boundary conditions, such as dispersion models [7, 8], empirical models [9, 10], or a film and side channel models [11, 12]. Some studies are scrutinizing the ideal boundary conditions and demonstrate that several apparatus, structure, and operation-related deviations from the ideal assumptions could occur. Inhomogeneous cake geometry [13], cake shrinkage [14], inhomogeneous dosage of wash liquid [15], compressibility, and other physically and/or chemically caused changes of cake structure [16] or the mixing of wash and mother liquid at the end of the cake formation zone as it happens on horizontal filters such as vacuum belt or pan filters should be exemplarily denominated here.

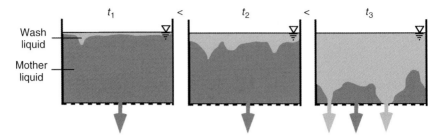

Figure 7.17 Fingering effect during permeation washing of filter cakes.

Of high influence on dispersion is the pore size distribution in a filter cake, which causes larger and smaller pore channels across the filter cake. The larger the channel, the lower the flow resistance and the faster the displacement of the mother liquor by the wash liquid. This effect of "fingering" is demonstrated in Figure 7.17 (cf. [9], p. 824) on the example of a filter experiment in a transparent lab filter cell.

Shortly after the start of the washing procedure at the time t_1, the penetration of the cake by the wash liquid takes place first, where the largest pores are located. Until this time, exclusive mother liquor appears at the filtrate outlet in perfect agreement with the ideal wash curve, although dispersion in the upper part of the cake starts obviously. Even at t_2, a breakthrough of the wash liquid through the largest pores has not happened, although the fingering effect becomes stronger. At t_3, the first large pores meet the filter medium, and wash liquid gets into the filtrate, although several locations in the cake are still fully saturated with mother liquid. From now on, wash liquid flows useless through the still washed pore channels, whereby mother liquid is removed from other smaller pore channels at the same time. As a consequence, more wash liquid is spent than necessary for ideal displacement of the mother liquid. The wash curve deviates from the ideal wash curve toward greater wash ratios. As had become clear in different investigations (cf. [9], p. 825 and [16], p. 112), the relation between mother and wash liquid viscosity plays a significant role on the fingering effect. Figure 7.18 illustrates this effect schematically.

The more viscous the wash liquid becomes in comparison to the mother liquor, the more piston like the displacement takes place. If the mother liquid has a considerable greater viscosity than the wash liquid, the influence of local inhomogeneities is increased because the length of the flow path, filled with mother

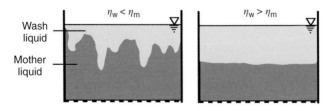

Figure 7.18 Influence of liquid viscosity on the permeation washing.

Figure 7.19 Influence of washing time on the washing result for different cake heights.

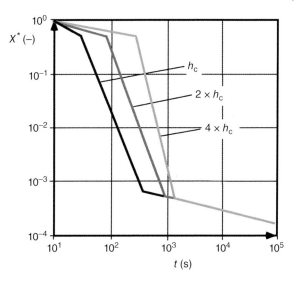

liquid, dominates nearly solely the hydraulic resistance. Contrary for considerably greater viscosity of the wash liquid, the influence of local inhomogeneities becomes less because a pore channel with an advancing wash liquid is appropriately slowed down.

If the filter cake height is varied, the flow velocity and residence time of the liquid in the porous system is changing, and a significant influence on the washing results can be observed. Figure 7.19 qualitatively shows the influence of the cake height on the kinetics of impurity reduction (cf. [9], p. 825).

At the beginning, the curves for different cake height are significantly differing. The thinner the cake becomes, the faster the impurity reduction takes place. This behavior can be observed during pure displacement at the beginning and during the transition zone of axial dispersion. Toward the end of the washing process, the curves are converging and the influence of washing time predominates. Then, thin or thick cakes need the same time. An asymptotic diffusion process predominates now.

Figure 7.20 illustrates, in addition to Figure 7.19, the necessary amount of wash liquid to get the respective washing results.

Here, the curves are placed very tight together at the beginning for pure displacement. This means that the wash ratio predominates the process. In the intermediate range and toward the final diffusion range, the curves are spreading. The thicker the cake becomes, the less wash liquid is necessary to get a certain purity. Thicker cakes need more time but consume less wash liquid for the same wash result. A secondary reason for this phenomenon is the increased influence of local inhomogeneities toward thinner cakes. Those are more neutralized toward thicker cakes.

The final diffusion range is independent of the cake height. Thin or thick cakes need the same time to reach a certain state of purity (cf. Figure 7.19) but a different amount of wash liquid.

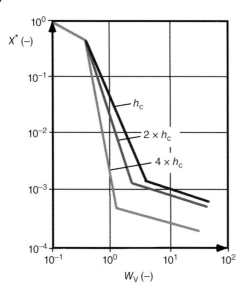

Figure 7.20 Influence of wash ratio on the washing result for different cake heights.

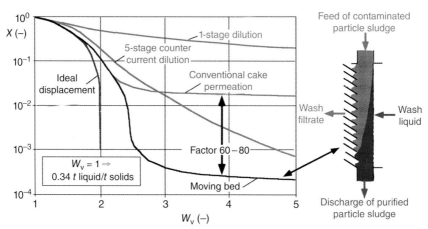

Figure 7.21 Moving bed washer.

Similar to the dilution washing procedure, permeation washing can be realized either as a slurry pretreatment process or as a filter cake posttreatment process. If the slurry should be purified before cake filtration by permeation, the so-called moving bed process can be used [17, 18]. This is again a gravity-driven process and requires highly concentrated slurries, which are just able to flow. Figure 7.21 illustrates the technical principle and compares the process with other possibilities of particle washing on the basis of experimental results (cf. [18], p. 130).

The liquid-filled particle bed moves downward through a channel, and wash liquid is pressed perpendicularly through the channel. To hold the situation permeable, at the wash liquid outlet, inclined channels allow the liquid to leave, but the particles are sliding back into the main particle flow. As can be observed clearly in Figure 7.18, a sharp front of wash liquid (dark) is permeating the moving bed nearly without any dispersion in the form of fingering. This effect is due to the relative movement of the particles to each other. The pore structure is permanently changing, and thus, an overall homogeneous porous system originates. In addition, all particle surfaces are optimally accessible in comparison to a fixed bed in a filter cake. The effectiveness of the principle can be seen in the wash curve on the left side of Figure 7.21. In comparison to the conventional permeation of a filter cake, the wash curve of the moving bed is much more close to the curve of ideal displacement, and in the equilibrium state, the particles are by the factor 60–80 better purified. In Figure 7.21, additionally, a calculated single-step cocurrent and a five-step counter current wash processes are documented. The moving bed wash process shows the comparatively best results.

Beside permeation washing as slurry pretreatment measure, permeation washing is normally realized as the filter cake posttreatment process. To get the best possible washing results, it is generally essential to apply the wash liquid directly after the end of the cake formation and not to allow gas intrusion into the filter cake pores before washing. At a first glance, it may be a good idea to displace as much mother liquor by desaturation first and to wash the cake afterward with less effort. Many investigations have proven that this procedure normally leads to poorer washing results. In addition, more deliquoring time must be provided in that case because at the end, low cake moisture content is required in most cases. If the filter cake is merely compressible, shrinkage cracks, like to be seen in Figure 7.22, can occur during the initial phase of cake deliquoring, and the wash liquid would flow mainly in a short circuit through the cracks without washing the cake.

Figure 7.22 Shrinkage cracks in a filter cake.

Figure 7.23 Vacuum belt filter for cocurrent or countercurrent cake washing. Source: Courtesy of BHS-Sonthofen GmbH.

A second point, which will be discussed in more detail in Section 8.2, is the fact that a rewetting of a desaturated filter cake never leads to complete saturation again and gas inclusions are remaining in the cakes structure. The wash liquid cannot reach the particles in such a gas inclusion, which is also including the interstitial mother liquor between these particles or adhering liquid at the particle surfaces anymore. The wash liquid flows around these hydraulically isolated zones. This is valid not only for filtration with gas differential pressure but also for filter centrifuges [19, 20].

In contrast to the dilution washing process, permeation washing is carried out principally on one single apparatus without remixing of the filter cake. This restricts the possibilities to realize countercurrent washing because a sharp separation between the wash filtrate of different wash stages is a necessary precondition for successful countercurrent operation. Horizontal vacuum belt filters can realize sharp separation of wash filtrates by a sufficient number of filtrate trays behind the filter medium. Figure 7.23 shows a vacuum belt filter inclusive wash pipes.

This principle of a high number of filter cells for sharp separation of filtrates is also realized for the most modern types of drum filters, as illustrated in Figure 7.24.

Although washing is more complicated on drum filters than on horizontal belt filters, a multistage countercurrent washing is principally possible. This type of filter can be operated as vacuum filter or, if installed in a vessel, as pressure filter (hyperbaric filter).

A further continuously operating filter, which is very well suited for intensive countercurrent permeation washing, is represented by the so-called FEST rotary pressure filter, as shown in Figure 7.25.

Figure 7.24 Drum filter for countercurrent washing. Source: Courtesy of BOKELA GmbH.

The filter area of the FEST rotary pressure filter is separated into several single filter cells, which are sealed against the ambient atmosphere by a narrow pressure vessel. The process zones are separated from each other by separating elements, which are pressed pneumatically on the filter cells. This arrangement allows separating several process zones and washing zones, respectively, from each other. A great amount of wash liquid can be pressed very effective through the filter cakes. In contrast to the hyperbaric filters, here no special cake discharge sluice is necessary because the whole filter apparatus resembles a kind of big star feeder as a solid sluice device.

In principle, possible but limited are the conditions for countercurrent washing on continuously operating vacuum pan filters, as can be seen in Figure 7.26.

The challenge here is the same, as discussed previously for filter cake formation (cf. Section 6.3.5). The wash liquid has to be distributed uniformly on the filter cake. Because of the circular and horizontal geometry of the filter area, the danger of wash liquid overdosage near the central shaft exists. In this case, the wash liquid would flow directly into the deliquoring zone and increases the residual cake moisture content. As a consequence, the dosage of wash liquid must be adjusted along the radius of the filter disc. Each location on the filter cake surface must get the same quantity of wash liquid.

All other types of discontinuous cake filtration apparatuses such as nutsche filters, candle, leaf, horizontal disc filters, filter presses, and all types of discontinuous and continuous filter centrifuges are only able to operate in the cocurrent mode. As an example for centrifuges, washing in a continuously operating two-stage pusher centrifuge is illustrated in Figure 7.27.

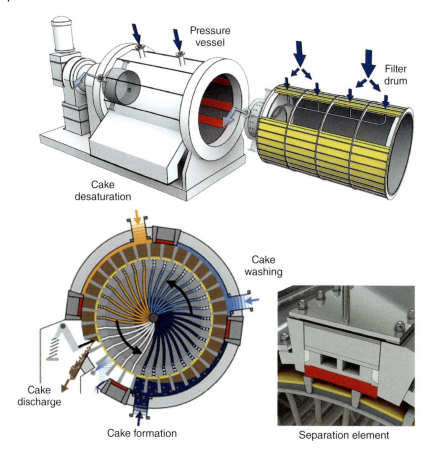

Figure 7.25 Washing on a FEST rotary pressure filter. Source: Courtesy of BHS-Sonthofen GmbH.

In the example of the moving bed washing process, it could be illustrated how undesired dispersion effects can be reduced by homogenizing the permeation conditions for the wash liquid. This issue should be discussed a little bit in more detail. To reduce dispersion due to size-distributed pore channels, the variance of the pore size distribution must be made smaller. Figure 7.28 shows the results from experiments with compressible filter cakes from very small particles on a lab-scale diaphragm filter press.

The filtration and washing conditions had been as follows:

$\Delta p_1 = 300$ kPa (cake formation);

$\Delta p_w = 400$ kPa (washing); and $\Delta p_{squ} = 800$ kPa (squeezing)

If the slurry was flocculated to increase the filtration velocity and no squeezing was realized at all, the washing results had been the poorest of all variations. The span of the pore size distribution was the greatest of all cases here and thus the dispersion effect had been most significant.

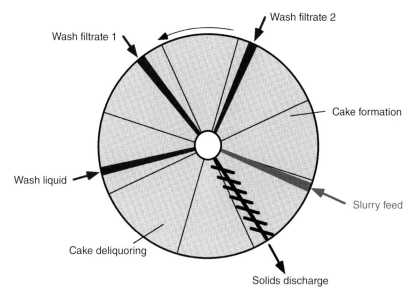

Figure 7.26 Washing on a vacuum pan filter.

Figure 7.27 Washing in pusher centrifuges.

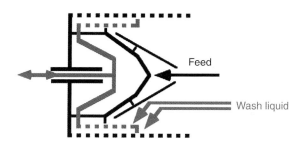

Figure 7.28 Permeation washing experiments on a lab-scale diaphragm filter press.

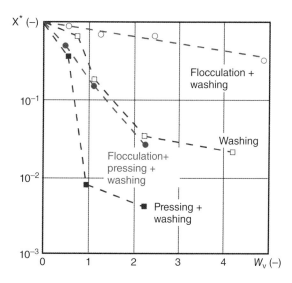

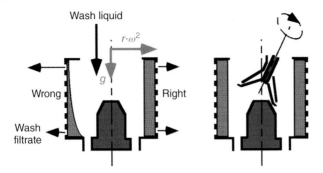

Figure 7.29 Washing in vertical filter centrifuges.

An improvement of washing efficiency could be obtained by flocculation and subsequent squeezing before washing or washing after cake formation without flocculation.

The most effective procedure was found for cake formation without flocculation, but subsequent squeezing. Here, the porosity gradient after the formation of the compressible cake was equalized by the squeezing procedure and thus the comparatively most homogeneous pore structure with the best washing conditions was produced. The renouncement of flocculation obviously leads to less cake production and the optimization of the whole solid–liquid separation process needs to consider all process steps.

As a result of the previous discussion, it can be stated that a porous layer of maximal homogeneity leads to minimal dispersion effects and thus to the best washing performance. Two more examples may illustrate this conclusion.

In vertical filter centrifuges, as shown in Figure 7.29, centrifugal force and gravity are perpendicular to each other.

If the rotation speed is not great enough, and if the slurry is not distributed evenly in the sieve basket, a hyperbolic shape of the cake surface can be expected. The wash liquid will permeate the cake most easily in the upper part of the basket, where the cake is the thinnest. In the worst case, the cake in the upper part of the basket no longer covers the filter media, and the wash liquid flows in a short circuit into the filtrate system. To avoid this situation, the rotation speed must be high enough and the slurry must be distributed evenly on the filter media. This is possible by installing an inclined rotating feed plate (cf. Section 6.5.2). The slurry is preaccelerated and evenly distributed across the basket height. Alternatives are special feed pipes with several adjusted openings.

The second example describes the cake washing in filter presses (cf. Section 6.4.3). In Figure 7.30, a horizontal chamber filter press and a vertical diaphragm press are depicted.

Horizontal chamber filter presses can be built for filter plates up to $20\,m^2$ and vertical diaphragm presses up to about $10\,m^2$. In the case of chamber filter presses, a volumetric pump conveys the slurry into the chambers between the filter plates. When the chambers are completely filled with the compressible filter cake, more slurry is fed under high pressure into the chambers for cake consolidation (cf. Section 6.4.4). As discussed before, the cake structure,

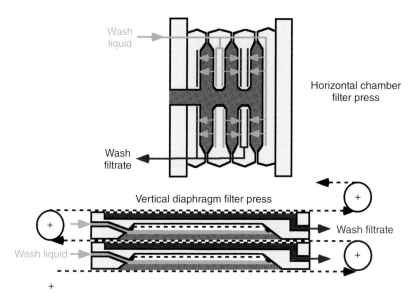

Figure 7.30 Washing on filter presses.

directly after the cake formation, is inhomogeneous because of the compressible behavior, which is not advantageous for the effectiveness of a washing process. During consolidation, the cake must be pushed across the entire plate into the corners of the plates, and because of high-friction forces, especially for large filter areas, the situation can become blocked more or less close to the feed channel. After stopping the feed, wash water is pumped through a filtrate channel from the backside through the filter cake. Again, the wash liquid prefers to flow through the more porous parts of the cake in the corners, instead of the highly compressed parts near the feed channel. This situation can be improved remarkably by choosing a diaphragm press, especially a vertical diaphragm press. The filter cake can be squeezed after formation very evenly by the diaphragm. After withdrawing of the diaphragm, wash liquid can be distributed all over the cake, as on a belt filter. Afterward, the wash liquid is pressed by the diaphragm through the homogeneous cake structure with much less dispersion than in the case of chamber filter presses. In addition, the horizontal arrangement of the filter plates reduces the danger of particle segregation during cake formation.

A further problem can be caused by the support system of the filter cloth in filter presses (cf. [[16], pp. 196–200]). This support system normally exhibits knops on the surface to guarantee easy filtrate drainage with minimal flow resistance behind the filter media. Unfortunately, the knops have a certain diameter, and according to Figure 7.31, blind areas are present, which hinder the filtrate from penetrating the filter cloth directly.

The filtrate has to drain first parallel to the filter area, until it can flow into the filtrate channels. This effect leads to zones above the knops, which are washed with less effectiveness than the rest of the cake. To minimize this deficiency drainage media, like screen fabrics, with relative coarse pores may be mounted behind the filter media to facilitate the filtrate drainage toward the filtrate channels.

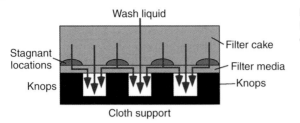

Figure 7.31 Stagnant locations near the filter media of filter presses.

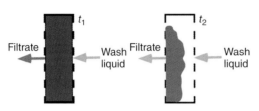

Figure 7.32 Shrinkage of filter cakes during washing.

A further measure to increase the effectiveness of the washing process in filter presses consists in a periodical change of the wash liquid flow direction [21]. This procedure equalizes a not perfect uniform permeation of the filter cake in one direction, and stagnant locations, as discussed before, can be reached more effective by the wash liquid (cf. [[16], pp. 121–123]). The flow of wash liquid takes place along different paths and thus hidden impurities could be removed.

As indicated in Figure 7.32, the physical and chemical properties of the solid/liquid system can cause filter cake shrinkage in the chambers of filter presses during the washing.

In extreme cases, parts of the filter media are not more covered by filter cake and can be permeated directly by the wash liquid in a short-circuit flow into the filtrate system. The reason for such a phenomenon is the formation of a highly porous cake from mother liquor with high ion content, which promotes agglomeration and a breakdown of these porous structures after removal of ions during washing. Also in this case, a diaphragm press is the appropriate solution for the problem because the diaphragm follows the shrinking cake and enforces a constant cake height across the entire filter area.

In some cases, the dispersion effect is significant because of fast flow of wash liquid through large pore channels and very slow diffusion out of small pores in agglomerates or micropores inside the particles. In such cases, a so-called wash break can save a lot of wash liquid. This procedure represents a kind of dilution washing during the permeation washing process. This means that after the breakthrough of wash liquid through the filter medium, the wash liquid feed and the wash filtrate discharge valves are closed. Now, the liquid in the filter cake pores is given time to homogenize the ion concentration by diffusion without further consumption of wash liquid. After reaching the concentration equilibrium, the wash liquid feed and wash filtrate discharge valves are opened again to continue the permeation wash process. This procedure can be repeated, if necessary. This wash break can be realized on discontinuous pressure nutsche filters or press filters such as frame, chamber, or diaphragm filter presses. In the centrifugal field, this is only possible in the case of horizontal or vertical siphon peeler centrifuges. Figure 7.33 illustrates the principle on the example of horizontal centrifuges.

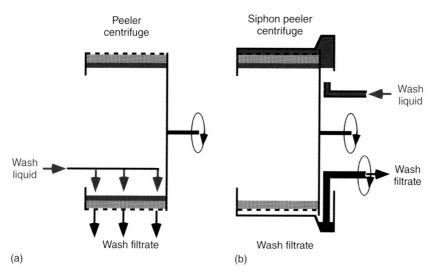

Figure 7.33 Washing in horizontal peeler centrifuges.

In the peeler centrifuge without siphon on Figure 7.33a, wash liquid can be applied either via a separate wash pipe or through the slurry feed pipe. A benefit of the latter procedure is the purging of, eventually in the feed pipe, remaining particles into the centrifuge basket and thus simultaneously cleaning the feed pipe. In this case, only a cocurrent permeation washing is possible.

In contrast to the peeler centrifuge with direct discharge of the filtrate into the centrifuge housing, the filtrate in a siphon peeler centrifuge is collected in a special filtrate-collecting room, which is part of the rotating basket. The filtrate is discharged from an open-ring channel of greater diameter than the basket at the backside of the basket by a pipe, which can be immersed into the liquid. The fast rotating liquid enters the fixed pipe and is discharged under pressure. As has been described previously in Section 6.5, because of the principle of communicating pipes, the liquid level in the ring channel and thus in the basket can be adjusted by positioning of the discharge pipe. If the overflow edge of the channel is high enough, the liquid level can be adjusted above the filter cake height and the centrifuge rotates during the wash break without discharging the filtrate (upper part of the siphon peeler centrifuge on Figure 7.33b). The wash liquid can be applied conventionally via wash pipe, slurry feed pipe, or from the backside via a wash liquid pipe and through the ring channel and filtrate-collecting room. Here, the latter is shown. An additional advantage of this procedure consists in the filter cloth cleaning by back flow of the wash liquid. Then, the discharge pipe is immersed into the ring channel and sucks off the wash filtrate (lower part of the siphon peeler centrifuge on Figure 7.33b). The process is functioning not only for siphon baskets but also for backwashing baskets with a ring channel of the same outer diameter than the basket. The possibility of washing in both directions from the cake surface toward the filter medium and vice versa can also lead to similar advantages as have been discussed for filter presses.

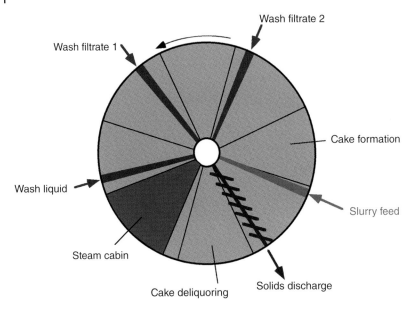

Figure 7.34 Vacuum pan filter with steam cabin.

A completely different but very efficient method of cake washing consists in the application of saturated steam. As an example shown in Figure 7.34 that on vacuum pan filters, the filter cake can be exposed to saturated steam in a special part of the cake-deliquoring zone, which is covered with a so-called steam cabin.

The steam is saturated and exhibits ambient pressure of about 100 kPa and thus a temperature of about 100 °C. The hot and saturated steam gets into contact with the cold cake surface, which enters the steam hood. According to Figure 7.35, the steam condenses and heats up the upper particle layers.

Then, the liquid is displaced out of the filter cake pores, if the applied pressure difference is greater than the capillary pressure in the pores. The condensate

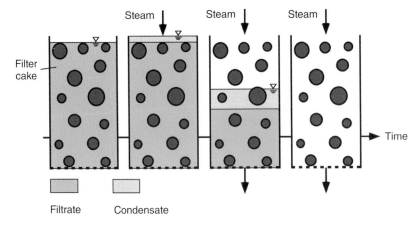

Figure 7.35 Steam washing and deliquoring.

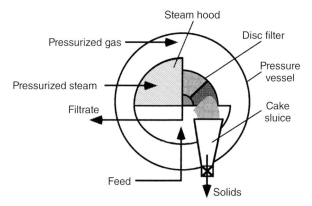

Figure 7.36 Steam pressure filtration on disc filters.

consists of pure hot water and is thus a perfect wash liquid. It originates exactly where it is needed at the contaminated surfaces of the particles. Very helpful is the phenomenon that the condensate front is migrating piston-like through the cake structure and the steaming can be finished, when the steam breaks through the filter medium (cf. Section 8.3.5). This steaming of filter cakes is not only applied for vacuum filters but also to gas overpressure filters, such as drum or disc filters. In Figure 7.36, a continuous operating disc filter in a pressurized vessel (hyperbaric filter) is shown as an example [22].

Also here, a steam hood covers a part of the cake-deliquoring zone. The vessel is loaded with pressurized gas (air). Slurry, filter apparatus, and vessel exhibit ambient temperature. The steam hood is loaded with overheated steam of the same pressure, such as the surrounding gas pressure. This avoids sealing problems between steam hood and vessel.

Beside gas pressure filters such as hyperbaric drum and disc filters, the Fest pressure filter can operate with pressurized steam and thus condensate washing. Last but not least, steam pressure filtration can be realized in discontinuous inverting filter centrifuges (cf. Section 6.5.2) because the housing of those centrifuges can be sealed relatively easy and pressurized up to about 600 kPa.

References

1 Wilkens, W. (2014). *Flushing - Entfernen von organischen Lösungsmitteln aus Haufwerken und Filterkuchen, Freiberger Forschungshefte, 912*. Freiberg: TU Bergakademie Freiberg.
2 Hoffner, B., Fuchs, B., and Heuser, J. (2004). Washing processes for disperse particulate systems – process spectrum and aspects for the process choice. *Chemical Engineering Technology* 24 (10): 1065–1071. https://doi.org/10.1002/ceat.200406142.
3 Wilkens, W. and Peuker, U. (2012). Grundlagen und aktuelle Entwicklungen der Filterkuchenwaschung. *Chemie Ingenieur Technik* 84 (11): 1873–1884. https://doi.org/10.1002/cite.201200095.

4 Anlauf, H. (2007). Overview and recent developments in effective particle decontamination by washing processes. *Filtration* 7 (1): 20–25.
5 Stahl, W. and Langeloh, T. (1998). The paddle washer – a mighty clean washer. *Aufbereitungs-Technik-Mineral Processing* 29 (6): 316–323.
6 Fried, J. and Combarnous, M. (1971). Dispersion in porous media. *Advances in Hydroscience* 7: 169–282.
7 Sherman, W. (1964). The movement of a soluble material during the washing of a bed of packed solids. *AIChE Journal* 10 (6): 855–860.
8 Wakeman, R. (1986). Transport equations for filter cake washing. *Chemical Engineering Research and Design* 64 (7): 308–318.
9 Bender, W. (1983). Das Auswaschen von Filterkuchen. *Chemie Ingenieur Technik* 55 (11): 823–829.
10 Heuser, J. and Stahl, W. (2000). Experimentally supported modeling of filtration cake washing performance. *Advances in Filtration and Separation Technology* 14: 447–454.
11 Kuo, M. (1960). Filter cake washing performance. *AIChE Journal* 6 (4): 566–568.
12 Wakeman, R. (1976). Diffusional extraction from hydrodynamically stagnant regions in porous media. *The Chemical Engineering Journal* 11: 39–56.
13 Neu, W. (1981). Untersuchung des Waschprozesses in Filterpressen. *Chemie Ingenieur Technik* 53 (12): 957–959.
14 Harderkopf, F. (1992). Experimentelle und theoretische Unterschungen zum Waschen von Filterkuchen in Filterpressen. *Chemie Ingenieur TechnikIng Technik* 64 (11): 1041–1044.
15 Wakeman, R. (1998). Washing thin and nonuniform filter cakes: effects of wash liquor maldistribution. *Filtration and Separation* 35 (March): 185–190.
16 Heuser, J. (2003). *Filterkuchenwaschprozesse unter besonderer Berücksichtigung physikalisch-chemischer Einflüsse*. Aachen: Shaker. ISBN: 3-8322-1369-4.
17 Hoffner, B. and Stahl, W. (2003). The effect of a relative particle motion during the washing process of granular materials. *The Transactions of the Filtration Society* 3: 156–161.
18 Hoffner, B. (2006). *Ein neuartiges Verfahren zur Feststoffwaschung - Die Anwendung des Wanderbetts zur hochgradigen Aufreinigung partikulär-disperser Systeme*. Aachen: Shaker. ISBN: 3-8322-4832-3.
19 Ruslim, F. (2008). *Flow Phenomena in Cake Washing Driven by Mass Forces*. Göttingen: Cuvillier. ISBN: 978-3-86727-788-4.
20 Ruslim, F., Hoffner, B., Nirschl, H., and Stahl, W. (2009). Evaluation of pathways for washing soluble solids. *Chemical Engineering Research and Design* 87 (8): 1075–1084.
21 Bernhard, P. and Gramlich, R. (1989). Fortschritte in der Kuchenwaschtechnik auf Filterpressen. *Chemie Ingenieur Technik* 61 (3): 239–240.
22 Bott, R., Langeloh, T., and Meck, F. (2002). Continuous steam pressure filtration of mass mineral products. *Aufbereitungstechnik-Mineral Processing* 43 (3): 19–30.

8

Filter Cake Deliquoring

8.1 Introduction

Desaturation and consolidation are the two principle physical possibilities to remove liquid from a filter cake. In the case of incompressible structures, capillarity is the decisive phenomenon. Here, surface tension of the liquid and wetting properties between liquid and solids are playing the most important role. According to Figure 8.1, liquid is displaced from the cake pores by gas, if the capillary pressure in the cake pores is smaller than the pressure difference, applied from outside.

This kind of deliquoring is called desaturation. The necessary gas pressure difference can be generated by gas or centrifugal pressure. Unfortunately, a complete drying of a filter cake by mechanical desaturation is not possible from the physical point of view because liquid bridges between the particles, adhering liquid at the particle surfaces, and eventually liquid in micropores of the particles themselves cannot be displaced. Although the overcoming of the capillary pressure is the dominating phenomenon for desaturation, it makes a significant difference, whether a cake is desaturated by a gas pressure difference (vacuum, overpressure) or in a centrifugal field. The exchange of gaslike air or nitrogen by pressurized steam as one of the latest developments leads to further interesting and advantageous effects. The time-dependent desaturation kinetics comes to a final end at the equilibrium state, when capillary pressure and external pressure become equal. Desaturation kinetics and equilibrium for filter cakes can be described with validated model equations, which nevertheless need some characteristic product data, which have to be measured experimentally. The combination of model equation and experiment guarantees realistic forecast data for the technical application, reduces experimental effort, and extends the possibilities of inter- and extrapolation of the included parameters. To carry out the experiments in a correct way, special guidelines, such as VDI guidelines, are available [1].

A complicated situation arises if slightly compressible filter cakes are desaturated by a gas pressure difference. Because of shrinkage of the cake, cracks can occur in the cake structure, similar to the phenomena in nature, as can be seen in Figure 8.2 for the example of a thermally dried harbor sludge and a filter cake from limestone.

Wet Cake Filtration: Fundamentals, Equipment, and Strategies, First Edition. Harald Anlauf.
© 2019 Wiley-VCH Verlag GmbH & Co. KGaA. Published 2019 by Wiley-VCH Verlag GmbH & Co. KGaA.

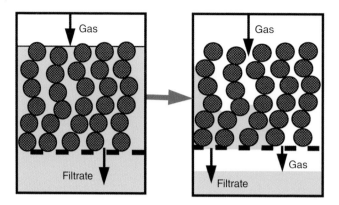

Figure 8.1 Incompressible cake desaturation by gas differential pressure.

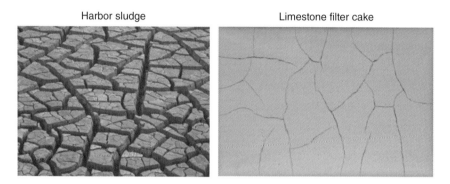

Figure 8.2 Shrinkage cracks in a dried harbor sludge and limestone filter cake.

In the case of a filter cake, gas breaks in a short circuit through the cake into the filtrate system. As a consequence, very often, the compressor cannot maintain the filtration pressure and the process fails. The available knowledge from research today can give the conditions to largely restrict or even avoid cake cracking. If the cake is consolidated to a certain extent, the structure is able to withstand the tensile strength, which occurs during desaturation and, as a result, no cracks appear. Unfortunately, the pressure, which is necessary to stabilize the cake, in most cases is too high to be realized on state-of-the-art filter equipment. In these cases, a rather undesired press filter or a centrifuge has to be chosen. Although in filter centrifuges the cake is desaturated, cracking can be observed very seldom, and if so, no gas throughput is present, which could cause a problem. Only in the case of cake washing, cracks would be unfavorable because the wash liquid would prefer to flow through the cracks and not through the cake structure. Recent research results, which are discussed later on (cf. Section 8.5.2) in more detail, could show a successful way out of the present dilemma.

Besides the desaturation of incompressible filter cakes by overcoming the capillary pressure, deliquoring can be realized by squeezing or consolidation, if the cake structure is compressible, as depicted in Figure 8.3.

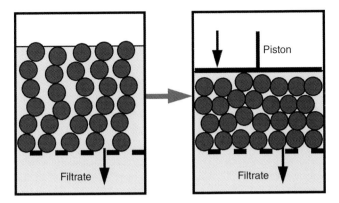

Figure 8.3 Deliquoring of compressible filter cakes by squeezing.

In general, the smaller the particles become and the more the physical and chemical properties of the slurry play an important or dominating role, the more compressible the cake structure will become. Normally, filter cakes with average particle sizes of less than 10 µm show this behavior notably. In contrast to that, also from particles in the nanometer range, nearly incompressible cake structures can be expected, if the physical and chemical slurry properties are adjusted adequately. If electrostatic repulsion between the particles is strong enough, agglomeration by van der Waals forces and thus a compressible cake structure cannot develop. Compressible filter cakes are processed normally in press filters. Compressible filter cakes and sediments show a porosity gradient after formation, but in contrast to sedimentation apparatuses in press filters, this gradient can be equalized during a cake posttreatment by different squeezing measures, such as piston, diaphragm, screw, double wire, or additional high-pressure slurry feed. The deliquoring mechanism here consists in the displacement of liquid by reducing the void volume between the particles. As in the case of desaturation, the consolidation (squeezing) undergoes a time-dependent kinetic process and ends at a consolidation equilibrium. Then, the cake structure is homogeneous with minimal possible and pressure-dependent cake porosity. The final moisture content corresponds to the compressibility of the structure and the applied pressure. However, the pores of the cake remain fully saturated.

The understanding of the deliquoring mechanisms and the proper characterization of the filter cake deliquoring behavior are central and decisive elements on the way to realize desired cake moisture contents in the technical practice. However, further aspects have to be considered, which are influencing the final result of the deliquoring process. The capillary pressure distribution in a filter cake depends on the cakes pore structure. This means, the conditions for later deliquoring are defined at the beginning of the separation process and depend on the slurry properties and the cake formation conditions. Slurry pretreatment, such as thickening, fractionation, agglomeration, or addition of filter aids, and the specific conditions during cake formation have to be considered to be able to get proper deliquoring conditions and desired cake moisture contents. In this context, the choice of well-suited filter media is also necessary. Filter media are the decisive interface between slurry, apparatus, and operational conditions. They can influence the cake formation and thus the cake deliquoring. In addition, cake washing (if relevant) can influence the subsequent deliquoring results. In the case of continuously operating filters, one part of the deliquoring zone is reserved for the washing procedure and the cake structure can change by removing soluble substances.

Last but not least, the apparatus operation and design exert strong influences on the final moisture content of the filter cake. If, for example, a peeler centrifuge is fed in the wrong way, layers of finest particles can form inside and/or on top of the filter cake. This increases, in comparison to a homogeneous cake structure, not only the batch time but also the final moisture content remarkably. If in the case of vacuum disc filters the filter cells are not completely emptied from filtrate before the air blowback detaches the cake, the filtrate is blown back into the filter cake and rewets it.

8.2 Characterization of Deliquoring Results

The filter cake liquid content during the deliquoring process can be characterized by different parameters. For incompressible filter cakes, the liquid saturation degree S is particularly well-suited because it gives a good imagination of the still with liquid-filled void volume and also varies, as a normalized parameter, within the fixed limits between 0 and 1. According to Eq. (8.1), the liquid saturation degree S is defined as the ratio of the actual liquid volume V_L in the filter cake and its total void volume V_{void}

$$S = \frac{V_L}{V_{void}} \tag{8.1}$$

To be able to quantify the void volume in the filter cake, the porosity ε as the ratio of void volume and the total volume of the filter cake V_{tot} have to be determined (cf. Section 3.2). However, besides the saturation degree, other definitions, such as residual moisture content MC, dry substance DS, or liquid–solid volume ratio θ, can be used to describe the liquid content of filter cakes.

The moisture content MC in Eq. (8.2) relates to the actual mass of liquid m_L to the total mass of the moist filter cake $(m_L + m_s)$

$$MC = \frac{m_L}{m_L + m_s} \tag{8.2}$$

The dry substance DS in Eq. (8.3) uses alternatively to the moisture content MC the relation of solid mass m_s to the total mass of the moist filter cake $(m_L + m_s)$

$$DS = \frac{m_s}{m_L + m_s} \tag{8.3}$$

Last but not least, the liquid–solid volume ratio B in Eq. (8.4) relates the volumes of liquid V_L and solids V_s in the cake to each other

$$B = \frac{V_L}{V_s} \tag{8.4}$$

Residual moisture content and dry substance are not descriptive as the saturation degree, but they are easier to measure because of the unnecessary porosity determination. On the other hand, it is not possible to compare different materials, if their densities are different. The same quantity of water results in these cases in different numbers of residual moisture.

In the case of compressible filter cakes, which are deliquored by squeezing, the saturation degree makes no sense because all pores of the cake are remaining fully saturated during the entire deliquoring process. For such cases, the liquid–solid volume ratio can be used alternatively to the residual moisture content or dry substance. Materials of different density can be compared well because the volume of liquid and solid is used and not the mass.

In any case, it would be meaningless to characterize the cake moisture content only by indicating a general percentage, if it is not clear, how this percentage was calculated. If the definition is clear, the moisture content can be transferred, according to Table 8.1, into each other residual cake moisture characterizing parameter.

In some cases, some further aspects must be considered to interpret the data correctly. If water is bound in the structure of crystals, the drying temperature determines which part of this water will remain as solid mass and which quantity of crystal water will be removed. In those cases, a defined upper limit of allowed drying temperature is necessary to avoid the removal of chemically bound liquid and a change of the solid structure.

The situation becomes even more difficult for biological particles, such as cells, because they contain inner liquid, which is separated from the ambience by a permeable membrane. If such a material is dried by thermal means, not only the free movable water around the cells but also water from inside will be removed. The problem here is not the determination of the remaining dry substance but the deformation of the particles during drying and to distinguish between inside-bound and outside-free movable liquid.

A third aspect consists in the eventual presence of dissolved substances in cake residual liquid. If such a moist filter cake is dried, the dissolved substances are becoming solid and are counted as solids. To avoid a mistake, when determining the residual cake moisture content MC, the mass of dissolved substances per

Table 8.1 Conversion of parameters for characterizing the cake moisture content.

	(S)	(MC)	(DS)	(B)
S	$= \dfrac{V_L}{V_{void}}$	$= \dfrac{\rho_s(1-\varepsilon)MC}{\rho_L\varepsilon(1-MC)}$	$= \dfrac{\rho_s(1-\varepsilon)(1-DS)}{\rho_L\varepsilon DS}$	$= \dfrac{(1-\varepsilon)B}{\varepsilon}$
MC	$= \dfrac{\rho_L\varepsilon S}{\rho_L\varepsilon S + \rho_s(1-\varepsilon)}$	$= \dfrac{m_L}{m_s + m_L}$	$= 1 - DS$	$= \dfrac{B\rho_L}{\rho_s + B\rho_L}$
DS	$= \dfrac{\rho_s(1-\varepsilon)}{\rho_L\varepsilon S + \rho_s(1-\varepsilon)}$	$= 1 - MC$	$= \dfrac{m_s}{m_s + m_L}$	$= \dfrac{\rho_s}{\rho_L B + \rho_s}$
B	$= \dfrac{\varepsilon S}{1-\varepsilon}$	$= \dfrac{\rho_s MC}{\rho_L(1-MC)}$	$= \dfrac{\rho_s(1-DS)}{\rho_L DS}$	$= \dfrac{V_L}{V_s}$

gram liquid $m_{s,L}$ [$g_{dissolved\ material}/g_{liquid}$] has to be measured independently in the liquid and then the residual liquid content of the cake can be corrected according to Eq. (8.5)

$$\mathrm{MC} = \frac{m_L}{m_L + m_s - (m_L \cdot m_{s,L})} \tag{8.5}$$

In analogous manner, the other moisture-characterizing parameters have to be corrected.

8.3 Desaturation of Filter Cakes

8.3.1 Boundary Surface and Surface Tension

Before the influences on the desaturation conditions of filter cakes will be discussed in detail, the physical mechanisms of mechanical desaturation have to be explained. Schubert [2] has given a most comprehensive fundamental description of capillary effects in porous systems of solid particles and the basic correlations, which are discussed in this chapter.

If two molecularly not miscible fluids, such as liquid and gas, are getting in contact, a boundary phase originates, which exhibits a certain volume. This volume has to be fixed by definition. Because of the fact that the dilation of the boundary phase parallel to the phase boundary is normally much greater than vertical to it; this phase is called boundary surface. For this concept, a model was formulated [3]. The two fluids in contact exhibit a common and freely movable boundary surface. The energy in the boundary surface is supposed to be greater than in the two bulk phases. According to Figure 8.4, the intermolecular forces of the molecules within the liquid phase are compensating each other, but not in the boundary surface in contact to the gas phase, where less interaction takes place.

At the boundary surface, a resulting force is directed into the liquid. There is work necessary to transport additional molecules into the boundary layer to enlarge it. A thermodynamic interpretation of the surface tension γ_L can be derived from the fundamental equation of Gibbs as formulated in Eq. (8.6)

$$dF = -S \cdot dT - p \cdot dV + \gamma_L \cdot dA + \sum_{i=1}^{j} \mu_i \cdot dn_i \tag{8.6}$$

F represents free energy, S entropy, T absolute temperature, p pressure, A boundary surface, μ chemical potential, n number of molecules, and γ_L surface tension. The surface tension results as the reversible, isothermal work to increase the

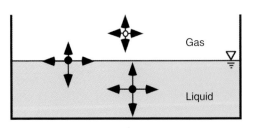

Figure 8.4 Boundary surface between liquid and gas.

Table 8.2 Surface tension of different liquids.

Liquid	Temperature (K)	Surface tension (N m^{-1})
Platinum	2273	1.8000
Mercury	293	0.4790
Water	298	0.0720
Ethanol	293	0.0226
Pentane	293	0.0160
Hydrogen	15	0.0028

boundary surface by a certain amount under the condition of constant volume and constant number of molecules, as indicated in Eq. (8.7)

$$\gamma_L = \left[\frac{dF}{dA}\right]_{T,V,n_i} \tag{8.7}$$

The dimension of γ_L is J m^{-2} or N m^{-1}. A mechanical interpretation of this equation considers the boundary surface as an arbitrarily thin membrane, parallel to the homogeneous phases. The boundary surface is plain if the pressure on both sides is equal and isotropic. Also, in this case, work is necessary to enlarge the boundary surface. The surface tension is a per length acting force, which is independent of the direction and parallel to the isotropic boundary surface [4].

The index of the surface tension normally denotes both involved phases, which means here the surface tension between liquid and gas $\gamma_{L,g}$, whereby the first index represents the phase of greater density. Because gas or vapor behaves similar to the vacuum, simplifying γ_L with only one index can be written normally. Surface and boundary surface are supposed as synonym here. Table 8.2 shows some numbers for the surface tension of different liquids.

The surface tension is depending on the kind of involved substances, the temperature, and on the presence of surfactants. The ambient pressure is normally of subordinate influence. Only for very high pressures beyond 1 MPa, some influences can be observed because of more intensive gas/liquid interaction and changing of the liquid composition by dissolution of greater amounts of gas.

If the temperature increases, the surface tension decreases because the molecular mobility and the distance between the molecules are increasing. Figure 8.5 shows the dependency between surface tension and temperature for the system water/air.

The function can be described by the formula in Eq. (8.8), which was originally formulated by van der Waals and approximated for water

$$\gamma_{H_2O} = 0.1179 \cdot \left(1 - \frac{T}{T_{cr}}\right)^{4/5} \tag{8.8}$$

The critical temperature T_{cr} is 374.1 °C for the critical pressure $p_{cr} = 2.26$ MPa and the critical specific volume $V_{cr} = 3.17$ dm^3 kg^{-1}.

The second measure to decrease the liquid surface tension consists in the dosage of surfactants [5, 6]. These are molecules with a polar and a nonpolar

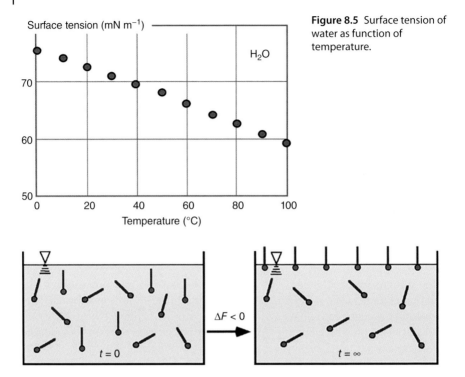

Figure 8.5 Surface tension of water as function of temperature.

Figure 8.6 Surfactant migration to the boundary surface.

part. The polar part often consists of a cationic, anionic, or amphoteric group, the hydrophobic part of a hydrocarbon chain. If such a molecule is dissolved in water as a polar fluid, high free energy F originates between the nonpolar part of the molecule and the surrounding polar water molecules. Nature tries to generate a state of minimal energy and thus the surfactants migrate to the liquid–gas boundary surface and orientate their nonpolar parts into the gas phase, as can be seen in Figure 8.6.

This decreases the energy of the system, and the cohesion between the water molecules in the surface is loosened. As a consequence, the surface tension decreases. The greater the concentration of surfactants becomes, the lower the surface tension is. This comes to an end at the critical micelle concentration (cmc), if surfactant molecules occupy all possible places in the liquid boundary surface. Additional dosage of surfactants does not reduce the surface tension more but leads to the formation of micelles. The qualitative development of surface tension in dependency of the surfactant concentration is drawn schematically in Figure 8.7.

In the slurry or a filter cake, the surfactants are interacting, according to Figure 8.8, not only with the liquid/gas interface but also with the particle surfaces, and an overall adsorption equilibrium is formed for constant conditions.

Depending on the kind of surfactant molecule, the surfactant concentration, the concentration gradient dc/dx near the boundary surface, and the diffusion

Figure 8.7 Surface tension as a function of surfactant concentration.

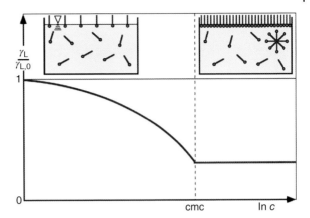

Figure 8.8 Adsorption equilibrium of surfactants in slurry.

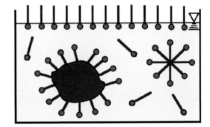

constant D ($D \approx 10^{-5}$ cm^2 s^{-1}) in the liquid, the equilibrium state is reached after times from hours to parts of a second [7, 8]

$$\frac{dn}{dt \cdot A} = D \cdot \frac{dc}{dx} \qquad (8.9)$$

The concentration gradient results from the reduction of surfactant molecules in the layer near the boundary surface, which is called subsurface, as illustrated in Figure 8.9.

An example for the dynamic change of surface tension with surfactant concentration variation is given in Figure 8.10 [9].

In a first step after the dosage of the surfactant, the system has to be homogenized. In a second step, after the end of mixing, the migration of surfactant molecules into the boundary surface starts. The greater the surfactant concentration at the beginning is, the faster the equilibrium state is reached.

During the desaturation of a filter cake, permanently fresh boundary surface is generated. This means that an efficient surfactant must migrate as fast as possible to the boundary surface and must be present in a sufficient amount in the slurry. Unfortunately, a part of the surfactants adsorb at the solid surface and is not more

Figure 8.9 Concentration gradient of surfactants near the boundary surface.

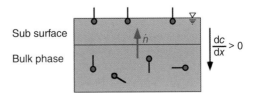

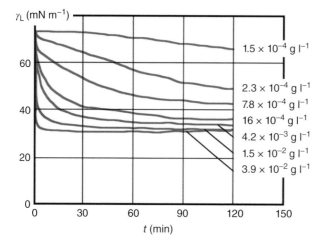

Figure 8.10 Dynamic surface tension $\gamma_L(t)$.

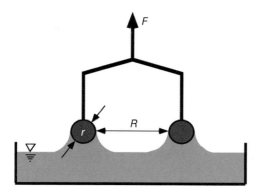

Figure 8.11 Ring method to measure surface tension.

available to lower the surface tension of the liquid. This effect becomes worse, the smaller the particles and thus the larger their surface becomes. The usage of surfactants as pretreatment measure before cake desaturation is principally possible, but not often practiced (cf. Section 5.7).

The surface tension between liquid and gas is one key parameter to describe the desaturation of filter cakes. It can be measured by different techniques and one can distinguish between static and dynamic measurement methods [10–12].

The ring method (Lecomte du Noüy) is based, according to Figure 8.11, on a ring of perfect wetting platinum/iridium (cf. Section 8.3.2), which is first submerged in and then pulled out of the liquid to be analyzed, whereby the force is registered.

The maximal force is reached when the lamella, which has formed underneath the ring, is breaking. Along the circumference of the ring, the surface tension γ_L is acting as an overall force F per length L, as indicated in Eq. (8.10)

$$\gamma_L = \frac{F}{L} \tag{8.10}$$

8.3 Desaturation of Filter Cakes

The force F consists, according to Eq. (8.11), of the ring mass F_m, the capillary force (cf. Section 8.3.3), and a correction factor f, which depends on the geometry

$$F = F_m + 4 \cdot \pi \cdot R \cdot \gamma_L \cdot f \tag{8.11}$$

Equation (8.12) describes the influencing parameters of the correction factor f

$$f = f\left[\frac{R^3}{V_L}, \frac{R}{r}\right] \tag{8.12}$$

Now, the surface tension can be calculated in Eq. (8.13)

$$\gamma_L = \frac{F - F_m}{4 \cdot \pi \cdot R \cdot f} \tag{8.13}$$

This method is denoted as a static method because only the static value of the surface tension can be measured and not its dynamic change. The factor 4 in Eq. (8.13) is due to the fact that there are two lamellae with a circumferential length for each lamella.

A second static force measuring method can be realized, according to Figure 8.12, with a plate (Wilhelmi), which also consists of platinum/iridium.

Here, the plate is driven near to the liquid surface, the liquid jumps to the plate, and pulls it down. The surface tension has to be calculated again from a force balance with the weight of the plate F_m, the buoyant force F_b, and the length of the so-called three-phase contact line (3pcl), as is formulated in Eq. (8.14)

$$\gamma_L = \frac{F - (F_m - F_b)}{L_{3pcl}} \tag{8.14}$$

A third static method consists in the measurement of the liquid height in a capillary, as shown in Figure 8.13.

In the equilibrium state, the capillary force equals the weight force of the liquid column in the capillary, as shown in Eq. (8.15)

$$(\pi \cdot r^2 \cdot h + V_M) \cdot \rho_L \cdot g = 2 \cdot \pi \cdot r \cdot \gamma_L \cdot \cos \delta \tag{8.15}$$

The volume of the meniscus V_M should be much smaller than the volume of the entire liquid column, as indicated in the Eq. (8.16)

$$r \ll h \Rightarrow V_M \ll \pi \cdot r^2 \cdot h \tag{8.16}$$

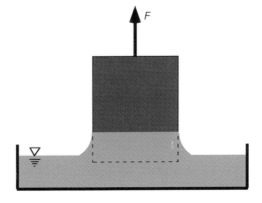

Figure 8.12 Plate method to measure surface tension.

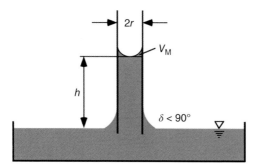

Figure 8.13 Capillary method to measure surface tension.

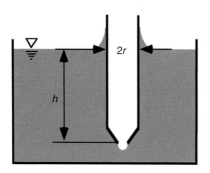

Figure 8.14 Bubble method for dynamic measurements of surface tension.

The wetting angle δ should be 0 because in that case, $\cos \delta = 1$ and the surface tension can be easily calculated, according to Eq. (8.17)

$$\gamma_L = \frac{1}{2} \cdot r \cdot \rho_L \cdot g \cdot h \tag{8.17}$$

To measure the dynamic behavior of the surface tension, the bubble method can be used. Figure 8.14 explicates the principle.

A wetting capillary is immersed into the liquid. Then, gas is pressed into the capillary, and if the gas pressure is greater than the capillary pressure of the capillary orifice plus the hydrostatic pressure at the orifice, gas flows through the capillary. The greater the gas flow becomes, the shorter the time for formation and displacement of the originating bubbles is. The age of the boundary surface in the moment of bubble detachment in combination with the surfactant concentration and mobility determines the surface tension. The surface tension then can be calculated, according to Eq. (8.18), from a pressure balance with the knowledge of the respective gas pressure in the capillary

$$p = \frac{2 \cdot \gamma_L}{r} + \rho_L \cdot g \cdot h \tag{8.18}$$

In Eq. (8.19), the pressure balance is resolved for the surface tension

$$\gamma_L = \frac{r \cdot (p - \rho_L \cdot g \cdot h)}{2} \tag{8.19}$$

Precondition here is again the complete wetting ($\cos \delta = 1$) of the capillary by the liquid.

Figure 8.15 Three-phase contact line.

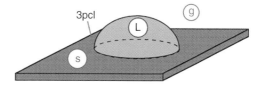

8.3.2 Three-Phase Contact Line, Contact Angle, and Wetting

The surface tension characterizes a boundary surface between two phases in a filter cake. During desaturation, three phases are in contact. These are solid, liquid, and gas. This situation can be described by the concept of 3pcl [13] and is depicted in Figure 8.15 in the example of a liquid drop on a plain solid surface in gaseous ambience.

The situation is of course also valid for a solid and two nonmolecularly mixable liquids. In the equilibrium state, according to Figure 8.16, a force balance can be formulated using the three involved surface tensions [14].

From this force balance, the wetting angle δ can be calculated, according to Eq. (8.20)

$$\cos \delta = \frac{\gamma_{s,g} - \gamma_{s,L}}{\gamma_{L,g}} \tag{8.20}$$

The Young equation is also valid for three-phase systems such as s–L_1–L_2. The equation is not valid if $\gamma_{s,g} > \gamma_{L,g} + \gamma_{s,L}$. In that case, the liquid is spreading and overlays the solid surface.

The question can be asked whether the surface tensions can cause a deformation of the solid body. According to Figure 8.17, there is γ_r, a reaction force per length from the solid body.

Figure 8.16 Force balance for three phases in contact.

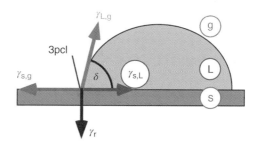

Figure 8.17 Deformation of a solid body near the three-phase contact line.

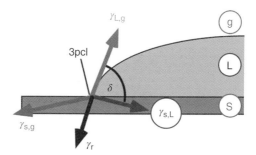

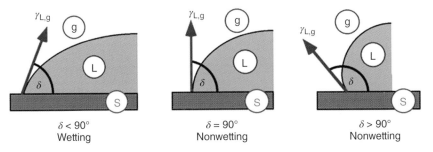

Figure 8.18 Definition of wetting and nonwetting.

Normally, this is not resulting in a deformation of the solid material, as indicated in Figure 8.17. A measurable effect can be observed for small E-moduli and high surface tensions $\gamma_{L,g}$, as indicated in Eq. (8.21) [15, 16]

$$\frac{E}{\gamma_{L,g}} < 10^{10}\ \text{m}^{-1} \tag{8.21}$$

Normally, $E > 10^9$ N m^{-1} and $\gamma_{L,g} < 0.1$ N m^{-1} (water at 20 °C: 0.072 N m^{-1}). In these cases, the solid body can be considered to be rigid and nondeformable. As a consequence, the contact angle δ is not affected. However, there are several other effects, which are affecting the contact angle and lead to deviations in comparison to the ideal situation in Figure 8.16.

Before discussing these deviations, the principle definition has to be given: what is denominated wetting and what is denominated nonwetting. This can be illustrated by the examples given in Figure 8.18 [17].

According to a general convention, the contact angle is measured in the fluid phase of greater density, which means here in the liquid. If the contact angle is smaller than 90°, the situation is denominated wetting and if the contact angle is equal or greater than 90°, the situation is nonwetting. If the contact angle is zero, the situation is denominated as completely wetting.

As mentioned before, several effects are influencing the contact angle such as surfactants, homogeneity, and roughness of the solid surface, geometry of particles and capillaries, or dynamic effects. It is obvious that the contact angle is changed in the presence of surfactants because they are influencing the surface tension. This is used, for example, in soil-washing processes, if oil should be separated from soil particles. Figure 8.19 illustrates the principle of such a process schematically.

The addition of water-soluble surfactants (t_1) changes the good wetting properties of the solids for oil at the beginning (t_0). This leads to a bulging of the oil drops because of a reduction of the solid wettability for oil. At the end (t_2), the oil drop can be relieved relatively easy and washed away. The surfactant should make $\gamma_{s,o}$ as great and $(\gamma_{s,w} + \gamma_{w,o})$ as small as possible.

The homogeneity and roughness of the solid surface lead to an effect that is denominated as contact angle hysteresis and explained in Figure 8.20.

If a liquid drop is put on a horizontal even and more or less smooth solid surface, the drop has a symmetric and spherical shape. The contact angle along the

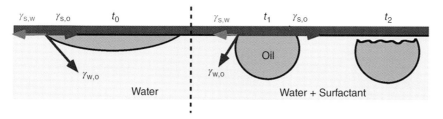

Figure 8.19 Soil-washing with the support of surfactants.

Figure 8.20 Contact angle hysteresis.

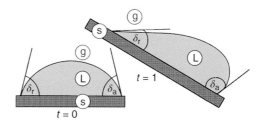

3pcl is constant. If the surface is declined, the advancing contact angle δ_a gets greater than the receding contact angle δ_r. If in a third step the surface is brought again in horizontal position, a remaining asymmetry of advancing and receding contact angle can be observed. Depending on the flow direction of the liquid, the contact angle is named receding angle or advancing angle. The asymmetry persists for repeated inclinations in the form of a hysteresis. It was found that a roughness of <0.1 µm has no longer influence on the wetting angle [18].

Besides this hysteresis effect, the roughness of the solid surface directly affects the contact angle. The Young equation in Eq. (8.22) results from an energy balance in the case of a differential shift of the three-phase contact line on an ideally smooth surface (cf. Figure 8.16)

$$dF = \gamma_{s,L} \cdot dA - \gamma_{s,g} \cdot dA + \gamma_{L,g} \cdot \cos\delta \cdot dA = 0 \quad (8.22)$$

In the case of nonideally smooth surface, a roughness factor r can be introduced, according to Eq. (8.23), and the energy balance can be formulated in Eq. (8.24) for that case again

$$r = \frac{A_{rough}}{A_{smooth}} \quad (8.23)$$

$$dF = \gamma_{s,L} \cdot r \cdot dA - \gamma_{s,g} \cdot r \cdot dA + \gamma_{L,g} \cdot \cos\delta \cdot dA \quad (8.24)$$

From this balance in Eq. (8.25) results the Wenzel equation for the contact angle on rough surfaces [19]

$$\cos\delta_{rough} = r \cdot \frac{\gamma_{s,g} - \gamma_{s,L}}{\gamma_{L,g}} = r \cdot \cos\delta_{smooth} \quad (8.25)$$

As a consequence, wettability of a wetting system becomes even better (Eq. (8.26)) and the nonwettability of a nonwetting system becomes even worse (Eq. (8.27)). The so-called lotus effect [20] is based on this phenomenon

$$\delta_{smooth} < 90° \Rightarrow \delta_{rough} < \delta_{smooth} \quad (8.26)$$

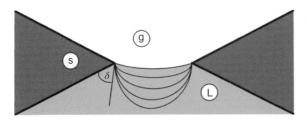

Figure 8.21 Contact angles for sharp edges and corners.

$$\delta_{smooth} > 90° \Rightarrow \delta_{rough} > \delta_{smooth} \qquad (8.27)$$

A further influence on the contact angle is given by the geometry of the wetted system. For broken particles or crystals, which exhibit sharp edges and/or corners, an infinite number of equilibrium situations can exist, as Figure 8.21 illustrates [21].

A second geometrical effect can be seen in Figure 8.22 for capillaries of different shape.

For cylindrical capillaries, the definition of wetting and nonwetting seems to be distinctly defined. If the shape of the capillary becomes noncylindrical, the situation becomes more complicated. If the capillary has conical shape, as on the right side of Figure 8.22, the contact angle is, according to the general definition, less than 90°. This is the condition for a wetting situation, although the situation is obviously nonwetting. In the nonwetting situation, the gas is the wetting phase and tries to displace the liquid out of the capillary.

A third example for geometrical effects regarding the contact angle is depicted in Figure 8.23 for liquid bridges between two spherical particles.

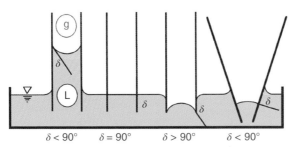

Figure 8.22 Wetting in case of different pore geometries.

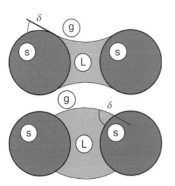

Figure 8.23 Liquid bridges between spherical particles.

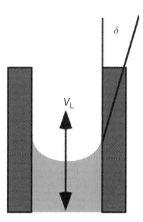

Figure 8.24 Dynamic contact angle.

Although the wetting angle is in both cases less than 90°, the liquid pressure inside the liquid bridge can be either smaller or greater in comparison to the ambient gas pressure. If the liquid pressure is smaller than the gas pressure, the liquid/gas boundary surface is bent inward. If the liquid pressure is greater than the gas pressure, the liquid/gas boundary surface is bent outward.

All the influences on the contact angle, which have been discussed before, refer to a static equilibrium situation. During a desaturation or an imbibition process of a filter cake, the liquid/gas boundary surface is moving, according to Figure 8.24, with a certain velocity through the void system.

This dynamic process influences the contact angle. If a capillary is desaturated and the liquid moves downward, the receding contact angle becomes smaller, the greater the velocity is. In the opposite case of imbibition, the liquid moves upward and the advancing contact angle becomes greater, the greater the velocity is. If the system is saturated and wetted, in most cases, the receding contact angle during the kinetic process of desaturation can be approximated as zero.

The measurement of the contact angle can be carried out with different methods, which are based on a visual or image analysis. If a drop of liquid is put on a horizontal smooth solid surface, as in Figure 8.15, the contact angle can be measured by microscopic observation. A further method is shown in Figure 8.25 [22].

A plate of the solid material in question is immersed first vertically into the liquid(s). Then, the plate is inclined, until the meniscus between liquid and plate disappears. This indicates the contact angle between liquid and solid. The plate

Figure 8.25 Plate method (Adam) to measure the contact angle.

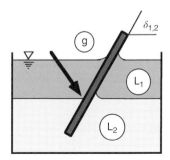

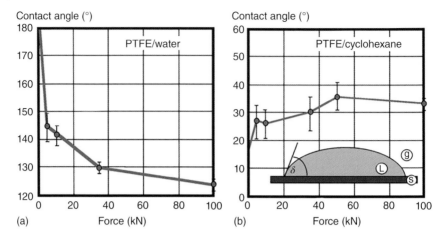

Figure 8.26 Influence of press force on contact angle (cf. [24], p. 220).

method exhibits in comparison to the drop method the additional option to measure not only the three-phase system s/L/g but also the system $s/L_1/L_2$.

If not a solid plate of a certain material is available, but only a powder, an impermeable tablet can be produced in a press apparatus and the contact angle again can be measured by putting a liquid drop on the tablet [23]. The week point of this method is the influence of surface roughness on the contact angle. Depending on the press force and the material, the surface of the tablet becomes more or less smooth and thus the contact angle varies. This effect has been measured extensively by Bröckel and Löffler [24] and is demonstrated in the example of PTFE-particles in Figure 8.26.

The system PTFE/water on Figure 8.26a is nonwetting and the contact angle is greater than 90°. The smaller the press force on the powder becomes, the more rough the surface is and the more the contact angle increases (cf. Eq. (8.27)). The system PTFE/cyclohexane on Figure 8.26b behaves wetting and thus the contact angle is smaller than 90°. The smaller the press force on the powder becomes, the more rough the surface is and the better the liquid wets the solid (cf. Eq. (8.26)).

By using a special measuring method, it is possible to become independent of this roughness effect and to determine the "true" contact angle of the system. This method is based on the preparation of the particles of interest on an object plate, which is covered with an adhesive film, as is illustrated in Figure 8.27.

The particles are randomly distributed on the plate. It is important that the adhesive film is thin enough to avoid too deep immersion of the particles. After preparation, the following steps of measurement have to be carried out. With the plate method, shown in Figure 8.25, the contact angle between pure adhesive film and liquid δ_{ad} and the contact angle between adhesive film plus particles $\delta_{ad,p}$ is measured. In addition, the area porosity of the plate plus particles ε_p has to be determined by image analysis. Now, the contact angle of the particles δ_p can be calculated from the porosity-weighed contributions of the single contact angles, as formulated in Eq. (8.28)

$$\cos \delta_p \cdot (1 - \varepsilon_p) = \cos \delta_{ad,p} - \varepsilon_p \cdot \cos \delta_{ad} \qquad (8.28)$$

Figure 8.27 Preparation of particles on an object plate.

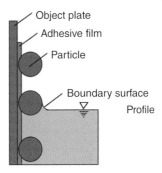

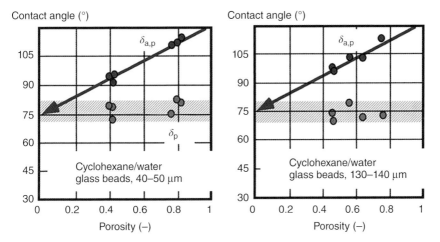

Figure 8.28 Contact angle determination for dispersed particles (cf. [24], p. 219).

As limits of application are described, a particle size ranges between 10 and 1000 μm, a porosity ranges between 40% and 70%, and a particle embedding in the adhesive film is less than 50%. Figure 8.28 demonstrates the results of this method in the example of a cyclohexane/water mixture and glass beads in a different size range.

For all parameter combinations, the same contact angle of the pure particles result and this value can be validated by a standard experiment with a drop of cyclohexane/water on a polished glass plate.

As has been discussed before, different methods are available to measure the contact angle, but it has also been demonstrated that there can be found several phenomena, which are leading to deviations from the simple situation of a rotationally symmetric drop on a smooth solid plate. Especially in the pore system of a filter cake of nonspherical and different shaped particles of often nonuniform material, an exact determination of a meaningful contact angle is not possible. Because of the fact that for a wetted system, already the static contact angle is smaller than 90° and the dynamic receding contact angle during desaturation becomes even smaller, in practice, the approximation of total wetting is allowed in most cases. In this case, the wetting angle can be set to 0 and the cosine of the wetting angle to 1.

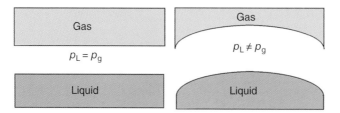

Figure 8.29 Even and bent boundary surface.

8.3.3 Capillary Pressure and Capillary Pressure Distribution

If the boundary surface between two not mixable fluids is even, the pressure in both homogeneous phases must be equal because the boundary surface is freely movable. If the boundary surface is bent, as shown in Figure 8.29, for the system of gas and liquid, the pressures p_g in the gaseous phase and in the liquid p_L must be different.

The higher pressure is related to the phase with outward and the lower pressure to the phase with inward bent boundary surface. The resulting pressure difference $(p_g - p_L)$ is defined as capillary pressure p_{cap}. Depending on the direction, in which the boundary surface is bent, the capillary pressure can be defined as positive or negative. According to a general agreement, the capillary pressure is defined as positive, if the pressure in the phase of higher density is lower than in the phase of lower density. Figure 8.30 gives examples for three-phase and two-phase systems.

On Figure 8.30a, the capillary pressure for the nonwetting capillary and the droplet is negative because the pressure in the liquid is greater than the pressure in the gas. On Figure 8.30b, the capillary pressure for the wetting capillary and a gas bubble is positive because in the liquid phase of higher density, the pressure is lower than in the gas phase.

The capillary pressure can be calculated from a force balance for an infinitesimal small part of the boundary surface, as depicted in Figure 8.31.

The two radii of curvature R_1 and R_2 characterize the shape of the boundary surface here. From this force balance, the Young–Laplace equation [14, 25], as expressed in Eq. (8.29)–(8.32), results

$$\Delta p_{cap} \cdot ds_1 \cdot ds_2 = \gamma_L \cdot (ds_1 \cdot d\alpha_2 + ds_2 \cdot d\alpha_1) \tag{8.29}$$

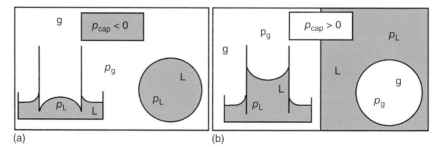

Figure 8.30 Positive and negative capillary pressures.

Figure 8.31 Force balance for a piece of boundary surface.

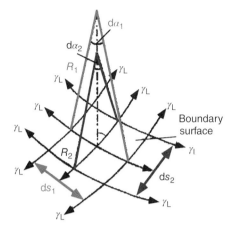

Figure 8.32 Deformation of a gas bubble in liquid.

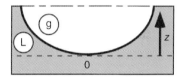

$$ds_1 = R_1 \cdot d\alpha_1 \tag{8.30}$$

$$ds_2 = R_2 \cdot d\alpha_1 \tag{8.31}$$

$$\Delta p_{cap} = \gamma_L \cdot \left[\frac{1}{R_1} + \frac{1}{R_2}\right] \tag{8.32}$$

Depending on the deformation of the boundary surface, the capillary pressure changes. This should be discussed on the example of a gas bubble of sufficient diameter, as shown in Figure 8.32, which is immersed in a liquid and influenced by an external field, as gravity, which acts against the direction of the coordinate z (cf. [2], p. 30).

If the deepest point of the bubble is chosen as a reference point, the hydrostatic pressure decreases in the upward direction and thus the pressure difference between gas and liquid, which means capillary pressure. In the equilibrium situation at each point on the bubble surface, the force balance leads to a different bent of the surface. This can be expressed in Eq. (8.33)

$$p_{cap} = \gamma_L \cdot \left(\frac{1}{R_1} + \frac{1}{R_2}\right) = p^0 + (\rho_L - \rho_g) \cdot g \cdot z \tag{8.33}$$

The reference point advantageously is positioned at the deepest point of the bubble, which is here zero. According to Eq. (8.34), the capillary pressure at this point can be calculated with the respective radii of curvature

$$p_{cap}^0 = \gamma_L \cdot \left(\frac{1}{R_1^0} + \frac{1}{R_2^0}\right) \tag{8.34}$$

Cake filtration leads to pore diameters in the micrometer range and capillary pressures, which are much greater than the hydrostatic pressure in the earth field. As a consequence, for cake filtration, this phenomenon plays no role and can be neglected in correspondence to Eq. (8.35)

$$p_{cap} \gg g \cdot z \cdot (\rho_L - \rho_g) \Rightarrow p_{cap} = p^0_{cap} \tag{8.35}$$

A further influence on the capillary pressure is given by its correlation with the liquids vapor pressure p_{vap}. This correlation is formulated in Eq. (8.36) and is known as Kelvin equation [26]

$$\ln \frac{p^0_{vap}}{p_{vap}} = \frac{V_{mol,L} \cdot p_{cap}}{R \cdot T} \tag{8.36}$$

The reference vapor pressure p^0_{vap} is defined for an even boundary surface and a capillary pressure of zero. $V_{mol,L}$ characterizes the molar volume of the liquid, R the universal gas constant (=8.31 J mol^{-1} K^{-1}), and T the absolute temperature. For even boundary surface, there is, according to Eq. (8.37), no influence because the capillary pressure is zero

$$p_{cap} = 0 \Rightarrow p_{vap} = p^0_{vap} \tag{8.37}$$

If the capillary pressure in Eq. (8.38) is positive, the vapor pressure is reduced

$$p_{cap} > 0 \Rightarrow \ln p_{vap} = \ln p^0_{vap} - \frac{V_{mol,L} \cdot p_{cap}}{R \cdot T} \tag{8.38}$$

If the capillary pressure in Eq. (8.39) is negative, the vapor pressure is increased

$$p_{cap} < 0 \Rightarrow \ln p_{vap} = \ln p^0_{vap} + \frac{V_{mol,L} \cdot p_{cap}}{R \cdot T} \tag{8.39}$$

These effects are relevant only for very small pore diameters. As discussed before, a capillary pressure is not necessarily bound to a three-phase system but can also be observed for two-phase fluid systems, such as gas bubbles or liquid drops. Table 8.3 illustrates for a temperature of 293 K that the effect for water drops of less than 1 μm in air is negligible.

A macroscopic consequence of the interaction of capillary and vapor pressure is certain residual moisture of wettable particles, if they exhibit micropores and are in contact with a gaseous atmosphere of certain humidity. In such a case, liquid molecules are condensing into the pore as long as the corresponding capillary pressure is reached. As a result, such a pore will remain fully saturated

Table 8.3 Influence of the capillary on the vapor pressure for different pore sizes.

x (nm)	p_{cap} (MPa)	p_{vap}/p^0_{vap}
5	72.2	0.5803
50	6.0	0.9567
100	2.9	0.9784
1000	0.3	0.9978

Figure 8.33 Circular model pore.

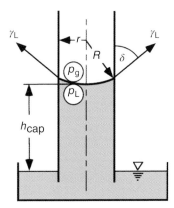

because of its high capillary pressure and little length. It will not be dehumidified by evaporation, if the ambience conditions are not changing.

A moist filter cake in gaseous environment represents a three-phase system of solid, liquid, and gas. Solid, liquid, and gas are meeting at the 3pcl and the liquid/gas boundary surface touches the solid wall of the pore with the contact angle δ. A circular–cylindrical model pore represents the simplest model of a porous filter cake. For positive capillary pressures, a certain equilibrium situation is shown in Figure 8.33.

The formulation of a force balance results in this case in a simplified form of the Young–Laplace equation (cf. eq (8.32)). Because of the circular shape of the pore, the curvature of the boundary surface can be described by one radius R. However, one is normally not interested to know the capillary pressure in dependence of R, but in dependence of the pore radius r. The correlation between R and r is given in Eq. (8.40), according to the rules of trigonometry, by the cosine of the contact angle δ

$$p_{cap} = p_g - p_L = \frac{2 \cdot \gamma_L}{R} = \frac{2 \cdot \gamma_L \cdot \cos \delta}{r} \tag{8.40}$$

If liquid rises up in a capillary to an equilibrium value, the capillary height h_{cap} can be calculated by a force balance between capillary and hydrostatic pressure, as shown in Eq. (8.41)

$$h_{cap} = \frac{p_{cap}}{\rho_L \cdot C \cdot g} = \frac{2 \cdot \gamma_L \cdot \cos \delta}{r \cdot \rho_L \cdot C \cdot g} \tag{8.41}$$

The characterization of the pore system of a filter cake by parallel cylindrical tubes of different diameter is the simplest model to explain some basic effects of deliquoring, but there are further effects in reality, which cannot be captured.

A little more realistic but nevertheless simple model can be seen in Figure 8.34 in the form of a pore channel with changing diameter.

Below the sketch of the pore channel, an equilibrium diagram for desaturation and imbibition is drawn. This diagram and the procedure to get it correspond in principle with the procedure for characterizing the capillary pressure distribution in a filter cake.

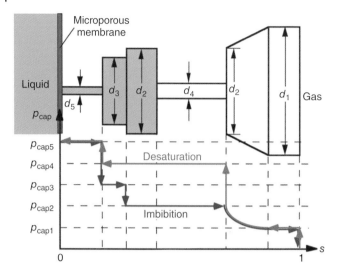

Figure 8.34 Model pore channel of changing diameter.

At first, the entire pore channel is completely filled with liquid, as a filter cake after formation, which is just emerging from the slurry. The capillary is connected at one side with a microporous membrane and a liquid reservoir and is on the other side open to the gas atmosphere. Then, the gas pressure is increased stepwise by an increment at a time. After each pressure increase, the equilibrium state is awaited. If the gas pressure becomes greater than the capillary pressure of the pore with the diameter "1," this part of the capillary is emptied and the gas pressure must be increased to desaturate more parts of the pore channel. If the gas pressure becomes greater than p_{cap4}, the small pore "4" is deliquored. After deliquoring of pore d_4, the following pores d_2 and d_3 are additionally deliquored automatically because the actual gas pressure "4" is greater than p_{cap2} and p_{cap3}. After increasing the gas pressure to more than p_{cap5}, the final smallest pore is deliquored, but no gas is able to get behind the filter medium, which consists of a microporous membrane of a capillary pressure greater than for the smallest pore in the capillary. Now, the model pore is emptied completely and the saturation degree becomes zero. The complete emptying of the pore channel represents one deficiency of this simple model because in reality, not the entire pore liquid can be removed by mechanical means. On the other hand, the model is nicely able to demonstrate that the imbibition curve is principally different from the desaturation curve, if the gas pressure is reduced stepwise again and the liquid is sucked back into the pore channel. This corresponds with experimental observations. However, in this simple model, the pore can be filled completely again, which is not the case in reality because some gas inclusion will remain principally in the pore structure and a saturation degree less than 1 results. There do exist different considerations to explain this phenomenon. One possibility is demonstrated in Figure 8.35 by desaturation and rewetting a pore channel, which includes a sharp edge.

Figure 8.35 Model to explain gas inclusions in filter cakes.

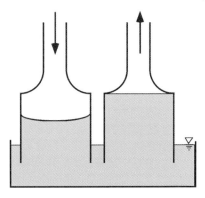

If the capillary pressure of the small capillary is exceeded, the capillary is emptied completely because the diameter in the lower part is increasing. In the opposite direction during imbibition, the liquid is sucked into the lower part of the capillary. At the transition point to the reduced diameter, a sharp edge is present. At a sharp edge, infinite equilibrium values of the wetting angle are possible (cf. Figure 8.35). As a consequence, the wetting angle here is increasing until it reaches 90°. For a horizontal boundary surface, the capillary pressure becomes zero and the imbibition stops.

One-dimensional models, as discussed before, are able to explain a certain number of capillary phenomena, which can also be observed for real filter cakes, but of course, they exhibit some limitations. In Figure 8.36, a two-dimensional model is presented, which was proposed by Schubert (cf. [2], p. 96).

This model is able to explain why in reality gas breaks through the filter media very early, whereby great parts of the cake are still fully saturated. This dispersion effect is called fingering and could already be observed for permeation washing processes (cf. Figure 7.17). The numbers between the particles denominate the respective pore throat diameter (cf. Figure 3.20). "1" characterizes the smallest pore diameter and "10" the largest pore diameter. They are distributed by chance in the pore system. As can be seen, a pressure difference of "8", which

Figure 8.36 Two-dimensional pore model.

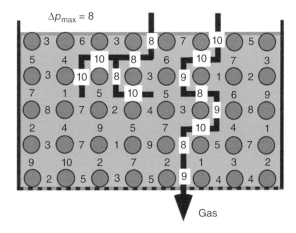

enables to desaturate pores with a diameter of at least "8," is enough to find a path throughout the whole pore system and leads to a gas breakthrough. In other parts of the cake, the desaturation is stopped due to too small pores, which are surrounding the desaturated pores. This fingering effect is very unfavorable because the compressor to maintain the necessary pressure difference must compensate the gas flow. This gas compression determines the main part of the operational costs of a cake filter. Unfortunately, the fingering effect cannot be avoided for conventional gas pressure-driven cake filtration processes. Unconventional solutions to improve this deficiency will be discussed later on (cf. Section 9.4).

After discussing some model configurations to explain principle effects, the real particle system should be examined. As a result of a stochastic arrangement of particles in a filter cake, a pore size distribution and thus a capillary pressure distribution arises. Even this would be the case if particles would be spherical and of the same size. The capillary pressure distribution can be described, according to the terminology of the particle size analysis, as a cumulative volume distribution of the hydraulic pore throat diameters. On the one hand, the pressure is measured, which leads to the diameter of the pore throat via the Young–Laplace equation, and on the other hand, the displaced liquid from the void volume is measured, which leads to the corresponding saturation degree. The capillary pressure distribution in a filter cake of an approximately incompressible, homogeneous, and isotropic structure can be characterized, according to Figure 8.37, by a diagram, which describes the correlation of applied pressure difference and saturation degree in the equilibrium state.

The resulting function depends on the history of desaturation and/or imbibition. The desaturation of the filter cake is described by the so-called capillary pressure curve. This function characterizes for mechanical equilibrium the relationship between capillary pressure p_{cap} and liquid saturation degree S for the entire desaturation process. At full cake saturation, the liquid covers the top layer of particles completely. For such a flat boundary surface, the capillary pressure, as a difference of the pressure in the gas and liquid bulk phase on both sides of the freely movable boundary surface, is zero. With a gradual increase of this

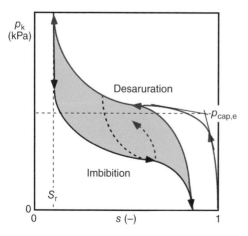

Figure 8.37 Capillary pressure distribution.

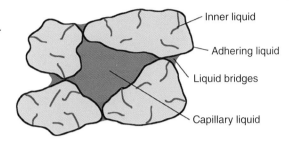

Figure 8.38 Different bound components of liquid in the cake.

differential pressure, the liquid between the particles is pressed first a little bit into the pores and forms menisci at the cake surface, which leads to a marginal reduction of saturation without real desaturation of pores. As can be seen from Figure 8.37, significant reduction of the saturation degree can be registered only after exceeding the capillary entry pressure $p_{cap,e}$ of the largest pore diameter. With increasing differential pressure, smaller and smaller pores can be emptied. The capillary entry pressure is fixed conventionally by intersection of two tangents, attached to both curve segments at the beginning of desaturation. In most cases of incompressible filter cakes, the first formation of menisci is not measurable exactly, and the desaturation starts from $S = 1$. If finally the complete capillary liquid, which is hydraulically connected with the filter medium, is displaced by gas, the mechanical limit of desaturation S_r is reached. Further increase of the differential pressure does no longer decrease the saturation. This is due to the fact that, according to Figure 8.38, the remaining liquid in the filter cake is now hydraulically isolated and bound at the contact points of the particles as liquid bridges, on the surface of the particles as adhering layer, and inside of micropores in the solid matter as internal liquid.

If the liquid can penetrate back into the filter cake due to stepwise reduction of the pressure difference, an imbibition curve results. This curve case is located in any case below the desaturation curve. A complete remoistening of the filter cake does not succeed because of unavoidable gas inclusions during the imbibition process. If desaturation is started again after maximal imbibition, the new desaturation curve meets the initial desaturation curve at a certain point and a hysteresis loop can be passed through [27]. Within this hysteresis loop between desaturation and imbibition curve, any point can be set in the equilibrium state by an appropriate change of desaturation and imbibition. For this reason and to have an unambiguous definition, the capillary pressure curve has to start from full saturation $S = 1$ and ends at the mechanical limit of desaturation $S = S_r$. The capillary pressure curve is independent of the filter cake dimensions and the procedure of desaturation. It represents a material characterizing function of the respective filter cake and has to be measured experimentally for each individual material.

However, the question can be asked whether it would be possible to find at least some correlations between the particle size distribution and the capillary pressure curve, as in the case of filter cake permeation (cf. Section 4.3). Practical background is the desire for enabling a forecast, how the cake moisture content of

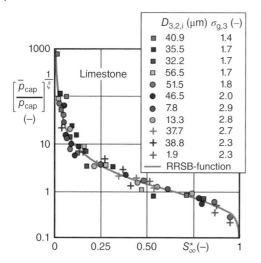

Figure 8.39 Representation of capillary pressure curve by RRSB-function.

a running filter would change in the case of some changes in upstream processes, such as crystallization, comminution, or fractionation.

For the estimation of the characteristic capillary entry pressure $p_{cap,e}$, several approaches can be found in the literature [28]. In Eq. (8.42), one of these approaches is formulated, which is partially validated by experiments

$$p_{cap,e} = C \cdot \frac{1-\varepsilon}{\varepsilon} \cdot \frac{\gamma_L \cdot \cos \delta}{x_{av,SV}^2} \quad (8.42)$$

The factor C yields for spheres to 6 [29].

Extended experiments with limestone particle fractions, characterized by different Sauter mean diameters $D_{3,2,i}$ and geometric variance $\sigma_{g,3}$, which had been dispersed in water and stabilized, have proved that all measured capillary pressure curves could be represented by one master curve [30]. For this purpose, the RRSB (Rosin–Rammler–Sperling–Bennet) [31, 32] distribution, well known from the particle size distribution characterization, turned out to be useful. In Eq. (8.43), this function is applied for cake desaturation and must be adjusted by three physically interpretable parameters

$$S_\infty^* = \frac{S_\infty - S_r}{1 - S_r} = 1 - \exp\left[-\left[\frac{\overline{p}_{cap}}{p_{cap}}\right]^{\frac{1}{\xi}}\right] \quad (8.43)$$

The parameter S_∞ represents the measured equilibrium saturation degree for the respective capillary pressure p_{cap}. The adjustment parameters depend on the respective particle size distribution and can be interpreted as a limit of mechanical desaturation S_r, mean characteristic capillary pressure $\overline{p}_{cap}$, and pore size distribution index ξ. A detailed investigation on how each of these three parameters is depending on the particle size distribution is unfortunately still missing. Figure 8.39 shows the result of the above-mentioned capillary pressure measurements for different limestone fractions (cf. [30], p. 1083).

Figure 8.40 Capillary pressure curve.

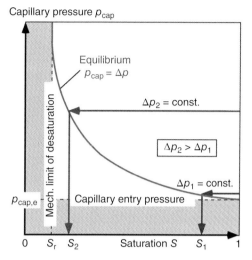

8.3.4 Desaturation of Incompressible Filter Cakes by Gas Pressure Difference

8.3.4.1 Equilibrium of Cake Desaturation with a Gas Pressure Difference

The mechanical limit of cake desaturation by a gas pressure difference is characterized by the capillary pressure curve. This function is an individual "fingerprint" of a filter cake and must be determined experimentally (cf. Section 8.3.4.4). The capillary pressure curve can be used as well for apparatus selection and design as for the analysis of process results. In Figure 8.40, the desaturation behavior of a filter cake for two gas pressure differences is demonstrated qualitatively.

Δp_1 should symbolize here the maximal limit for vacuum filtration, and the capillary pressure curve provides the information that from the physical point of view at best, the saturation degree S_1 can be reached in the equilibrium state. The desaturation starts at saturation $S = 1$ and desaturation time $t_2 = 0$ and ends in the equilibrium state at $S = S_1$ for $t_2 \to \infty$. In the technical reality and cake heights of only some mm to a few cm, the equilibrium is reached approximately after a few minutes. If the saturation degree S_1 does not meet the process specifications and a lower saturation degree is required, the capillary curve tells that vacuum filtration has no chance to undergo the saturation degree S_1. If a lower saturation degree of S_2 must be reached, only a change of the technical system to an overpressure process with $\Delta p_2 > \Delta p_1$ can fulfill this requirement. In this case, the principle decision about a successful filtration technique can be derived from the capillary curve, but it can also be used as a tool for process diagnosis. This should be explained in the example shown in Figure 8.41.

If in the laboratory, the capillary curve is measured carefully for a homogeneous filter cake of a certain material, it can be observed clearly in Figure 8.41 that in this case, a relatively low residual cake moisture can be realized by vacuum filtration. If the message is sent afterward from the production site that the vacuum drum filter is not working properly, but wet mud is discharged, although the same

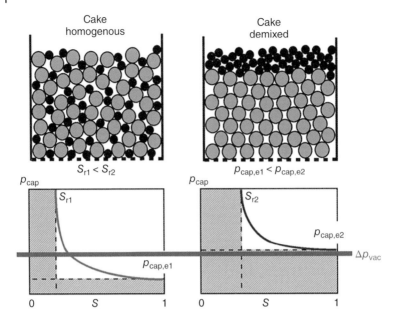

Figure 8.41 Capillary curve for homogeneous and inhomogeneous filter cakes.

material, as previously tested in the laboratory, was filtered with the same vacuum pressure difference, the filters operation cannot be correct. One possibility for the poor filter performance could be an insufficient slurry homogenization in the filter trough. In such a case, a particle size gradient is forming across the height of the filter trough. The filter area immerses at first into a zone of fine particles near the slurry surface. Increased filtrate pollution and in the worst case a clogging of the filter cloth can happen. In the deeper regions of the slurry, coarse particles are filtered, and shortly before the cake appears to the atmosphere, a sealing layer of small particles is finally formed at the cakes surface. The capillary curve of such a segregated filter cake is shown on the right side of Figure 8.41. Because of the covering layer of fine particles on top of the filter cake, the capillary entry pressure rises up dramatically. If this capillary entry pressure in the worst case becomes greater than the applied vacuum pressure difference, no desaturation takes place at all, the cake remains fully saturated, and wet mud is discharged from the filter apparatus. By the way, not only the capillary entry pressure rises in the case of particle segregation but also the remanent saturation. The reason for that is the spontaneous spreading of the intruding gas below the fine particle layer, if the capillary entry pressure is overcome locally at some places of the cakes surface. This results in hydraulically isolated "wet islands" of fine particles at the cake surface, which are surrounded by emptied greater pores. This discussion is again a hint to produce filter cakes of maximal homogeneity for getting the best filter performance.

The segregation tendency of particles in the slurry during cake formation is influenced by the slurry concentration and the particle sedimentation, which takes place parallel to the filtration. The greater the particle concentration

Figure 8.42 Residual cake moisture profile as a function of the slurry concentration.

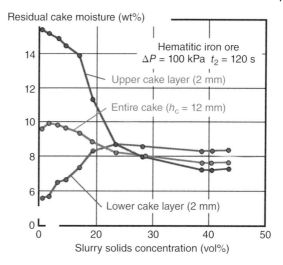

becomes, the less segregation occurs because the particles are not more settling individually and undisturbed by other particles, but the hindered settling and swarm sedimentation effect becomes more and more dominant. If particle segregation can take place, each layer of different particle size in the resulting filter cake exhibits a different capillary pressure distribution. This leads to a residual moisture profile across the cake height, as shown exemplarily in Figure 8.42 for experiments in a laboratory pressure filter cell [33].

After reaching the desaturation equilibrium, the filter cakes, which had been formed from slurries of different particle concentrations, had been cut into slices of 2 mm thickness and analyzed with regard to the residual cake moisture content. In the diagram, the moisture of the upper and lower layer of the cake and the average moisture of the entire cake are plotted. In the case of low slurry concentration and thus dominant single particle sedimentation, maximal segregation takes place. The upper cake layer of the finest particles exhibits the greatest moisture, and the lower layer of the coarsest particles exhibits the minimal cake moisture. The more the slurry concentration rises, the less particle segregation takes place, and for a concentration of approximately 23 vol%, the filter cake is built up homogeneously as far as possible. Then, the lower layer is slightly moister than the upper layer because there might have been, after pressure release in the filter cell, a little rewetting from the filter media.

If the capillary pressure distribution for more or less homogeneously formed filter cakes is measured, Figure 8.43 confirms the previously discussed effects exemplary for filter cakes, which had been formed from slurries of two different solid concentrations.

It becomes clear that the capillary pressure distribution of the more inhomogeneous cake is shifted to the right side, greater residual moistures for comparable pressures are achieved, and the capillary entry pressure, as well as the remanent cake moisture, is increased. Here, not the saturation degree but the residual cake moisture content is plotted on the abscissa to demonstrate the different liquid content of both cakes at the beginning, when both cakes are saturated completely

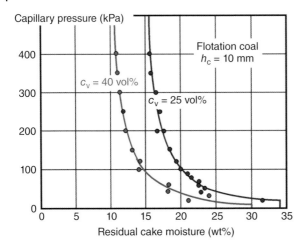

Figure 8.43 Capillary pressure curve for filter cakes of different homogeneities.

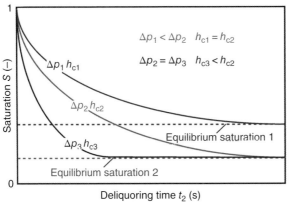

Figure 8.44 Influence of Δp and h_c on the cake desaturation kinetics.

($S = 1$). This is due to the fact that the porosity of a homogeneous cake is smaller than the porosity for segregated cakes. In the case of homogeneous cakes, smaller particles are located in the space between greater particles, whereby in a segregated cake, the particle fractions are more isolated from each other.

8.3.4.2 Kinetics of Filter Cake Desaturation with Gas Pressure Difference

For vacuum and gas overpressure filters, as they are realized in the form of continuously operating drum, disc, belt, and pan filters or discontinuously operating stirred nutsche filters, candle, and sheet filters, the gas pressure difference for cake deliquoring is generally held constant. Thus, in the case of sufficient time to reach the equilibrium state, all pores hydraulically connected with the filter medium, which have a capillary pressure lower than the applied gas pressure difference, can be desaturated. This correlates for the respective applied pressure difference with a point on the capillary pressure curve. In the case of filtration with gas differential pressure, the equilibrium saturation is depending on the pressure difference, but not on the cake height. This of course is different for the desaturation kinetics. The kinetics of desaturation is depending on the pressure difference and the cake thickness, as Figure 8.44 exemplifies at a glance.

8.3 Desaturation of Filter Cakes

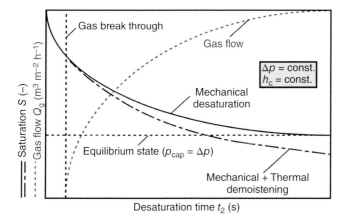

Figure 8.45 Desaturation kinetics for gas pressure deliquoring.

The discussion should start with the curve for Δp_1 and h_{c1}. After some time, this curve approaches the equilibrium saturation 1. If the pressure difference is increased from Δp_1 to Δp_2, but the cake thickness is held constant, the desaturation takes place faster and approaches to a lower equilibrium saturation 2 because even smaller pores can be desaturated by the greater pressure difference. If finally the pressure difference is held constant, but the cake is made thinner, the desaturation velocity is increased even more, whereby the equilibrium saturation 2 remains the same.

For a quantitative description of the desaturation kinetics, a deeper insight in this process is necessary. Figure 8.45 shows in detail the time-dependent cake desaturation process for constant gas pressure difference.

The desaturation process starts with emersion of the cake surface from the slurry to the gas atmosphere at $S = 1$ and ends at the equilibrium saturation S_∞, when the applied pressure difference equals the capillary pressure. At the beginning, the deliquoring of all involved pores lead to a steep decrease of the saturation degree with time. Then, the saturation decrease is flattening because less and less smaller pores are still partly filled with liquid. As soon as the first through-going pore in the filter cake is emptied, gas flows through the filter medium into the filtrate system. The more pores become open, the more gas is flowing. It is important to realize that the gas flow is not the reason but only the consequence of pore desaturation. The vacuum pump or the compressor must compensate the gas flow in order to hold the gas pressure difference constant. This determines the main part of the operating costs for gas pressure filters.

A mechanical influence of the gas flow on the deliquoring by induced shear forces into the liquid can be completely neglected. Normally, the gas flow through the micropores behaves laminar and the shear forces, which could be transferred from gas to liquid, are very small in comparison to the capillary forces. As a consequence, liquid and gas are flowing independently from each other. A superimposed thermal deliquoring by evaporation, as indicated in Figure 8.45, normally for technical deliquoring times of a few minutes likewise

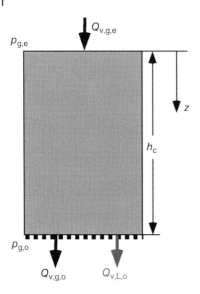

Figure 8.46 Principle situation during cake desaturation.

can be neglected. If in the case of high-permeable and thin filter cakes in combination with increased temperatures and volatile liquids thermal deliquoring seems to be no longer negligible, the effect can be estimated by the amount and absorbing capacity of the gas flow and included into the calculation of the mechanical deliquoring result. Decisive for the desaturation from the physical point of view is the absolute value of the gas pressure difference applied to the filter cake, which can be realized either by a vacuum below the filter medium or gas overpressure above the filter cake. In industrial processes eventually from the pressure level depending secondary effects like precipitation, gas desorption, or chemical reaction can occur and then have to be considered. Depending on the specific cake behavior, the technical upper limit of gas pressure difference is reached at approximately 1 MPa. The smaller the particles become, the smaller the pores in the cake become and thus the higher the capillary pressures are, which have to be overcome by the gas pressure difference. The limitation of pressure increase in combination with smaller particles depends on the increasing compressibility of the cake and the tendency to form shrinkage cracks, which lead to noncontrollable high gas throughputs or unacceptable high gas compression costs. A further aspect consists in the increasing wall thickness of the pressure vessel around the filter, which increases the weight and costs and thus limits economical feasibility.

The time-dependent course of desaturation can be described under the precondition of homogeneous cake structure and uniform liquid displacement from all locations in the filter cake. The latter condition is not valid for the very first moment of desaturation, but already after a short time, there is nearly no saturation gradient present across the cake height. This is due to the strong fingering effect during desaturation. Figure 8.46 explains the principle situation.

A gas pressure difference between cake entry "e" and cake outlet "o" causes a liquid volume flow $Q_{v,L,o}$ out of the cake and a gas volume flow Q_g, which is changing from the cake entry to the cake outlet due to the changing pressure.

8.3 Desaturation of Filter Cakes

In comparison to the cake formation phase, not a single-phase liquid flow but a two-phase flow of liquid and gas takes place, whereby both fluids are flowing independently from each other.

The basis for the modeling is, as for all process steps of cake filtration, the Darcy equation (cf. Eq. (4.14)), which has to be adapted to the respective special conditions here. The characteristic for the fluid flow during desaturation is a two-phase flow of liquid and gas, which is changing with time because of the changing saturation degree. According to Eq. (8.44) and Eq. (8.45), the Darcy equation has to be formulated for liquid and gas flow separately

$$w_L(S) = \frac{p_{c,L}(S) \cdot \Delta p}{\eta_L \cdot h_c} \tag{8.44}$$

$$w_g(S) = \frac{p_{c,g}(S) \cdot \Delta p}{\eta_L \cdot h_c} \tag{8.45}$$

For cake formation, a single-phase liquid flow through a growing porous layer is characteristic. When the cake is formed, the cake height remains constant, but now a two-phase flow of liquid and gas is originating during desaturation. Liquid and gas are flowing parallel in the same direction, but independently from each other. The liquid flow decreases and the gas flow increases during desaturation time. This oppositional behavior can be described by the concept of saturation-dependent liquid and gas permeability [34]. Wyckoff and Botset investigated the parallel and independent flow of liquid and gas through a sand-filled column. The permeability of the cake as a function of the cake saturation for liquid $p_{c,L}(S)$ or gas flow $p_{c,g}(S)$ could be described, according to Eqs. (8.46) and (8.47), by the product of single-phase permeability p_c and relative permeability $p_{c,\text{rel}}(S)$, which depends on the saturation degree

$$p_{c,L}(S) = p_c \cdot p_{c,\text{rel},L}(S) \tag{8.46}$$

$$p_{c,g}(S) = p_c \cdot p_{c,\text{rel},g}(S) \tag{8.47}$$

The single phase and relative permeability have to be measured as a material function experimentally. Figure 8.47 shows the results qualitatively.

The relative specific liquid permeability starts at 1 (100%) and $S = 1$ (100%), when all pores in the filter cake are still fully saturated. This correlates with single-phase liquid flow. The function becomes zero at the mechanical limit of desaturation S_r, where no more liquid movement is possible. The relative specific gas permeability starts at zero and $S = 1$. It ends at 100%, when the saturation becomes zero. At $S = S_r$, the curve shows an inflection because the mechanism of deliquoring changes from mechanical pore displacement to thermal evaporation of the remaining liquid bridges at the particle contact points and the adhesive liquid on the particle surfaces. This leads to an additional small increase of the relative gas permeability in comparison to the steep increase during the mechanical opening of pore channels.

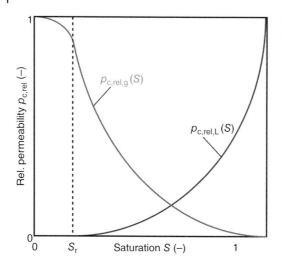

Figure 8.47 Relative liquid and gas permeability.

The function for the specific relative liquid permeability is approximated in Eq. (8.48) by an exponential function [35, 36]

$$p_{c,rel,L}(S) = \left[\frac{S - S_r}{1 - S_r}\right]^n \qquad (8.48)$$

The exponent n is positive and characteristic of the particle system. As previously supposed, the liquid is displaced all over the filter cake and it is supposed that the desaturation process can be described by one-dimensional flow, as formulated in Eq. (8.49)

$$w_L = \frac{Q_{v,L,a} \cdot z}{A \cdot h_c} = -\frac{p_c \cdot p_{c,rel,L}(S)}{\eta_L} \cdot \frac{dp}{dz} \qquad (8.49)$$

The influence of the filter media resistance on the flow velocity of the liquid can be considered by the addition of the real cake height h_c and the equivalent cake height h_{eq} (cf. Eq. (6.11)), as can be seen in Eq. (8.50)

$$w_L = \frac{Q_{v,L,a} \cdot z}{A \cdot (h_c + h_{eq})} = -\frac{p_c \cdot p_{c,rel,L}(S)}{\eta_L} \cdot \frac{dp}{dz} \qquad (8.50)$$

Similarly, as in the case of filter cake formation in practice, normally, the filter media resistance should be negligible in comparison to the filter cake resistance. For this reason, h_{eq} is neglected further on, but this has to be validated in practice from case to case. Now, Eq. (8.49) has to be integrated from $z = 0$ to $z = h_c$ and from $p = p_{g,e} - p_{cap}(S)$ at the cake surface to $p = p_{g,o}$ ($p_{cap} \approx 0$) behind the filter medium. At the cake surface, the gas pressure is reduced by the capillary pressure, whereby behind the filter media the filtrate drops down with $p_{cap} \approx 0$. The result of the integration can be seen in Eq. (8.51)

$$Q_{v,L,o} = 2 \cdot \frac{p_c \cdot p_{c,rel,L}(S)}{\eta_L} \cdot A \cdot \frac{\Delta p - p_{cap}(S)}{h_c} \qquad (8.51)$$

Between the saturation degree S and the filtrate volume flow at the cake outlet $Q_{v,L,o}$, the correlation in Eq. (8.52) for constant porosity ε and liquid density ρ_L is formulated

$$A \cdot h_c \cdot \varepsilon \cdot \frac{dS}{dt} = -Q_{v,L,o} \tag{8.52}$$

Now, Eq. (8.51) has to be interrelated to Eqs. (8.52) and (8.48). To make the following integration more uncomplicated, the capillary pressure function $p_{cap}(S)$ in Eq. (8.53) is replaced by a characteristic average capillary pressure $p_{cap,av}$

$$\frac{d(S-S_r)}{(S-S_r)^n} = \frac{-2 \cdot p_c \cdot (\Delta p - p_{cap,av})}{\eta_L \cdot \varepsilon \cdot h_c^2 \cdot (1-S_r)^n} \cdot dt \tag{8.53}$$

This differential equation has to be integrated for the initial condition of $t_2 = 0$ and $S = 1$ and results in Eq. (8.54)

$$\frac{S-S_r}{1-S_r} = \left[1 + \frac{2 \cdot (n-1) \cdot p_c \cdot (\Delta p - p_{cap,av}) \cdot t_2}{\eta_L \cdot \varepsilon \cdot h_c^2 \cdot (1-S_r)}\right]^{\frac{1}{1-n}} \tag{8.54}$$

Precondition for the validity of Eq. (8.54) is the possibility to reach $S = S_r$ with the applied pressure difference. To fulfill this condition, Δp has to be much greater than $p_{cap,av}$. Unfortunately, these conditions are in most cases not given in real applications of vacuum or gas overpressure filtration because normally the mechanical limit of desaturation cannot be achieved by the applied pressure differences. To solve this problem, Eq. (8.54) is not more generally related to the mechanical limit of desaturation S_r, but for each particular pressure difference to the corresponding equilibrium saturation S_∞. The capillary entry pressure $p_{cap,e}$ can be used as the characteristic capillary pressure, and two material-dependent fitting parameters a and b, as well as a kinetics parameter K, are implemented in Eq. (8.55) [37–39]

$$\frac{S-S_\infty}{1-S_\infty} = [1 + a \cdot K]^{-b} \tag{8.55}$$

The kinetic parameter K in Eq. (8.56) contains all relevant influencing parameters

$$K = \frac{p_c \cdot (\Delta p - p_{cap,e})}{\varepsilon \cdot \eta_L \cdot h_c^2} \cdot t_2 \tag{8.56}$$

The capillary pressure curve inclusive of the capillary entry pressure, the specific cake permeability, the cake porosity, and the liquid viscosity must be measured in any case experimentally as material parameters, and thus, all necessary parameters are quantitatively available to describe the desaturation kinetics for practical use.

Now, it is possible to describe the filter cake desaturation in dependence of all relevant material and operational parameters. Figure 8.48 demonstrates this graphically for one selected desaturation time, the pressure difference, and the cake thickness.

In this way, for example, the question can be answered, which parameter combinations of Δp, h_c, and t_2 are resulting in a certain saturation degree, which is necessary to fulfill the requirements of the next process step, that is, pelletizing.

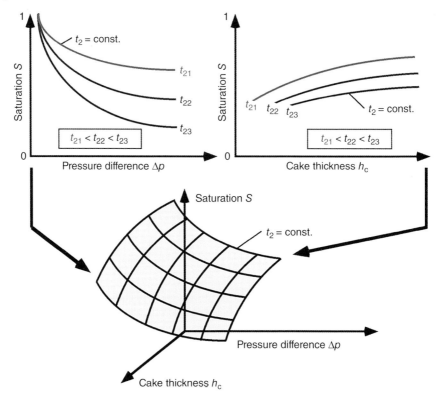

Figure 8.48 Filter cake desaturation matrix.

As an alternative to the calculation via Eq. (8.55), it would also be possible to follow a completely empirical strategy. In the interesting field of parameters, the time-dependent cake desaturation has to be measured experimentally for at least nine parameter combinations of pressure difference and cake height. All interesting information is then available by graphical or mathematical interpolation.

8.3.4.3 Kinetics of Gas Flow through Filter Cakes and Energetic Considerations

Parallel to the decrease of the saturation degree, the gas flow rate through the already emptied pores rises (cf. Figure 8.45). Normally, the gas volume flow rate is related to the filter area and indicated for normal conditions ($T_N = 273.15\,\text{K} = 0\,°\text{C}$ and $p_N = 101.325\,\text{kPa}$). Instead of the time-dependent specific gas volume flow rate $q_{v,g,N}(t_2)$ in $\text{m}^3\,\text{m}^{-2}\,\text{h}^{-1}$, the integrated gas flow rate in the form of the time-dependent specific gas consumption $v_{g,N}(t_2)$ in $\text{m}^3\,\text{m}^{-2}$ can also be used. As an example, Figure 8.49 shows experimental results for the time-dependent gas flow rate and gas consumption for different filtration pressure differences.

Analogues to the liquid flow rate during desaturation, the gas flow rate can also be described on the basis of the adjusted Darcy equation. Now, it has to be considered that gas is a compressible fluid, which expands from the cake surface to

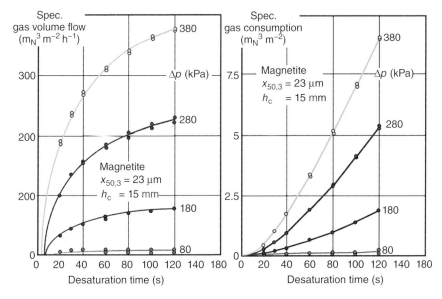

Figure 8.49 Time-dependent gas flow rate and gas consumption.

Figure 8.50 Principle situation for the gas flow during cake desaturation.

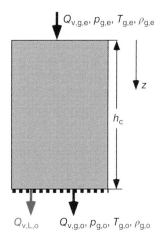

the filter medium because of the pressure drop. Figure 8.50 explains the principle situation and relevant parameters.

In a first step, the one-dimensional gas flow rate through a homogeneous and isotropic dry cake of constant thickness should be examined. Stationary conditions are considered for the gas flow rate. This means that the gas mass flow remains constant. Assuming no heat exchange with the cake and relative low flow velocity of less than 30 m s^{-1}, the enthalpy of the gas is not changing. The enthalpy of ideal gases is only depending on the temperature, and therefore, isothermal throttling process takes place here, as expressed in Eq. (8.57)

$$h_{g,e} = h_{g,o} = \text{const.} \Rightarrow T_{g,e} = T_{g,o} = \text{const.} \tag{8.57}$$

As a consequence, in Eq. (8.58), the dynamic viscosity of the gas also remains constant

$$\eta_{g,e} = \eta_{g,o} = \eta_g = \text{const.} \tag{8.58}$$

For a stationary flow in Eq. (8.59), it is formulated that the product of pressure and velocity remains constant

$$p_{g,e} \cdot w_{g,e} = p_{g,o} \cdot w_{g,o} = p_g \cdot w_g = \text{const.} \tag{8.59}$$

Because of the changes of state for the gas, from the cake surface to the filter media, the Darcy equation must not be used in its integral but in its differential form, as shown in Eq. (8.60)

$$w_g = -\frac{p_c}{\eta_g} \cdot \frac{dp}{dz} \tag{8.60}$$

The minus sign indicates that the pressure decreases in the direction of flow. The flow velocity w is considered as a constant for the differential length dz. Substituting Eq. (8.59) into Eq. (8.60) leads to the differential equation in Eq. (8.61), which is integrated in Eq. (8.62) within the limits of $z = 0$ to $z = h_c$ and $p = p_e$ to $p = p_o$

$$p \cdot dp = -\frac{p_{g,e} \cdot w_{g,e} \cdot \eta_g}{p_c} \cdot dz \tag{8.61}$$

$$p_{g,e} \cdot w_{g,e} \cdot h_c = \frac{p_c}{\eta_g} \cdot \frac{1}{2} \cdot (p_{g,e}^2 - p_{g,o}^2) \tag{8.62}$$

If the influence of the filter medium resistance cannot be neglected, it can be approximated by the addition of the equivalent cake height $h_{c,e}$ to the real cake height h_c. As shown in Eq. (8.63), the pressure term of Eq. (8.62) can be transformed using the average pressure p_{av} and the pressure difference Δp

$$\frac{1}{2} \cdot (p_{g,e}^2 - p_{g,o}^2) = \frac{1}{2} \cdot (p_{g,e} + p_{g,o}) \cdot (p_{g,e} - p_{g,o}) = p_{av} \cdot \Delta p \tag{8.63}$$

In a final step, the gas volume flow rate through the dry cake can be formulated in Eq. (8.64)

$$Q_{v,g,e} = \frac{p_c}{\eta_g} \cdot \frac{p_{av} \cdot \Delta p}{p_{g,e}} \cdot \frac{A}{h_c} \tag{8.64}$$

Equation (8.64) has to be modified for the description of the gas flow during the desaturation process with consecutive decrease of the saturation degree. This succeeds by implementing the relative gas permeability $p_{c,rel,g}(S)$ in Eq. (8.65) (cf. Eq. (8.47) and Figure 8.47)

$$Q_{v,g,e} = \frac{p_c \cdot p_{c,rel,g}(S)}{\eta_g} \cdot \frac{p_{av} \cdot \Delta p}{p_{g,e}} \cdot \frac{A}{h_c} \tag{8.65}$$

Figure 8.51 shows experimental data for the measured relative gas permeability function.

To get this function, experiments for different pressure differences, cake heights, and desaturation times had been carried out, and the relative gas permeability was calculated using Eq. (8.65).

Figure 8.51 Relative gas permeability as a function of the saturation degree.

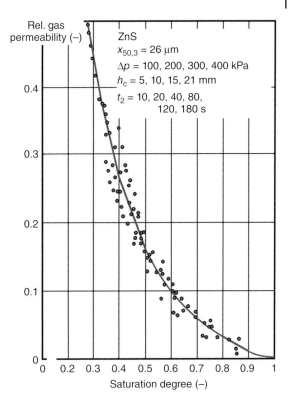

Similar to the saturation degree, it would also be possible for the description of the gas consumption to follow the completely empirical strategy, as shown qualitatively in Figure 8.52.

Simultaneous to the measurement of the optional parameter field for the cake desaturation, the gas consumption has to be measured for those at least nine parameter combinations of pressure difference and cake height. All data are then available for a graphical or mathematical interpolation of gas consumptions.

The gas flow rate is reacting much more sensitive to little changes of the cake structure than the liquid flow. Shrinkage cracks make calculations impossible and additional aspects, such as evacuation of the filtrate piping system after aeration or leakages in the system, cannot be identified and quantified by an ideal filter in the laboratory. As a consequence, the experiment in the lab scale can give only one component of the real gas flow rate in the industrial practice, which means the gas permeation of the ideal homogeneous filter cake.

As mentioned before, the energy for the compression of the gas flow rate through the opened pores of the cake to maintain the applied gas pressure difference represents the main part of the filters' energy consumption. To estimate the necessary mechanical energy consumption W_g due to the gas consumption, the air is supposed to be an ideal gas and its change of state may be adiabatic or rather isentropic with an isentropic coefficient of $\kappa = 1.4$. From these preconditions, the mechanical compression energy W_{mech} can be

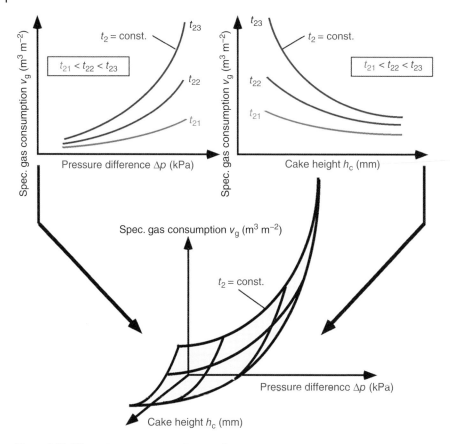

Figure 8.52 Filter cake gas consumption matrix.

calculated in Eq. (8.66)

$$W_{mech} = p_e \cdot V_e \cdot \frac{\kappa}{\kappa - 1} \cdot \left(1 - \left[\frac{p_o}{p_e}\right]^{\frac{\kappa-1}{\kappa}}\right) \qquad (8.66)$$

To get the specific gas compression energy w, the absolute compression energy W has to be related to the solid mass of the filter cake, according to Eq. (8.67)

$$w_{mech} = \frac{W_{mech}}{m_s} = \frac{W_{mech}}{A \cdot h_c \cdot (1 - \varepsilon) \cdot \rho_s} \qquad (8.67)$$

If vacuum and overpressure filtration are compared to each other, it is interesting to recognize that it is more energy consuming to produce vacuum than to produce overpressure. From Figure 8.53, it becomes clear that the compression of 1 m³ air from 20 to 100 kPa needs roughly the same energy than to compress 1 m³ air from 100 to 500 kPa.

To calculate the real compression energy, the efficiency factor η of the compressor must be included in addition for vacuum (Eq. (8.68)) and overpressure

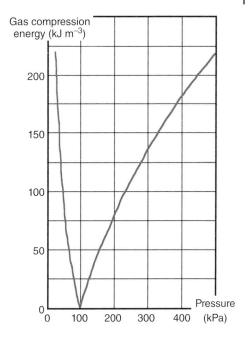

Figure 8.53 Compression energy for vacuum and overpressure.

(Eq. (8.69))

$$w_{vac} = \frac{p_N \cdot V_N}{\eta_{vac} \cdot m_s} \cdot \frac{\kappa}{\kappa - 1} \cdot \left(1 - \left[\frac{p_{vac}}{p_N}\right]^{\frac{\kappa-1}{\kappa}}\right) \quad (8.68)$$

$$w_{pr} = \frac{p_N \cdot V_N}{\eta_{pr} \cdot m_s} \cdot \frac{\kappa}{\kappa - 1} \cdot \left(1 - \left[\frac{p_N}{p_{pr}}\right]^{\frac{\kappa-1}{\kappa}}\right) \quad (8.69)$$

The necessary specific energy to desaturate a filter cake can be realized by mechanical or thermal means. In the case of thermal drying, the filter cake must be heated up first and afterward the liquid must be evaporated. Equivalent to the specific gas compression energy, according to Eq. (8.70), a specific thermal drying energy w_{therm} can be calculated

$$w_{therm} = \frac{W_{therm}}{m_s} = c_s \cdot \Delta\vartheta + \frac{m_L}{m_s} \cdot c_L \cdot \Delta\vartheta + \frac{h_{vap,L}}{m_s} \cdot \Delta m_L \quad (8.70)$$

To calculate w_{therm}, the specific heat of solids and liquid c_s and c_L, as well as the temperature difference $\Delta\vartheta$, and the evaporation enthalpy of the liquid $h_{vap,L}$ must be known. On the basis of known w_{mech} and w_{therm}, a comparison between mechanical and thermal deliquoring of filter cakes can be carried out. In Figure 8.54, an example for such a comparison is given in qualitative form.

The thermal energy is directly proportional to the quantity of evaporated liquid and in total much greater than that for mechanical desaturation. The energy demand for mechanical desaturation increases with decreasing saturation degree because the gas flow increases due to more and more open pores. The greater the pressure difference becomes, the lower the saturation degree is, which can be

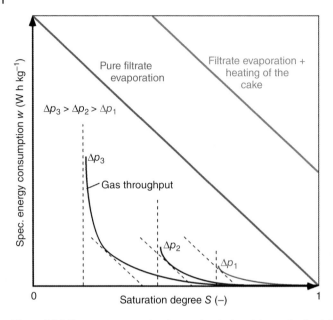

Figure 8.54 Energy consumption for mechanical and thermal cake deliquoring.

achieved. Near the mechanical equilibrium of desaturation, the energy consumption for each pressure difference approaches a vertical asymptote because energy is still consumed, although no further desaturation takes place. The hand-over point for a moist filter cake from the mechanical to the final thermal drying should be set not before the inclination of the mechanical energy consumption curve is still smaller than the function for the thermal energy. In every case, the handling properties of the moist filter cake must fit to the following dryer.

8.3.4.4 Measurement of Cake Desaturation Equilibrium and Kinetics

In principle, the same lab-scale pressure filtration setup can be used to measure the cake desaturation and the capillary pressure curve, as was described in Section 6.3.2. The filter cake, which is formed in the pressure filter cell, afterward can be directly desaturated without pressure release. By registration of the filtrate on the electronic balance during the desaturation phase, the kinetics of the process can be determined. After measuring the residual cake moisture after the end of the experiment, the time-dependent course of desaturation can be recalculated from the registered filtrate accumulation.

For the measurement of the capillary pressure curve, two methods are available, which are both based on the pressure filter cell. The procedures are described in the VDI guideline 2762-3 for cake desaturation (cf. [1]). Each procedure needs a special lower part of the filter cell for the filtrate discharge, as can be seen in Figure 8.55.

The first method enables to measure the capillary pressure curve during one single experiment. However, this needs some hours of time. Shortly before the end of cake formation, the pressure difference has to be reduced to zero. Then, the gas pressure has to be increased carefully but must remain slightly below the

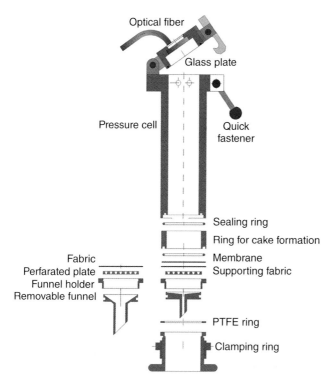

Figure 8.55 Pressure filter cell with special filtrate discharge systems.

capillary entry pressure. The little rest of free liquid on top of the cake surface is displaced in this way. Now, the filter cake exhibits a saturation degree of $S = 1$ and the measurement of the capillary pressure curve can start. The pressure difference is increased stepwise with increments of about 10 kPa at the beginning to find the capillary entry pressure as precise as possible and about 20 kPa or more afterward. The displaced filtrate of each pressure level is registered on a scale. As soon as no more filtrate displacement takes place, according to a criterion to be determined (e.g. no drop after five minutes), the pressure is increased by an increment. This continues until the upper limit of the measuring range is reached or further pressure increase leads to no more filtrate. The saturation degree, corresponding to the respective pressure level, can be calculated from recording the time-dependent filtrate flow. Because this measurement needs hours, no gas flows through the already deliquored pores and the filter media is allowed to avoid a thermal drying of the cake. Superimposed thermal drying would falsify the measurement, which should only register the mechanical deliquoring. This is achieved by using a microporous filter medium with a capillary entry pressure higher than the maximal adjustable gas pressure difference. The filter media must be wettable with the filtrate, which means hydrophilic in the case of aqueous slurries. The large measurement times make it useful to automate the measurement process. Because of the small steps of pressure increase, one speaks in this case of a static capillary pressure curve. Because of the semipermeable behavior of

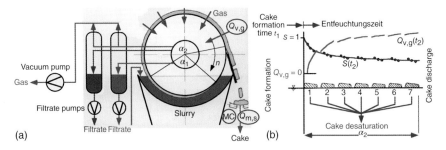

Figure 8.56 Cake desaturation on a vacuum drum filter.

the filter membrane, the small designed space behind the filter medium remains completely filled with liquid. The evaporation of filtrate out of the filtrate reservoir on the scale is restricted by its covering. There is of course no friction allowed between the cover and the filtrate pipe.

The second method is related very close to the technical practice and is known as dynamic method. Here, several independent filter experiments with different filtration pressures are carried out, and conventional filter fabrics are used instead of microporous membranes. Each experiment is stopped after some minutes, when no more filtrate flow can be observed. This enables to measure approximately the complete time-dependent deliquoring process for each pressure difference. The equilibrium saturation, measured by this procedure, is representing points on the capillary pressure curve, which can then be interconnected by an appropriate mathematical approximation function. The dynamic capillary pressure curve may have only little deviations compared with the static version. The filtrate discharge system is here relatively open designed, not to hinder the free flow of liquid and gas. For relative high gas flow, an influence of the gas pulse on the scale has to be excluded. This is supported by a greater diameter of the filtrate discharge pipe. Just before the experiment is shut down, the filtrate funnel has to be pulled out and the backside of the perforated plate for supporting the filter media has to be wiped to exclude remoistening of the cake after pressure release. To determine the kinetics of the desaturation process, portions of the filtrate, which was registered time dependent on the scale, are added to this final moisture content and the cake moisture content has to be recalculated.

8.3.4.5 Transfer of Desaturation Results from Bench Scale to Rotary Filters

The principle process of cake deliquoring is shown in Figure 8.56 in the example of a vacuum drum filter, including a more detailed view to the desaturation zone.

The cake deliquoring starts as soon as the cake appears from the slurry and ends at the moment of cake discharge. The expenditure for the filtration originates mainly in the form of gas volume flow $Q_{v,g}$, which increases with desaturation time and decreasing cake moisture. On Figure 8.56b, the development of cake saturation and gas throughput is itemized for the single cells, which are staying in the desaturation zone of the rotary filter. Because of the slightly smaller cake height at the leading edge of the filter cells in comparison to the trailing edge, the cake moisture decreases from the trailing to the leading edge a little bit. This

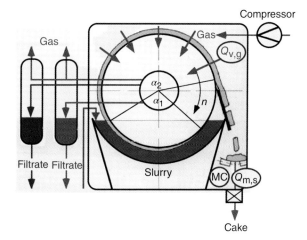

Figure 8.57 Cake deliquoring on a hyperbaric drum filter.

effect is more significant for the filter cells of disc filters (cf. Figure 6.37) but can be observed in reduced form also for drum filters. The final cake moisture content is normally registered at the discharge point. The gas flow rate is measured as average gas flow rate of all filter cells because all involved filter cells are connected with one joint filtrate pipe, which is connected with the filtrate receiver.

The same is valid if such filters are installed in a pressurized vessel and then denominated as hyperbaric filters, as shown in Figure 8.57.

In this case, the pressurized air is delivered by a compressor and fed into the pressure vessel, whereby the pressure in the filtrate receiver is ambient pressure and no filtrate pumps are necessary. In comparison to vacuum filters, the filter cake must be discharged from the pressure vessel via a special gastight sluice system.

Section 6.3.4 discusses on how the bench-scale experiments for cake formation can be transferred to a rotary filter. The rotation speed n of the filter can be calculated, according to Eq. (8.71), from the cake formation time t_1, which leads to a certain cake thickness and the cake formation angle α_1 of the rotary filter

$$n = \frac{\alpha_1}{360°} \cdot \frac{1}{t_1} \qquad (8.71)$$

The cake desaturation time t_2 has to be calculated in Eq. (8.72) for the same rotation speed and the deliquoring angle α_2

$$t_2 = \frac{\alpha_2}{360°} \cdot \frac{1}{n} \qquad (8.72)$$

As discussed in the previous chapter, the desaturation kinetics is measured during the same experiment directly after the cake formation in the laboratory pressure filter cell. This experiment results the development of the cake moisture content as a function of the deliquoring time for a certain pressure difference and cake height, as shown in Figure 8.58.

Here, the cake moisture content instead of saturation degree was chosen because it respects the more practical aspects of the following discussion, but likewise, the saturation degree or an alternative cake moisture characterizing

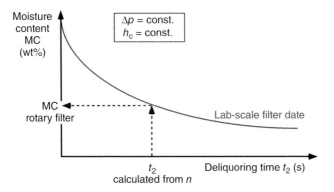

Figure 8.58 Cake moisture content as a function of deliquoring time.

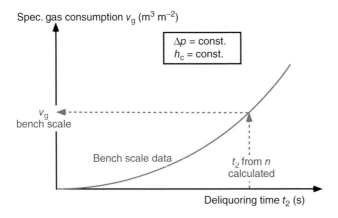

Figure 8.59 Specific gas consumption as a function of deliquoring time.

parameter could be used. After calculation of the desaturation time on the rotary filter, according to Eq. (8.72), the resulting cake moisture content or saturation degree can be read from the diagram of the bench-scale experiment results.

The determination of the spent specific gas volume during desaturation is carried out similarly. At first, the specific gas volume flow rate $q_{v,g,bs}(t_2)$ during cake desaturation has to be measured with the laboratory filter cell. In a second step, the volume flow rate must be integrated to get the specific gas volume consumption $v_{g,bs}(t_2)$. Equation (8.73) shows the result

$$v_{g,bs} = \int_0^{t_2} q_{v,g,bs} \cdot dt \tag{8.73}$$

This result of integration is indicated schematically in Figure 8.59.

The average specific gas flow rate of the rotary filter $q_{v,g,rf}$ can be calculated from the rotation speed n and the specific gas consumption of the bench scale experiment $v_{g,bs}$, as indicated in Eq. (8.74)

$$q_{v,g,rf} = \frac{v_{g,bs}}{t_2} \tag{8.74}$$

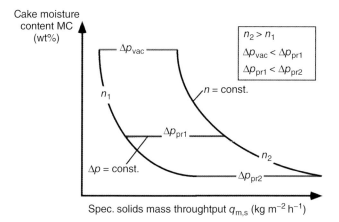

Figure 8.60 Throughput and cake moisture of rotary filters.

This gas volume flow rate is taken for further considerations as representative for a filter, but it must be realized that in reality, additional components of the gas consumption have to be added to design the required vacuum pump. After each cake discharge, the aerated filter cell and the filtrate piping system have to be evacuated again. Furthermore, the filter apparatus has to be sealed between the rotating and the fixed part at the control valve, which may not be realized perfectly with the consequence of some leakage air. Last but not least, some shrinkage cracks in the filter cake can increase the gas flow rate dramatically. This component of the entire flow rate can hardly be predicted by lab-scale tests, and under circumstances, this effect makes it impossible to run the process at all. Nevertheless, the pure gas flow rate through the filter cake gives a useful hint for estimating the minimum size of the vacuum pump and the economic feasibility of the process.

8.3.4.6 Interrelation of Throughput, Cake Moisture, and Gas Consumption for Rotary Filters

The rotary filter-specific solid mass throughput $q_{m,s}$, as a function of product, operation, and design parameters, has been derived in Section 6.3.4 and is given here again in Eq. (8.75)

$$q_{m,s} = \frac{Q_s}{A} = \rho_s \cdot (1-\varepsilon) \cdot \sqrt{\frac{2 \cdot p_c}{\eta_L}} \cdot \sqrt{\kappa \cdot \Delta p \cdot n} \cdot \sqrt{\frac{\alpha_1}{360°}} \qquad (8.75)$$

The question is how the cake moisture content correlates with the solid throughput. Figure 8.60 gives a principle answer to this question in a qualitative form, which is based on theory and extensive experimental experiences [40–42].

A first result from this diagram is that an increase of the rotation speed from n_1 to n_2 leads, according to Eq. (8.75), to an increase of solid throughput, but the cake moisture content remains constant. This phenomenon can be expected from the theory. In Section 8.3.4.2, it has been derived that the time-dependent decrease of the saturation degree for incompressible and isotropic filter cakes generally can

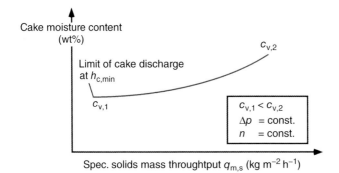

Figure 8.61 Influence of slurry prethickening on throughput and cake moisture.

be expressed by a deliquoring function, as shown in Eq. (8.76)

$$S(t_2) = f\left[\frac{r_c \cdot \varepsilon \cdot \eta_L}{(\Delta p - p_{cap,e})} \cdot \frac{h_c^2}{t_2}\right] \quad (8.76)$$

The square of the cake height h_c is proportional to the cake formation time t_1, which replaces h_c in Eq. (8.77)

$$S(t_2) = f\left[\frac{r_c \cdot \varepsilon \cdot \eta_L}{(\Delta p - p_{cap,e})} \cdot \frac{2 \cdot \kappa \cdot \Delta p}{r_c \cdot \eta_L} \cdot \frac{t_1}{t_2}\right] \quad (8.77)$$

The ratio of cake forming and cake deliquoring time is directly proportional to the ratio of the cake forming and cake deliquoring angle α_1/α_2 because the rotation speed n can be canceled and no more influence of the rotation speed is present. The ratio of the times t_1/t_2 is replaced by the ratio of the angles α_1/α_2 in Eq. (8.78)

$$S(t_2) = f\left[\frac{r_c \cdot \varepsilon \cdot \eta_L}{(\Delta p - p_{cap,e})} \cdot \frac{2 \cdot \kappa \cdot \Delta p}{r_c \cdot \eta_L} \cdot \frac{\alpha_1}{\alpha_2}\right] \quad (8.78)$$

If this function is held constant for a variation of the contained parameters, the deliquoring result should remain constant too. If the filter rotates faster, the deliquoring time decreases, but also the cake formation time does. This leads to a thinner cake, which deliquores faster. Both effects are compensating each other and the resulting cake moisture content remains constant. The maximum rotation speed is reached, if the filter cake cannot be discharged anymore because of its small height, as has been discussed in Section 6.3.4.

If the gas pressure difference is increased from Δp_{vac} to $\Delta p_{pr,1}$ or $\Delta p_{pr,2}$, not only the solid throughput is increased for a certain rotation speed, according to Eq. (8.75), but also smaller pores in the cake can be desaturated, and thus, the cake moisture content is decreasing, according to Eq. (8.78). This is especially beneficial if the filter feed slurry concentration has increased by a prethickening procedure [33]. For constant pressure difference and for constant rotation speed, the cake moisture would rise, according to Figure 8.61, for increased slurry concentration because the thicker cake gets no longer deliquoring time.

If for the greater slurry concentration and thus thicker cake and greater residual cake moisture the rotation speed of the filter would be increased, of course the

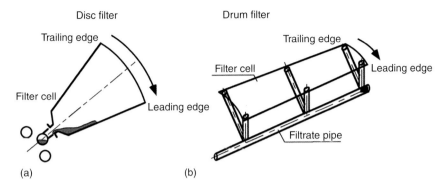

Figure 8.62 Filtrate discharge from disc and drum filter cells.

filter throughput would increase, but the residual cake moisture would remain constant at the higher level. Only by increasing the filtration pressure difference, this negative effect can be compensated or even overcompensated.

In practice, it can be sometimes observed that the cake saturation nevertheless increases with increasing rotation speed, although it should remain constant. Knowing from theory that this should not be possible, something with the filter must be wrong. A reason for this phenomenon could be a bottleneck in the filtrate piping system or an unfavorable design of the filter cell. Figure 8.62 shows a filter cell of a disc and a drum filter with filtrate discharge system.

If the rotation speed of such a filter is increasing, the cake thickness decreases and due to the greater throughput, the filtrate flow increases. A gas blow back supports the filter cake discharge, and if the filter cell was not completely emptied before, the remaining filtrate is blown back from the cell and rewets the cake. The greater the rotation speed, the thinner the cake, and the greater the portion of filtrate becomes, which is blown back, the more serious this effect becomes. To avoid this deficiency, no hydraulic limitations should be present in the filtrate piping system. If necessary, the capacity of the filtrate piping system must be increased by a filter revamping. A hint for the design may be that the recommended suction ratio (cross-sectional area of filtrate pipe/filter area) for disc filter cells should be greater than 5‰.

An additional problem originates if there are no filtrate pipes present at the leading and trailing edges of drum filter cells, but only one centrally positioned pipe. In the case of only one pipe in the middle of the cell, not all the filtrate is able to leave the cell in the discharge position. The best option is to have one filtrate pipe at the trailing and one at the leading edge of the cell, as can be seen in Figure 8.62b.

Besides the correct hydraulic design of the filtrate piping system, the homogeneity of the cakes on the filter cells sensitively influences the residual moisture content of the discharged solids. In Section 8.3.4.1, it has been discussed how seriously a layer of finest particles on top of the filter cake influences the residual cake moisture because of its increased capillary entry pressure. Especially for disc filters, it is relevant to get a filter cake of constant height on the filter cell

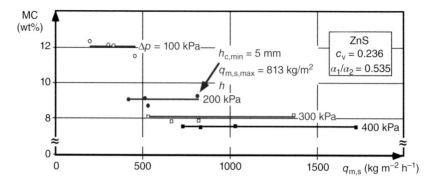

Figure 8.63 Residual cake moisture as a function of specific solid throughput.

and no cake height distribution, which affects the average residual cake moisture negatively. Measures to avoid those inhomogeneities have been discussed in Section 6.3.5.

After discussion of throughput and residual cake moisture content on rotary filters, the expenditure or operating costs, respectively, to get these results have to be considered. Most interesting in this connection is the specific gas throughput through the filter cake for standard conditions in relation to the specific solid mass throughput and the achieved residual cake moisture content. The question should be answered whether certain residual cake moisture content can be achieved particularly energy saving by a special parameter combination. The ratio between cake forming and cake deliquoring angle in the control head of the filter, the rotation speed of the filter, the gas pressure difference, and the solid volume concentration of the slurry are chosen as variable parameters. The following discussion is based on the previously introduced theoretical considerations and on extensive experimental parameter studies in the lab scale for slurry of zinc sulfide ore [43]. The lab-scale data had been scaled up to the operation conditions of rotary filters. Figure 8.63 shows the results for residual cake moisture content and specific solid throughput.

The specific solid mass throughput $q_{m,s}$ has been defined in eq. (8.75). It is proportional to the square root of the rotation speed, as indicated in Eq. (8.79)

$$q_{m,s} \propto \sqrt{n} \tag{8.79}$$

The residual cake moisture content or the cake saturation degree, respectively, which can be achieved on a rotary filter, is defined in Eq. (8.78). If all other parameters are held constant, the saturation degree is not depending on the rotation speed. To achieve a constant saturation degree on a rotary filter, the change of one parameter must be compensated by other parameters to hold the desaturation function constant. If, for example, the ratio of cake formation and desaturation angle is changed, a compensation is possible by changing the slurry feed concentration, which is implemented in the concentration parameter κ (cf. Eq. (2.39)). In the same way, a change of the pressure difference can be treated. However, the targeted residual cake moisture must be achievable by overcoming of the acting capillary pressure in the filter cakes pores. Principally,

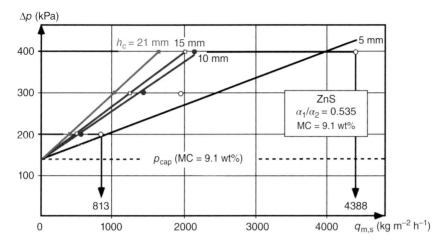

Figure 8.64 Pressure dependency of the filter performance for targeted cake moisture.

every theoretical change of parameters must be checked, whether it can be still realized technically on a rotary filter. However, certain target moisture can be reached not only by means of a certain pressure difference. That allows further possibilities of optimization of the separation problem with regard to an increased specific solid mass throughput or a minimization of the necessary filter area. If exemplarily the target moisture content should be 9.1 wt%, from Figure 8.63, it can be seen that this can be achieved with a pressure difference of 200 kPa. The maximal possible rotation speed is limited by the minimal allowed cake thickness of 5 mm and leads, in this special case, to a specific solid throughput of 813 kg m^{-2} h^{-1}. The question is now, which parameter combination also leads to the target cake moisture of 9.1 wt% and could be achieved a greater throughput than 813 kg m^{-2} h^{-1}. Figure 8.64 gives an answer to this question.

Pressure difference, slurry concentration, and rotation speed must be adjusted in a way to get for a certain ratio of cake forming and cake deliquoring angle a cake moisture content of 9.1 wt%. Respective cake height and solid throughput are resulting. To get the right slurry concentration, at first, the deliquoring times are taken from the experimental data, which lead to the targeted residual cake moisture for the measured pressure differences and cake heights. With these deliquoring times, the rotation speed of the filter is calculated according to Eq. (8.71). From the rotation speed, according to Eq. (8.72), the related cake formation time has to be determined. From Eq. (8.75), the associated concentration parameter and finally from Eq. (2.39), the necessary solid volume concentration of the feed slurry can be calculated. Under these conditions, the correlation between pressure difference and solid mass throughput becomes linear for each cake height in Figure 8.64. Interesting here is especially the function for the minimal allowed cake height of 5 mm and thus the maximal throughput. If the pressure difference is increased from 200 to 400 kPa, the specific solid throughput can be increased from 813 to 4388 kg m^{-2} h^{-1}. The straight lines are starting all from the same point on the ordinate, where the specific solid throughput approximates zero for infinite large dewatering time. This point is according to Figure 8.65 characterized by

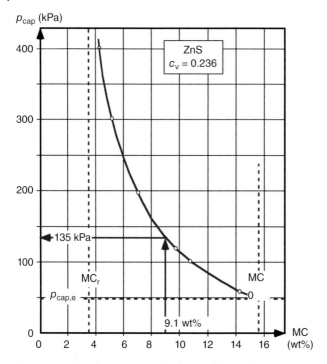

Figure 8.65 Capillary pressure distribution for ZnS filter cakes.

the capillary pressure of $p_{cap} = 135$ kPa, which enables to reach a cake moisture content of MC = 9.1 wt% in the equilibrium state.

The next step after these considerations is to determine how the results for the solid throughput at fixed cake moisture content are corresponding with the required gas throughput. As a characteristic number to describe the effectiveness of separation, the relation of specific gas and solid throughput or the consumed m³ of gas (standard conditions) per kg produced dry solids is used in Eq. (8.80)

$$\frac{q_{v,g,N}}{q_{m,s}} = \frac{V_{g,N}}{m_s} \left[\frac{m_N^3 \, m^{-2} \, h^{-1}}{kg \, m^{-2} \, h^{-1}} \right] = \left[\frac{m_N^3}{kg} \right] \tag{8.80}$$

From Eq. (8.65), it is evident that the specific gas throughput for constant cake saturation degree and thus constant relative gas permeability is, according to the Darcy law, inversely proportional to the cake thickness. The cake thickness itself is inversely proportional to the square root of the rotational speed of the filters. As a consequence, the specific gas throughput of the rotary filter should be proportional to the square root of the rotational speed, as indicated in Eq. (8.81)

$$q_{v,g,N} \propto \sqrt{n} \tag{8.81}$$

The specific solid mass throughput is proportional to the square root of the rotational speed too. Both functions are considered for the precondition of constant cake moisture content. This leads to the conclusion that, according to Eq. (8.82), the ratio of specific solid mass throughput and gas flow rate should be

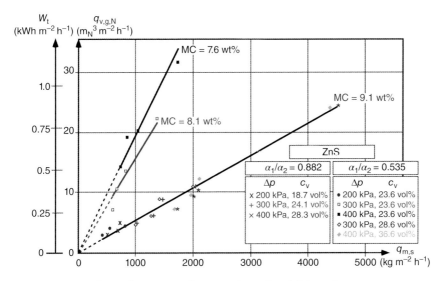

Figure 8.66 Specific gas flow rate as a function of specific solid throughput.

constant for a constant desaturation degree (cf. Eq. (8.78))

$$\frac{q_{v,g,N}}{q_{m,s}} \propto \frac{\sqrt{n}}{\sqrt{n}} = 1 \tag{8.82}$$

Figure 8.66 confirms this theoretical conclusion for different cake moisture contents by experimental data.

As general result, it can be observed that for each value of residual cake moisture, a linear correlation exists between specific gas throughput and specific solid throughput. As can be seen from the second ordinate, this linear correlation is valid in an analogous way for the technical gas compression energy, which was formulated in Eq. (8.66). This means that for constant cake moisture, the ratio of spent gas volume and solid mass unit remains constant. In the example of a targeted residual cake moisture of $MC = 9.1$ wt%, it is shown that the specific gas consume remains constant for a variation of rotation speed, pressure difference, slurry concentration, and the ratio of process angles. Only in the case of changing cake moisture, the specific gas consume is also changing. Thus, the solid throughput can be increased up to the technical limit without increase of the specific gas consumption. If the throughput is given, the smallest possible filter area can be installed without increasing the costs for the compression of the originating gas volume flow.

8.3.5 Desaturation of Incompressible Filter Cakes by Steam Pressure Difference

The fundamentals of steam pressure filtration with gas differential pressure [44–48] and in the centrifugal field [49] have been investigated intensively in bench scale experiments. For vacuum pan filters, rotary pressure filters, and

258 | 8 Filter Cake Deliquoring

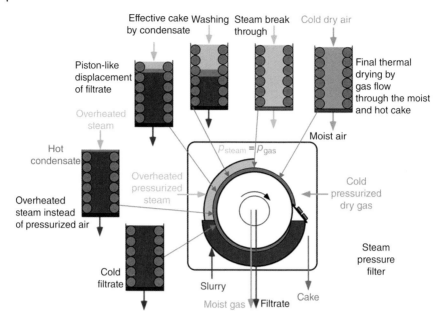

Figure 8.67 Principle process of steam pressure filtration.

inverting filter centrifuges, this process has been transferred to real technical applications. Besides desaturation of filter cakes, steam pressure filtration can be used for effective permeation washing of filter cakes with the necessarily originating condensate (cf. Section 7.6). Now, the desaturation mechanisms and results should be discussed in more detail. Figure 8.67 illustrates the process in the example of a pressure drum filter with steam hood.

The gas atmosphere in the pressure vessel exhibits ambient temperature and the same pressure as the heated steam in the steam cabin. This makes it easy to seal the cabin, to avoid steam, and thus condensate in the pressure vessel. Pressure and temperature of the steam are correlated by the steam pressure function, as exemplarily shown for water in Figure 8.68.

For example, a pressure difference of 0.5 MPa leads to a steam temperature of about 150 °C, which is in the range of technical applications. The cold filter cake enters the steam cabin directly after emerging from the slurry, and the cold cake surface is confronted with hot steam. Some condensate is forming, and the upper particle layers of the cake are heated up. Then, because of the pressure difference between cake surface and filtrate system behind the filter medium, the steam acts like other gases would do and displaces the liquid of the cake pores, if the capillary pressure is overcome. In contrast to the desaturation process with gases, such as air or nitrogen, where a fingering effect (dispersion) can be observed, during cake desaturation with steam, the liquid is displaced piston-like. An even condensate front migrates through the cake. If a coarse pore is desaturating faster than others, steam gets in contact to cold particles and the pore is partly refilled with the condensate. This leads to an overall even condensate front in the cake, which can be observed very well, for example, by magnetic resonance imaging (MRI) [50]. In any case, the thickness of the condensate layer grows much slower than

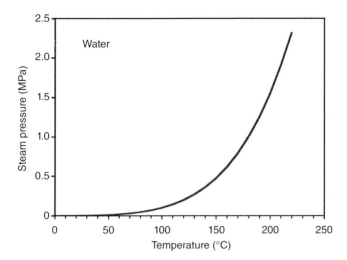

Figure 8.68 Steam pressure function.

the condensate/steam front migrates through the cake and is comparatively thin. Otherwise, a desaturation would not be possible. The steam breakthrough should take place, when the cake leaves the steam hood. This moment is important for the design of the process. After the steam breakthrough, a further steam flow would not be advantageous or effective. To calculate the position of the steam breakthrough on the rotary filter for correct positioning of the steam hood, a piston-like displacement of the liquid by the steam is presupposed with sufficient accuracy. The capillary entry pressure, which tries to hold the liquid in the cakes pores, reduces the applied pressure difference for the liquid displacement, as can be seen in Eq. (8.83)

$$\frac{dh}{dt} = \frac{(\Delta p - p_{cap,e})}{\eta_L \cdot (r_c \cdot h_c + R_m)} \tag{8.83}$$

The pressure loss due to the steam flow is neglected, as it was neglected for the cake desaturation by air or other gases. Integration of Eq. (8.83) and neglecting of the filter media resistance lead, in Eq. (8.84), to the steam breakthrough time t_{br}

$$t_{br} = \frac{\eta_L \cdot r_c \cdot h_c^2}{2 \cdot (\Delta p - p_{cap,e})} \tag{8.84}$$

In this calculation, the additional amount of liquid, which originates in the cake because of condensation, is likewise neglected. For the dynamic viscosity of the liquid, the ambient temperature outside of the steam hood is relevant. Normally, the displacement of the liquid is so fast that the liquid underneath the liquid front of displacement is not heated up on the whole that it would change the viscosity noticeable. Gerl found these coherences by measuring the temperature profile across the cake height during liquid displacement, as is illustrated exemplarily in Figure 8.69 [51].

The temperature, registered by the thermocouple direct at the filter medium, rises after 25 seconds steeply to 100 °C, which corresponds to the steam

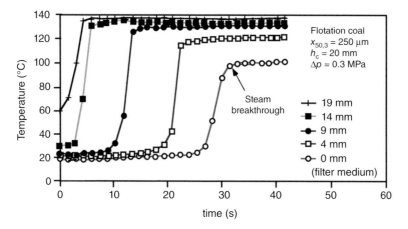

Figure 8.69 Temperature profile across the cake height during steam desaturation.

temperature for an absolute pressure of 100 kPa. Here, the steam breakthrough takes place after about 30 seconds. Furthermore, the temperature profiles show clearly that the steam is progressing piston-like through the cake and the filtrate is not heated up remarkably during displacement.

For the application on drum, disc, or pan filters, the steam breakthrough time t_{br} has to be transferred to the conditions of the rotary filter to design the position of the steam hood correctly. According to Eq. (8.85), cake thickness and steam breakthrough are depending on the parameters of the cake formation and have to be considered not independently at the rotary filter

$$h_c = \sqrt{\frac{2 \cdot p_c}{\eta_L}} \cdot \sqrt{\kappa \cdot \Delta p \cdot \frac{1}{n} \cdot \sqrt{\frac{\alpha_1}{360°}}} = \sqrt{\frac{t_{br} \cdot 2 \cdot (\Delta p - p_{cap,e})}{\eta_L \cdot r_c}} \tag{8.85}$$

Now, the breakthrough time t_{br} can be calculated in Eq. (8.86)

$$t_{br} = \frac{\Delta p}{(\Delta p - p_{cap,e})} \cdot \frac{\alpha_1}{360°} \cdot \frac{\kappa}{n} \tag{8.86}$$

The steam breakthrough time t_{br} and the rotary speed n are correlated by the steam breakthrough angle α_{br}, as shown in Eq. (8.87)

$$\alpha_{br} = t_{br} \cdot 360° \cdot n = \frac{\Delta p}{(\Delta p - p_{cap,e})} \cdot \frac{\alpha_1}{360°} \cdot \frac{\kappa}{n} \cdot 360° \cdot n = \frac{\Delta p \cdot \alpha_1 \cdot \kappa}{(\Delta p - p_{cap,e})} \tag{8.87}$$

As a result, the breakthrough angle is slightly dependent on the pressure difference and not dependent on the rotation speed. This is very advantageous for the practical filter operation and could be validated by experiments [52].

When the still efficient desaturated filter cake leaves the steam hood and enters the air-filled pressure vessel, it is hot and exhibits a large moist inner surface. The pressurized gas, which is now flowing through the cake, provokes an intensive evaporation and dries the cake thermally. The drying process ends after some seconds because the temperature in the cake and thus the drying velocity are

Table 8.4 Comparison of steam pressure with combined mechanical/thermal deliquoring.

Product (Data from BOKELA GmbH)	Limestone	Fine coal	Gypsum
Cake moisture pressure filtration	37%	20%	11%
Cake moisture steam pressure filtration	24%	10%	3%
Spec. energy consumption steam generator (efficiency 75%)	469 MJ t^{-1}	131 MJ t^{-1}	150 MJ t^{-1}
Spec. energy consumption thermal dryer (0.11 oil / 1 kg steam)	1140 MJ t^{-1} (37% → 24%)	583 MJ t^{-1} (20% → 10%)	389 MJ t^{-1} (11% → 3%)

slowing down remarkably. At the end, a filter cake of very low residual moisture is discharged.

The question may be asked whether it is more economical to produce overheated steam to get a certain cake moisture content instead of a pure mechanical cake desaturation with gas and following thermal drying to the same final residual cake moisture. The numbers in Table 8.4 for three different products give an answer to this question [53].

First, the residual cake moisture contents for pure gas pressure filtration are given with 37 wt% for limestone, 20 wt% for fine coal, and 11 wt% for gypsum.

For the same differential pressure, these cake moistures could be dropped remarkably for steam pressure filtration and subsequent gas blowing to 24 wt% for limestone, 10 wt% for fine coal, and only 3 wt% for gypsum.

The specific energy consumption for the steam generator with efficiency of 75% amounts to 469 MJ t^{-1} for limestone, 131 MJ t^{-1} for fine coal, and 150 MJ t^{-1} for gypsum.

The specific energy consumption for the thermal dryer was calculated with 0.1 l oil/kg steam. In the case of limestone, the cake had to be dried from 37 wt% after the pressure filtration to the final 24 wt%, which was reached by steam pressure filtration. This consumed energy of 1140 MJ t^{-1}, which means the factor 2.43 more than for steam pressure filtration. For fine coal and gypsum, 4.45 and 2.59 times, respectively. more energy was needed for the independent mechanical and thermal cake deliquoring.

This clearly demonstrates the effectiveness of the steam pressure process and can be explained by the synergy, which is produced with the hybrid process.

8.3.6 Desaturation of Incompressible Filter Cakes in the Centrifugal Field

8.3.6.1 Equilibrium of Filter Cake Desaturation in the Centrifugal Field

In principle, the same capillary pressure distribution is valid for cake desaturation by an external field like a centrifugal field or for cake desaturation by gas

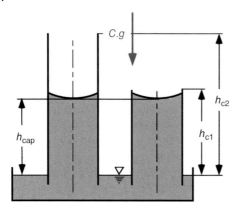

Figure 8.70 Equilibrium cake moisture in the centrifugal field.

differential pressure. Nevertheless, some differences can be observed regarding the distribution of the remaining liquid in the filter cake. This is due to the fact that the centrifugal pressure is depending on the actual liquid height in the centrifuge or in a capillary, respectively, as had been already discussed in Section 6.5.1 for the cake formation in the centrifugal field. Now, the desaturation of filter cakes should be discussed. This is first demonstrated in the simple example of two capillaries of different length, as shown in Figure 8.70.

The capillary pressure in both capillaries of the same diameter is identical. As a consequence, the capillary height in the centrifugal field is the same in the equilibrium state (cf. Eq. (8.41)). The capillary height is a typical phenomenon for desaturation by mass forces in contrast to the desaturation by a gas pressure difference. The gas pressure difference remains constant during the desaturation process, and if the capillary pressure is overcome, the capillary is emptied completely without any capillary height. In the centrifugal field, the centrifugal pressure decreases with decreasing liquid height in the capillary, and an equilibrium state is reached at a certain height of the liquid column. The greater the centrifugal pressure or the rotation speed of the centrifuge, respectively, becomes, the smaller the capillary height is. In Figure 8.70, two capillaries of different height, but the same diameter, are desaturated in the centrifugal field. The rotation speed should also be the same. The average residual moisture of the longer capillary is less than the average moisture of the shorter capillary, which remains completely saturated because its length corresponds, in this case, exactly with the capillary height.

Now, the simple scenario of circular and cylindrical capillaries has to be transferred to a filter cake. Experiments to desaturate filter cakes in the centrifugal field can be carried out in a lab-scale beaker centrifuge, as shown in Figure 8.71.

If a load cell and a transmitter are installed in the bottom of the filtrate-collecting beaker, not only the cake moisture after shut down of the centrifuge but also the time-dependent yield of the filtrate and thus desaturation kinetics can be registered and sent wireless to a computer for data collection and further analysis. If telemetric data transfer is not available, the data transfer can be realized in the classic manner by slip rings, which enable sliding contacts even at high rotational speeds. In the most simple case, the time-dependent decrease

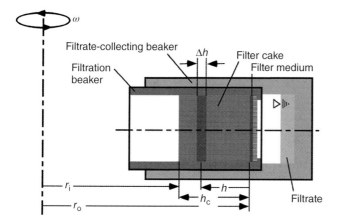

Figure 8.71 Cake desaturation in a centrifuge beaker.

of cake desaturation must be measured by performing several experiments for constant conditions, which are stopped after different desaturation times.

The centrifugal pressure in the filter cake depends on the distance of the boundary surface between liquid and gas and the filter medium. As a consequence, a local capillary pressure can be calculated, according to Eq. (8.88)

$$p_{cap} = \Delta \rho \cdot h \cdot a_{cent} = \rho_L \cdot h \cdot \left(r_o - \frac{h}{2}\right) \cdot \omega^2 \quad (8.88)$$

The cake slice thickness Δh is assumed to be infinitesimal thin, the specific weight of the gas above the two phase boundary surface can be neglected in comparison to the liquid, and the centrifugal acceleration is calculated, according to the long-arm approximation, as a constant average value.

When the kinetic process of desaturation comes to an end, the remaining capillary height looks in reality of course not like an even front as can be seen in Figure 8.70 for a constant pore diameter. As a consequence for a real pore size distribution in a homogeneous cake and therefore a distribution of capillary heights, according to Figure 8.72, a certain saturation gradient across the cake height must be existent in the equilibrium state.

This gradient is plotted for different centrifugal values in the diagram and should be explained in the example of two slices of the filter cake, which should have different distances to the filter medium or the filter cake surface, respectively. Both slices exhibit the same pore structure because the filter cake should be homogeneous and isotropic. The slice closer to the filter cake surface is exposed to a centrifugal pressure, which is determined by the liquid column of the greater length h_1. The slice closer to the filter medium is exposed to the lower centrifugal pressure, determined by the length of the liquid column h_2. For a const. C-value, the greater centrifugal pressure is able to desaturate more pores in the upper slice than the lower centrifugal pressure in the lower slice. Thus, the saturation degree rises up toward the filter media. Depending on the C-value at a certain cake height, all pores remain fully saturated because the centrifugal pressure does no longer overcome the capillary pressure of the largest pore. This

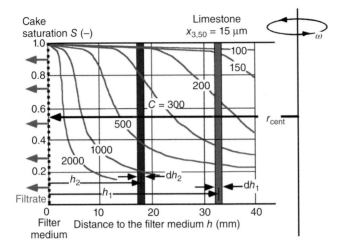

Figure 8.72 Equilibrium moisture profiles in centrifuge cakes (cf. [2]).

can be denominated as the real capillary height. The greater the rotation speed of the centrifuge and thus the centrifugal force (C-value) becomes, the more the capillary height is shifted toward the filter medium and the more homogeneous the total filter cake is desaturated. The capillary pressure curve, as presented in Figure 8.40, can also be measured in the centrifugal field, whereby one can distinguish between an incremental and a cumulative method.

The incremental method needs at least one single desaturation experiment with constant rotation speed until the equilibrium state is reached. After the desaturation experiment, the centrifuge is stopped, the cake must be taken out quickly, and cut into thin slices across the total cake height. Each slice represents one point on the capillary pressure curve $p_{cap}(S)$ because there is given an explicit correlation between capillary pressure and saturation. The saturation degree of each thin cake slice is assumed to be approximately constant. This method is comparatively quick. For safety, it is recommended to carry out the experiment at least for one additional rotational speed. Dismantling and preparation of the cake should be done as fast as possible to avoid drainage of remaining capillary liquid from more moist to more dry cake regions because the measurement would be falsified otherwise.

To realize the cumulative method, several independent experiments with constant cake height, but varied rotation speed, have to be carried out with the lab-scale beaker centrifuge. In each case, the integral average saturation degree S_{av} of the entire filter cake has to be determined and the centrifugal pressure has to be calculated for $h = h_c$. The equilibrium curve, determined by this method, is not the true capillary pressure curve, as defined before. This "capillary curve" is located above the true capillary pressure curve and allows no longer an explicit correlation of capillary pressure and saturation degree due to the averaging of the saturation gradient in the cake and the correlation of this average saturation degree with the average centrifugal or capillary pressure, respectively. Nevertheless, it is possible to get the true capillary pressure curve

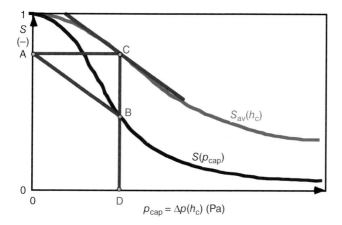

Figure 8.73 Cumulative centrifugal method and real capillary pressure curve.

by this cumulative method. Figure 8.73 illustrates how the true capillary pressure curve $S(p_{cap})$ can be found from the experimentally determined function $S_{av}(h_c)$ by a graphical method, which Schubert derived from theoretical considerations (cf. [2], p. 187–192).

At point C on the measured curve, a tangent is put to. For this purpose, it is recommended in a first step to approximate the function mathematically and in a second step to differentiate this function at point C. To the point C belongs point A on the ordinate and point D on the abscissa. A parallel shift of the tangent from point C to point A leads to an intersection point B on the straight line CD. Point B is finally located on the true capillary pressure curve.

Capillary pressure curves, which are measured by gas differential or centrifugal pressure, in principle should be identical. However, some differences may occur in detail (cf. [2], p. 195). For example, little differences in the cake structure can develop during cake formation in the pressure filter cell or in the beaker centrifuge. Centrifuge cakes are often more densely packed because the vibration of the centrifuge motor causes some consolidation during cake formation. In the case of gas pressure methods, under certain conditions, little hydraulically isolated liquid inclusions can remain in the cake, which are not more exposed to the pressure difference. In a centrifuge filter cake, such a phenomenon is to be expected less because on the one hand, the hydraulic connection of the liquid is better secured by the increasing saturation toward the filter medium and on the other hand mass forces are acting, which are generated directly by the liquid mass itself.

For users of centrifuges, it is common for reasons of easy use to represent the integral equilibrium saturation S_{av} as the function of the dimensionless Bond number Bo [54–56], which is also often denominated as capillary number. This number corresponds with a centrifugal force F_{cent}, which is made dimensionless by a capillary force F_{cap}. F_{cent} is calculated with the cake height h_c in the centrifugal field, as can be seen in Eq. (8.89)

$$Bo = \frac{4 \cdot F_{cent}}{F_{cap}} = \frac{4 \cdot h_c}{h_{cap}} = \frac{\rho_L \cdot g \cdot C \cdot d_h \cdot h_c}{\gamma_L \cdot \cos \delta} \tag{8.89}$$

The filter cake is considered here simplifying as a bundle of parallel circular capillaries of the same diameter in the direction of the centrifugal field and thus of uniform capillary height. Above the capillary height, the saturation degree is represented by the mechanical limit of desaturation S_r and below the capillary height the pore system is completely saturated. The pore size distribution of the filter cake is characterized by a mean hydraulic pore diameter d_h, as shown in Eq. (8.90)

$$d_h = \sqrt{\frac{32}{\varepsilon \cdot r_c}} \tag{8.90}$$

This pore diameter is calculated from a combination of the Darcy equation to describe the flow through a porous packed bed and the Hagen–Poiseuille equation to describe the flow through a circular pipe. As material function, the experimentally determined specific filter cake resistance r_c (cf. Eq. (6.28)) and the porosity ε (cf. Eq. (3.2)) are implemented into the equation.

The equilibrium saturation degree S_∞ is defined in Eq. (8.91), according to the above-mentioned conditions, by the ratio of capillary height h_{cap} and cake thickness h_c

$$S_\infty = \frac{h_{cap}}{h_c} = \frac{4}{Bo} \tag{8.91}$$

Figure 8.74 shows the representation of the Bond curve.

In range (I), the capillary entry pressure is still not overcome. According to Eq. (8.89), desaturation starts at Bo number $Bo = 4$, if $h_c = h_{cap}$. In the declining range (II) of the Bond curve, a capillary height remains in the cake, which becomes smaller for increasing Bo numbers. In range (III) of the Bond curve, the hydraulically interconnected pores of the cake are desaturated to the greatest possible extent, the capillary height approaches the value zero, and the saturation reaches the mechanical limit of desaturation. For very great Bo numbers in the range (IV) under circumstances, a very small part of the liquid bridges, which are bound very strongly by high capillary forces to the contact points of the particles or liquid adhering at the particle surfaces, can be removed. This can be observed

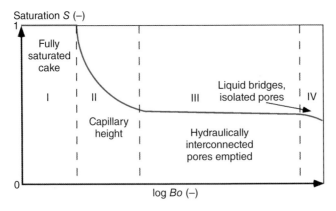

Figure 8.74 Bond curve.

especially for vibrating or sliding centrifuges, which are applied for relative coarse particles of several 100 μm.

8.3.6.2 Kinetics of Filter Cake Desaturation in the Centrifugal Field

The time-dependent decrease of cake saturation looks principally, as shown in Figure 8.45, for gas pressure deliquoring with the difference that in the centrifugal field, no gas flow through the emptied pores in the cake exists. As a consequence, a filter centrifuge could be an alternative to a vacuum or gas pressure filter in the case of shrinkage cracks in the filter cake. The modeling of the desaturation kinetics in the centrifugal field, which is presented here, is different from the model discussed for desaturation with gas differential pressure. This piston/film model was inspired by the observation of the filtrate flow from batchwise operating filter centrifuges, as peeler centrifuges [57]. In a first relative short period, the filtrate flow is very intensive than a relatively long drainage period of much less filtrate flow follows. This has led to the assumption that the filtrate drainage can be separated into a fast piston-like filtrate flow and a following slower trickle film drainage, as depicted in Figure 8.75.

Thus, the total saturation of a filter cake S_{tot} can be divided, according to Eq. (8.92), into several components, which have to be added

$$S_{tot}(t) = S_{pi}(t) + S_{fi}(t) + S_r \quad (8.92)$$

$S_{pi}(t)$ represents the time-dependent liquid piston, $S_{fi}(t)$ the time-dependent liquid film, and S_r the constant portion of remanent liquid.

Both time-dependent components of the total saturation degree contain a deliquoring number [58] or its inverse value, which is denominated as kinetic parameter λ [59] in Eq. (8.93)

$$\lambda(t) = \frac{4 \cdot \eta_L \cdot h_c}{\rho_L \cdot g \cdot C \cdot d_h^2 \cdot t_2} \quad (8.93)$$

The local time-dependent position of the piston $h_{pi}(t)$ can be described in an implicit form as a function of a kinetic parameter λ [60, 61], as formulated in Eq. (8.94)

$$\frac{1}{8 \cdot \lambda(t)} = \left[1 - \frac{h_{pi}(t)}{h_c}\right] + \frac{h_{cap}}{h_c} \cdot \ln\left[\frac{h_c - h_{cap}}{h_{pi}(t) - h_{cap}}\right] \quad (8.94)$$

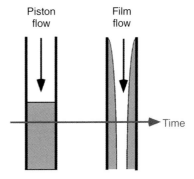

Figure 8.75 Piston-film model.

To get the time-dependent height of the piston, this equation must be solved numerically or by an approximation function (cf. [60], p. 41 or [62], p. 40). After that, the saturation, caused by the liquid piston in the cake, can be calculated in Eq. (8.95)

$$S_{pi} = \frac{h_{pi}(t)}{h_c} \tag{8.95}$$

In addition to the piston flow, the film flow must be formulated. The description of the film flow was inspired by experiments, where a plate was withdrawn from a bath of oil [63]. If this plate is converted to a capillary, this situation can be formulated mathematically (cf. [59], p. 53). The result is a power function with two material-dependent parameters a and b. Again, the kinetic parameter λ is included, as can be seen in Eq. (8.96)

$$S_{fi}(t) = a \cdot \left[\left(1 - \frac{h_{pi}(t)}{h_c}\right) \cdot \lambda(t)\right]^b \tag{8.96}$$

The term $(1 - h_{pi}(t)/h_c)$ represents the part of the cake above the fully saturated pore system, which is available for the film flow.

Now, the complete saturation $S_{tot}(t)$ can be formulated by summing up all components in Eq. (8.97)

$$S_{tot}(t) = \frac{h_{pi}(t)}{h_c} + \left[1 - \frac{h_{pi}(t)}{h_c}\right] \cdot \left[S_r + (1 - S_r) \cdot a \cdot \left[\left(1 - \frac{h_{pi}(t)}{h_c}\right) \cdot \lambda(t)\right]^b\right] \tag{8.97}$$

Figure 8.76 demonstrates by experimental and model results of Reif for different centrifuge types that the model could describe the desaturation results very well [64].

8.3.6.3 Aspects of Centrifuge Design and Operation Regarding Cake Deliquoring

The already discussed Bo curve can be taken to characterize the general deliquoring behavior of the different filter centrifuges, as depicted in Figure 8.77.

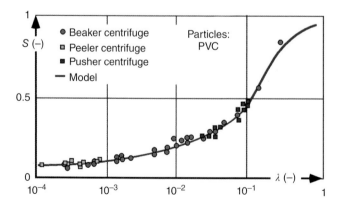

Figure 8.76 Centrifugal desaturation as a function of the kinetic parameter λ.

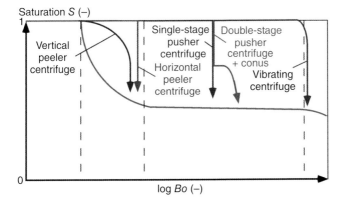

Figure 8.77 Representation of filter centrifuges in the Bo diagram.

A vertical peeler centrifuge very often increases speed during the cake-deliquoring phase, which can be seen by the increase of the Bo number. Because of the fact that in peeler centrifuges relatively small particles are separated, the curve is positioned in the Bo range of existing capillary height. The same is valid for the horizontal peeler centrifuge with the exception that this type of machine is operated with constant and in the optimal case maximal speed, which is normally slightly higher than for the vertical type. This leads to a constant Bo number in the diagram and a little lower cake moisture than in the case of vertical peeler centrifuges.

Pusher centrifuges are separating more coarse particles at comparatively greater Bo numbers. Here, in the best case, the mechanical limit of desaturation can be reached. The double-stage pusher centrifuge differs from the single-stage machine by a second screen basket of larger diameter. This causes a little step toward greater Bo numbers because of greater centrifugal acceleration. If in addition the second screen basket is conical shaped, the Bo number increases until the maximal radius has been reached.

Last but not least, the vibrating centrifuge separates still coarser particles in a conical basket of slight inclination. For these coarse particles in the range of several 100 μm to some mm, the capillary forces become very low and eventually the mass forces can remove a small part of the remanent liquid, which is represented by Sr.

Looking a bit more into the details of filter centrifuge operation, some more aspects should be mentioned except the general representation in the Bo curve, as demonstrated above. In the case of peeler centrifuges, it is advantageous to push down the capillary height into the remaining residual particle heel to avoid discharge of fully saturated cake by the peeler knife and to get in danger to plug the discharge chute by fully saturated and thus sticky cake.

Not only in the case of peeler centrifuges, but also generally for all filter centrifuges, a homogeneous cake formation is important for cake formation and cake desaturation. The formation of a sealing layer of the smallest particles on the filter cake surface increases the capillary entry pressure of the cake. This increases the residual cake moisture in any case and in the worst case no desaturation at all takes places. Measures to avoid particle segregation have been discussed in Section 6.5.2.

Peeler centrifuges with a rotating siphon (cf. Section 6.5.2) are able to provide an additional vacuum behind the filter medium, which not only increases the cake formation velocity but also improves the cake desaturation. Without siphon, a certain capillary height remains in the cake because of the discussed physical reasons. The additional gas differential pressure is able to desaturate this capillary height to a certain extent, if its capillary pressure is overcome. Unfortunately, the pressure difference breaks down instantly, when the first pores from the cake surface down to the filter media are emptied completely and pressure compensation takes place on both sides of the filter cake. In any case, this may be a possibility to avoid peeling into a fully saturated zone of the filter cake.

To be not limited by the vapor pressure of the liquid, which is limited by amount and time, the application of a permanent gas overpressure in the centrifuge basket would enable greater pressure differences and to maintain the gas pressure difference during the entire desaturation time. This is realized for inverting filter centrifuges. The pressurized gas of up to 600 kPa is fed into the process room via the feed pipe (cf. Figure 6.92). The process room of course must be made not only pressure tight but must fulfill all requirements of a pressure vessel for the intended gas pressure. This makes the principle more costly than the standard design and is used only for special applications. Especially, if shrinkage cracks during the cake desaturation with gas pressure filters are occurring, either exclusive centrifugation or combined centrifugal/pressure filtration could be an alternative [62]. By additional gas pressure difference, the kinetics of cake formation and desaturation are accelerated, the capillary height can be eliminated and the cake can be desaturated more homogeneously. Because of the reduced or completely avoided crack formation in the cake, the gas flow can be drastically reduced in comparison to pure gas pressure filtration. The principle enables beyond the standard overpressure filtration the application of saturated steam or the application of hot gas for thermal drying. In the last-mentioned case, the centrifuge operates as a filter dryer.

All continuously operating filter centrifuges with a conically opening basket toward the solid outlet, such as pusher, worm screen, vibrating, or sliding centrifuges, exhibit the advantage not only to facilitate the solid transport by the component of the centrifugal force parallel to the screen but also the absolute centrifugal pressure increases additionally because of the increasing basket diameter. This helps to decrease the filter cake residual moisture content. The problem of the conical screen basket shape, as discussed in Section 6.5.2, is the changing friction between cake and screen during the deliquoring phase. This is a limiting factor for the application of especially the self-transporting centrifuges, as the sliding centrifuge.

Centrifuges with an abrupt increase of the screen basket diameter, such as multistage pusher centrifuges, modern screen bowl decanters, or sliding centrifuges with cascade cone, exhibit as further advantage the reorganization of the cake structure at the transition point from stage to stage. This can lead to a liberation of liquid inclusions and to an intensification of the deliquoring by the impact of the particles on the next screen.

8.4 Consolidation of Compressible Filter Cakes by Squeezing

8.4.1 Fundamental Considerations Regarding the Consolidation Process

After the formation, a compressible filter cake exhibits a porosity gradient (cf. Section 6.4), which makes it possible to displace liquid from the cake by a subsequent squeezing or consolidation process. The pressure difference, necessary for the further deliquoring, is realized either hydraulically by the feed pressure of the slurry pump or mechanically by a press tool. Besides the uniaxial squeezing of the filter cake in batchwise operating chamber, diaphragm, piston, and tube filter presses by a pressure p, an additional shear stress τ is induced into the cake structure in continuously operating screw and double-wire presses. As can be illustrated in Figure 8.78, combined uniaxial pressing and shearing of the cake can lead under circumstances to a further compaction.

Uniaxial pressing leads to certain consolidation equilibrium, as soon as the cake structure becomes stable against the deforming force. This configuration may include greater voids, which still can be destroyed, if shear forces perpendicular to the press force destabilize the cake structure again.

A deliquoring of the filter cake takes place exclusively by reducing the pore volume and not by desaturation. For this reason, the quantification of the remaining liquid in the cake by the saturation degree S (cf. Eq. (8.1)), which is used to characterize the desaturation of incompressible filter cakes, is meaningless for squeezing. All pores of the filter cake are remaining fully saturated during the squeezing process, whereby the porosity is reduced simultaneously. However, the residual moisture content MC (cf. Eq. (8.2)), dry substance DS (cf. Eq. (8.3)), or the liquid–solid volume ratio B (cf. Eq. (8.4)) can be used. The liquid–solid volume ratio B in this special case directly correlates, according with Eq. (8.98), with the pore number e because the whole void volume in the cake is completely filled with liquid

$$B = \frac{V_L}{V_S} = \frac{\varepsilon}{(1-\varepsilon)} = e \tag{8.98}$$

Principally, a cake consolidation means automatically a reduction of moisture content. As discussed in Section 6.4.1, a compressible filter cake exhibits a strong

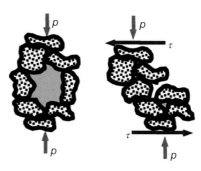

Figure 8.78 Consolidation by uniaxial pressing and shearing.

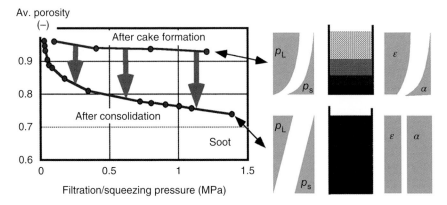

Figure 8.79 Average cake porosity before and after consolidation.

porosity gradient across the cake height (cf. Figure 6.50) after its formation. The porosity of the cake near the filter medium already corresponds approximately to the consolidation equilibrium for the applied pressure, whereas the cake surface is significantly more porous. This means that the cake layer near the filter medium is much more consolidated for higher pressures than the cake near the surface. This is the reason why the average porosity differs only a little for very different filtration pressures during the cake formation. Figure 8.79 demonstrates this observation again in a more general form (cf. [65], p. 112).

The difference between the average porosity directly after cake formation and the equilibrium porosity can be denominated consolidation potential. The cake directly after formation exhibits the mentioned porosity gradient, whereby the cake after consolidation is homogeneous [66, 67]. A first hint for an estimation of the consolidation potential is given by the porosity of a particle layer after sedimentation experiments. High compressibility and thus a high consolidation potential may be expected for high porosity values [68].

During the time-dependent consolidation process, the porosity gradient for a constant pressure becomes, according to Figure 8.80, more and more flat and the cake height decreases correspondingly.

At the end, the cake structure is homogeneous and the porosity is constant. The progress of consolidation with time is described in Eq. (8.99) by a so-called consolidation level CL

$$\mathrm{CL} = \frac{h_{c,0} - h_c(t)}{h_{c,0} - h_\infty} = \frac{e_0 - e(t)}{e_0 - e_\infty} \tag{8.99}$$

The actual cake height decrease is related to the maximal possible cake height decrease. This value can also be calculated from the respective pore numbers e. The consolidation time t is characterized in Eq. (8.100) by the dimensionless number TC

$$\mathrm{TC} = C_e \cdot \frac{i^2}{w_s^2} \cdot t \tag{8.100}$$

The consolidation time t is made dimensionless with a pressure-dependent material parameter C_e, the area-related solid mass w_s, which corresponds to the

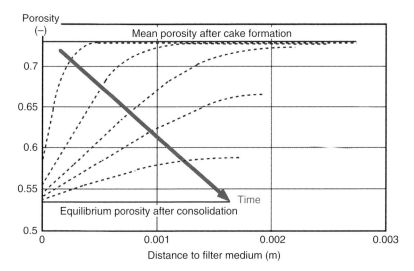

Figure 8.80 Porosity gradient during consolidation.

cake height, and the number of filter areas i. For example, in the chamber of a filter press, two filter areas on both sides of the chamber are present. From Eq. (8.100), it becomes clear that many filter areas and thin filter cakes (small numbers of w_s) are reducing the needed consolidation time. From this point of view, a vertical diaphragm press is more advantageous than a horizontal chamber filter press. Tubular piston presses with several drainage filter media are much more effective (cf. Figure 6.64). The pressure-dependent material parameter C_e can be calculated from experiments with the C-P-cell, as formulated in Eq. (8.101)

$$C_e = \frac{\rho_s}{\eta_L \cdot \alpha \cdot (de/dp_s)} \tag{8.101}$$

Experiments and simulations have proven that the consolidation level CL is, according to Eq. (8.102), more or less proportional to the square root of the dimensionless consolidation time TC

$$CL \propto TC^{0.5} \tag{8.102}$$

As a rough estimation, it can be stated that the filtrate of highly compressible filter cakes during the consolidation at first is flowing with a similar velocity as during the cake formation process. Only for consolidation levels of more than about 80%, the filtrate flow decreases distinctly (cf. [68]).

The final pore number of the consolidation equilibrium depends on the applied pressure and on the kind of the material. The consolidation mechanisms during increased pressure can also vary for different products and within certain ranges of pressure. In Figure 8.81, a simple and qualitative example for such product behavior is given, which has observed many times in various studies [69–71].

The pore number e (cf. Eq. (3.4)) in the equilibrium state is plotted against the applied pressure load p. In the range of low compaction, the particles still have a relatively great mobility and can easily be rearranged. In the second range of greater pressure, already agglomerate structures are destroyed, which had been

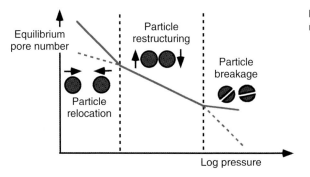

Figure 8.81 Consolidation mechanisms.

formed under the influence of adhesive forces. In the third range, the applied forces are as high as the particles themselves are destroyed by breakage. The dotted lines indicate that the inclination of the functions can vary depending on the specific material.

8.4.2 Aspects of Filter Design and Operation Regarding Cake Consolidation

In Section 7.6, with reference to permeation washing in frame and chamber filter presses, the problem had been discussed that the specific consolidation procedure of pressing additional slurry in the already cake-filled chambers normally leads to a porosity distribution across the plate area. This is due to the friction-caused constraints of cake movement far to the feed channel. If the cake sticks between the walls of the chamber exemplarily for a central feed pipe, it cannot be shifted longer to the outer regions and corners of the filter plate. Thus, the cake is highly compacted near the feed pipe and at the edges of the plates, far away from the feed pipe, the cake remains relatively porous. This means that the consolidation potential for the applied pressure often cannot be fully utilized. Because all pores of the cake remain fully saturated, this situation of course influences the residual cake moisture content negatively. In contrast to that deficiency, diaphragm filter presses exhibit the advantage of consolidating the filter cake homogeneously across the entire filter area by the diaphragm. This normally leads to comparatively lower cake moistures and shorter consolidation time. The vertical type of diaphragm presses is particularly advantageous because of the possibility of relatively small cake height and avoidance of segregation effects in the case of particle size distributions with a broad span.

Especially for diaphragm filter presses, a thermal vacuum contact drying of the cake after reaching the mechanical equilibrium of deliquoring can be realized. In this case, heat transfer plates are installed in the filter press, and at the filtrate outlet pipe, a vacuum is applied. Figure 8.82 illustrates the principle.

The applied vacuum in the chambers facilitates the evaporation of the liquid. The diaphragm must follow the shrinking cake to maintain the heat transfer. There are different heating systems possible. The liquid to press the diaphragm toward the cake can be heated or heating and filter plates are alternating. As a heat transfer medium, normally water or oil of up to 120 °C is used. Filter plates are often made of plastics such as polypropylene. To protect these thermoplastic

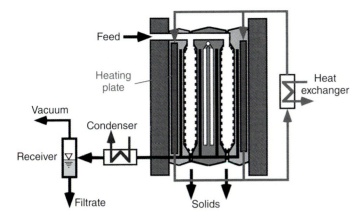

Figure 8.82 Diaphragm filter press with integrated thermal drying.

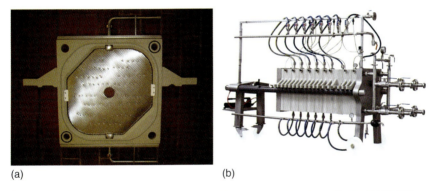

(a) (b)

Figure 8.83 Heat transfer plate for diaphragm filter presses (Courtesy of STRASSBURGER FILTER GmbH+Co.KG).

materials from high temperatures, special designed plates have been developed, as Figure 8.83 demonstrates in an example.

On Figure 8.83a, the filter plate is shown. A metallic heating plate is mounted floating in a polypropylene frame. This avoids direct contact of the hot surface with the heat-sensitive plastic. The metallic plate consists of two waved and by 90° against each other twisted steel sheets. On Figure 8.83b, the complete filter press installation is represented.

As discussed in Section 6.4.3, continuously operating press filters, such as screw or double-wire presses, exhibit on the one hand the deficiency of only moderate pressure, but on the other hand, the advantage of combined pressing and shearing the cake (cf. Figure 8.78). This enhances the deliquoring because of low cake porosities. In screw presses (cf. Figure 6.67), the shearing takes place only in one direction, whereby in double wire presses, the cake is sheared alternating in two directions. Double-wire presses or belt presses, respectively, can be designed in different configurations regarding the press roller system (cf. Figure 6.60). In Figure 8.84, an industrial example of a belt press is presented.

Figure 8.84 Belt press (Courtesy of FLOTTWEG SE).

After reaching the gel point in the first drainage zone, the sludge is transported into the wedge zone and carefully consolidated until traveling into the zone of medium pressure up to 150 kPa maximally and in technical practice frequently up to 100 kPa. The pressure is limited by the strength of the filter belts and the roller diameter. The pressure is applied on the entire area of the cake and increases consecutively by reducing the roller diameter. The pressure is generated by enlacement of the rollers by the filter belts, which are put under tension. The pressure on the filter cake p is proportional to the tensile force F_t and inversely proportional to the roller radius r, as in Eq. (8.103) is formulated

$$p = \frac{F_t}{r} \tag{8.103}$$

Figure 8.85 illustrates this situation.

The maximal pressure is held constant during a certain time, which can be adjusted by the number of consecutive rollers and the belt velocity.

Near the cake discharge, a zone of high-pressure deliquoring up to about 400 kPa can be installed. This can be realized either by squeezing the entire cake via an outer press belt or by applying a line pressure using press nips, which generate line pressures up to about $250\,\text{N}\,\text{m}^{-1}$. In the first case, an additional

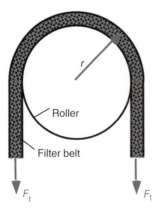

Figure 8.85 Generation of pressure at the press rollers.

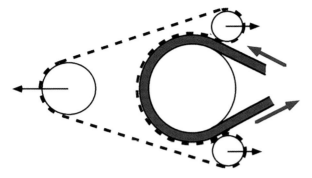

Figure 8.86 From outside applied press belt.

Figure 8.87 Press nips.

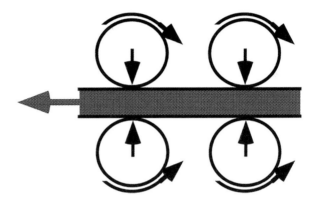

and extra stable belt is pressed from the outside on a press roller, as is illustrated in Figure 8.86.

Press nips can be seen at the end of the deliquoring zone, as depicted in Figure 8.87.

Here, the pressure rises up shortly to a maximum and then drops down again. The cake must be transported in any case through the press zone and must not be pressed back.

As mentioned before, the cake undergoes not only a squeezing but also a shearing of its structure. In contrast to the screw press, the shearing takes place not only in one direction but also because of the enlacement of the rollers forward and backward. The S-shaped enlacement of belts and rollers leads to an alternating parallel shift of both belts because of their different distance to the roller. This is illustrated in Figure 8.88.

Figure 8.88 Alternating cake shearing between press rollers.

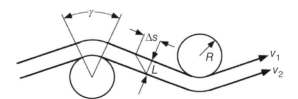

The parallel shift length Δs can be calculated, according to Eq. (8.104), from the geometric parameters of the roller system, whereby R represents the radius of the roller, L the distance between the two belts, γ the angle of enlacement, and v the velocity of the belts

$$\Delta s = \gamma \cdot (R + L) - \gamma \cdot R = \gamma \cdot L \tag{8.104}$$

For a successful operation, first the originating tensile strength between belt and cake must be transferred. Otherwise, slippage occurs. Secondly, the tensile strength inside the cake must be transferred. Otherwise, the cake would be fractured.

8.5 Consolidation/Desaturation of Compressible Filter Cakes by Gas Differential Pressure

8.5.1 Equilibrium of Filter Cake Consolidation/Desaturation

Not only nearly incompressible filter cakes but also filter cakes of a slight compressibility can be exposed to a gas differential pressure on a vacuum or pressure filter. If the applied gas pressure difference is greater than the capillary entry pressure, consolidation and desaturation are taking place simultaneously. For gas pressure differences below the capillary entry pressure, a pure consolidation of the cake without desaturation will happen. This process of consolidation is called shrinkage and lasts as long as the mobility of the particles is greater than the inner deformation resistance of the cake. For the example of a compressible barite filter cake, which consists of very small particles of about 2.6 µm, in Figure 8.89, the capillary pressure curve, measured for stepwise pressure increase, is demonstrated.

The two different mechanisms of cake deliquoring become visible, if both saturation degree and residual cake moisture content are plotted. It can be observed for pressure differences below the capillary entry pressure of about 230 kPa that there occurs some deliquoring but no desaturation. The deliquoring mechanism here is exclusively consolidation. The filtrate removal is much more than induced only by the initial forming of menisci, as discussed previously (cf. Figure 8.37). The pressurized gas above the cake surface acts like a piston and really squeezes the cake. After overcoming of the capillary entry pressure and emptying of the first pore channels, the inner deformation resistance of the cake increases steeply and the further deliquoring is mainly induced by desaturation. A second example is given in Figure 8.90 for filter cakes of fine-grained precipitated calcium carbonate, which had been consolidated or not before desaturation.

The not consolidated filter cakes had been exposed directly after the filter cake formation to the gas pressure difference, when a certain porosity gradient was present across the filter cake height. The first consolidated filter cakes had been pressed mechanically by a piston directly after the formation until the consolidation equilibrium for the applied pressure was reached. The consolidation pressure in both cases was the same as for the respective cake formation. For the not consolidated filter cakes, again a more or less linear decrease of moisture content can be observed with increasing gas pressure difference.

8.5 Consolidation/Desaturation of Compressible Filter Cakes by Gas Differential Pressure | 279

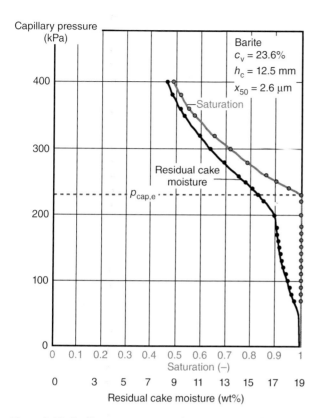

Figure 8.89 Capillary pressure curve for compressible filter cakes.

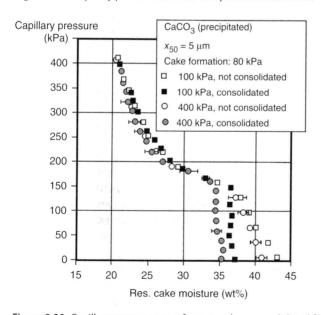

Figure 8.90 Capillary pressure curve for not and preconsolidated filter cakes [72].

The capillary entry pressure is not yet reached and the deliquoring mechanism is pure consolidation. The preconsolidated filter cakes of course cannot not be consolidated more during the pressure increase up to the capillary entry pressure and thus the residual cake moisture content remains the first constant The filter cakes of smaller cake formation pressure exhibit of course a bit more moisture content at the beginning in comparison to the cakes of greater cake formation pressure because their mean porosity is slightly greater. After overcoming the capillary entry pressure, which was very similar here for all filter cakes, the course of further desaturation was nearly identical within the limits of stochastic deviations. The cakes of smaller porosity have slightly smaller pores and thus greater capillary pressure but start at lower residual moisture and vice versa. These contrary effects are compensating more or less each other. The deliquoring mechanism beyond the capillary entry pressure is nearly exclusively desaturation, as discussed above for the barite filter cakes. After overcoming the capillary entry pressure, the inner deformation resistance of the filter cake increases rapidly and stabilizes the pore structure. Parallel to the cake stabilization, the tensile strength of the cake σ_t increases and reaches its maximum for a cake saturation degree of approximately 0.8–0.9. These by Schubert experimentally found results are illustrated qualitatively in Figure 8.91.

The function can be subdivided into three ranges. In range I, the liquid in the cake structure is represented mainly by liquid bridges between the particles. Range III is characterized by more or less filled capillaries, and in range II, bridges and filled capillaries are both present. The limits of the respective ranges are not fixed exactly to a special saturation degree but depend on the particle size distribution, the wetting behavior of the particles, and the surface tension of the liquid. Different models can describe the different ranges of the curve.

Range I is denominated by the liquid bridge range [72, 73] and can be described according to Eq. (8.105)

$$\sigma_{t,I} = \frac{(1-\varepsilon)}{\varepsilon} \frac{F_{ad,I}}{x^2_{av,SV}} \qquad (8.105)$$

The adhesion force F_{ad} can be calculated by addition of a line force F_1 and a capillary force F_{cap}, as formulated in Eq. (8.106)

$$F_{ad,I} = F_1 + F_{cap} = L \cdot \gamma_L \cdot \cos\delta + \pi \cdot R_1^2 \cdot p_{cap} \qquad (8.106)$$

Equation (8.107) describes the capillary pressure p_{cap}

$$p_{cap} = \gamma_L \cdot \left[\frac{1}{R_1} + \frac{1}{R_2}\right] \qquad (8.107)$$

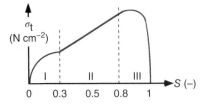

Figure 8.91 Transferable tensile strength in filter cakes (cf. [2]).

Figure 8.92 Liquid bridge between two particles.

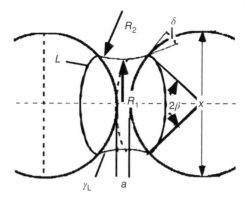

Figure 8.92 explains here the relevant parameters for a rotationally symmetric liquid bridge between two spheres of the same size.

The liquid bridge can be described by the radii of curvature R_1 and R_2, the 3pcl L, the bridge-filling angle β, the diameter of the particles x, the distance between the particles a, the wetting angle δ, and the surface tension of the liquid γ_L. Schubert has shown that for $\delta = 0$, $\beta = 10°–40°$, and $(a/x) \approx 10^{-2}$ as boundary conditions, which are often valid in practice, the adhesion force can be estimated by the simple correlation, given in Eq. (8.108)

$$F_{ad,I} \approx (2.2 - 2.8) \cdot \gamma_L \cdot x \tag{8.108}$$

Figure 8.93 illustrates calculative results of Schubert, from which Eq. (8.108) can be derived.

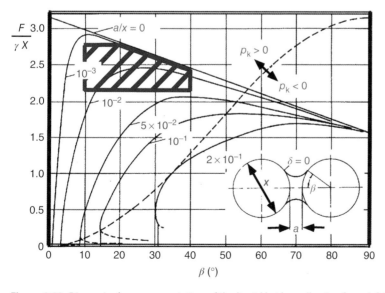

Figure 8.93 Dimensionless representation of the liquid bridge adhesion force (cf. [2]).

The shaded area in the diagram is given by the previously mentioned boundary conditions and the numbers in Eq. (8.108) can be read from the ordinate at the upper and lower limit of this area.

Range III represents the capillary range. The tensile strength in range III is formulated in Eq. (8.109)

$$\sigma_{t,III} = S \cdot p_{cap} \qquad (8.109)$$

As a consequence, the tensile strength $\sigma_{t,III}$ can be determined directly from the measured capillary pressure curve [29].

In range II, which can also be denominated as transition range, it was proposed to superpose the adhesion mechanisms because of bridge forces and capillary pressure [74] according to Eq. (8.110):

$$\sigma_{t,II} = \sigma_{t,I}^* + \sigma_{t,III}^* \qquad (8.110)$$

Consequences of these considerations are not only the correlation of cake cohesion and saturation but also the adhesion of the filter cake at the filter medium. This is important for the filter cake discharge from technical filter apparatuses (cf. Section 6.3.5). The higher the residual cake moisture content becomes, the more difficult the cake discharge is because of the higher adhesion between cake and filter cloth.

The relationship between consolidation and desaturation in the equilibrium state can also be demonstrated alternatively to Figures. 8.89 and 8.90, as demonstrated according to a proposal of Wiedemann in Figure 8.94 [75].

However, no direct information about the capillary pressure is included here. The pore number e is plotted against the liquid–solid volume ratio B. Using these both parameters, one can distinguish distinctly between consolidation and desaturation. If consolidation exclusively takes place, any reduction of the liquid–solid volume ratio corresponds with an equivalent reduction of the pore volume, and all measured points are located on a straight line through the origin, which represents the saturation degree $S = 1$. If the measured points deviate upward from the straight line, the capillary entry pressure is exceeded and the

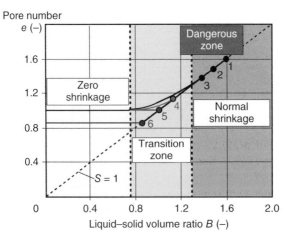

Figure 8.94 Deliquoring of compressible filter cakes by gas differential pressure.

cake is consolidated and desaturated simultaneously. Within the range, marked in the diagram as "dangerous zone," shrinkage cracks can occur in the filter cake (cf. Section 8.5.2). Finally, toward low-moisture content, the curve ends in a horizontal straight line. This means that in this case, liquid is removed from the cake only by displacement with gas, but no reduction of the pore volume itself is occurring even more. The cake behaves in the range of zero shrinkage incompressible.

Different starting points for the cake deliquoring in Figure 8.92 mean that the filter cake is preconsolidated by squeezing first and afterward desaturated by stepwise increased gas pressure. The numbers from "1" to "6" indicate in each case higher pressure during the preconsolidation phase. At the pressure "5," the cake already has reached the limit of shrinkage and is afterward exclusively desaturated. The same is valid for pressure "6," where the cake is preconsolidated below the limit of shrinkage.

To quantify the shrinkage behavior of a filter cake, a so-called shrinkage potential Δe can be defined. The maximal possible shrinkage for a certain particle system Δe_{max} is given by the difference between the pore number e_0 at the beginning and e_{min} as minimal possible pore number after shrinkage. The parameter e_{min} represents the limit of shrinkage. The course of the shrinkage curve is principally determined by the specific properties of the respective filter cake. The cake saturation, from which downward nearly no more shrinkage can be observed, amounts generally to about $S = 0.75$. This is beyond the maximum of tensile strength, and the inner cake deformation resistance has become already high. If the filter cake did not crack before, it will crack no longer. The diaphragm filter press represents a technical example for combined squeezing and following desaturation by gas differential pressure.

The diagram in Figure 8.94 was determined in its original form for the ideal case of homogeneous cake structures, which had been generated experimentally by pressing of a compressible filter cake to the equilibrium state and then drying by thermal means, to get the information about possible further shrinkage. The thermal drying can be used alternatively to the mechanical stepwise deliquoring by gas differential pressure [76]. The equivalence of stepwise mechanical desaturation and thermal drying can be explained by the pore size distribution and as a consequence the capillary pressure distribution, as is illustrated in Figure 8.95.

During stepwise increase of the gas pressure, the liquid is displaced from the largest pores first to the smallest pores at the end. If the liquid is evaporated thermally from the cake surface, further liquid is sucked via capillary forces to the surface. The smaller the capillaries become, the greater the capillary pressure is. As a consequence, smaller pores empty the greater pores and the result is similar to the mechanical process of desaturation, if the gas pressure is increased stepwise.

As discussed before, the course of desaturation for preconsolidated and not consolidated filter cakes is identical, if the gas pressure is increased stepwise. This can be understood well because the cake has time to consolidate from pressure step to pressure step. Also after overcoming the capillary entry pressure, only some pores are emptied for each pressure step and the cake deformation resistance increases only slowly. However, in the technical practice, the not

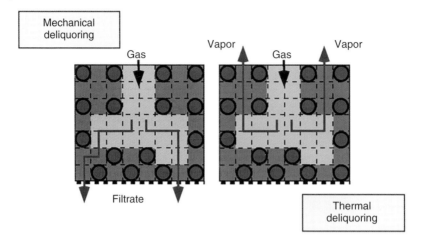

Figure 8.95 Mechanical and thermal cake deliquoring.

consolidated cake appears from the slurry and is faced abruptly to the full gas differential pressure. The capillary pressure of a broad spectrum of pore sizes is overcome now simultaneously, the solidification takes place very quickly, and the cake structure perhaps could be "freezed." This means less consolidation but more desaturation. Because of the mutual influence of consolidation and desaturation, it seems to be in question whether the final porosity and residual moisture content of compressible filter cakes, which have been deliquored according to the described two different deliquoring strategies, are comparable. Some experimental results seem to validate the expected different deliquoring results (cf. [77], p. 949). From Figure 8.96, it can be observed that preconsolidation and thus stabilization of the filter cake structure and subsequent desaturation has led to the same porosity as for pure squeezing.

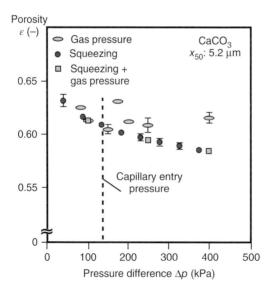

Figure 8.96 Cake porosity for different deliquoring procedures.

Figure 8.97 Residual cake moisture content for different deliquoring procedures.

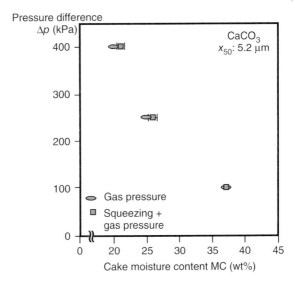

The subsequent desaturation had no further effect on the cake structure. However, the direct desaturation of the unconsolidated cake with constant gas pressure caused a certain "freezing" of the quick solidifying cake structure and thus a slightly greater cake porosity. As a consequence, the residual cake moisture content of the directly desaturated filter cakes had been slightly less than the preconsolidated cakes as can be seen in Figure 8.97.

The greater the gas pressure difference became, the greater the difference of the residual cake moisture content for the two procedures was. For the pressure difference of 100 kPa, the capillary entry pressure was not overcome and thus the cake moisture was identical. These experimental findings demonstrate that the smaller cake porosity in the case of desaturation after preconsolidation cannot balance completely the direct desaturation with slightly greater porosity.

Considering the aspect of optimal transferability of lab-scale data to a technical gas pressure filter, it would be recommendable to carry out the lab-scale experiments as similar as possible in comparison to the technical process. The capillary pressure curve would have to be determined in this case, according to the dynamic method (cf. Section 8.3.4.4), from several experiments and different filtration pressures. Cake formation and deliquoring pressure would have to be held constant for each experiment. The cake moisture content has to be given in wt% and not as saturation degree.

8.5.2 Cake Shrinkage and Shrinkage Cracking

Shrinkage cracks during desaturation of more or less compressible filter cakes are representing a serious and still not sufficiently solved problem for cake filtration apparatuses, such as nutsche filters or continuously operating rotary filters. Figure 8.98 shows a cracked filter cake from a circular pressure filter cell (a) and from a rectangular vacuum filter plate (b).

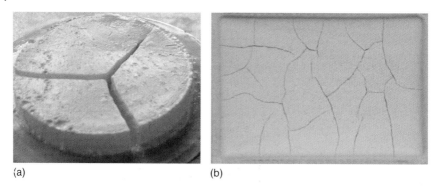

(a) (b)

Figure 8.98 Shrinkage cracks in filter cakes.

Although a rotary disc filter for large throughputs often would be the best solution for a certain separation task, shrinkage cracks can prevent the application. Shrinkage cracks lead to an abrupt and uncontrollable increase of gas throughput because of the shortcut flow through the cracks into the filtrate system. The vacuum pump or the compressor is often not more able to maintain the gas pressure difference, and the filtration process breaks down. If the filter cake has to be washed by permeation, the washing result will become very poor and inefficient because the wash liquid mainly flows through the cracks and not through the pore structure of the filter cake (cf. Section 7.6).

As discussed previously, compressible filter cakes, which are exposed to a gas pressure difference, are consolidated by rearrangement of the particles and eventually desaturated, if at least the capillary entry pressure is overcome by the gas pressure difference. If liquid is withdrawn from a fully saturated filter cake, a tensile strength originates in the cake structure, and a pressure difference exists between the gas atmosphere above and the liquid below the surface of the cake. These tensions originate already before capillaries are emptied and cracks can occur in still fully or nearly fully saturated cakes. The tensions can be released as long as the particles are movable enough. The liquid withdrawal leads to shrinkage of the cake structure and the inner deformation resistance of the cake increases with proceeding consolidation. During the shrinkage process, a crack develops at that location, where the tensile strength surpasses the adhesion between the particles. The danger of cracking is increased by inhomogeneities, such as porosity gradients, particle segregation, or defects in the cake structure, such as gas bubbles and outsized particles. This makes it nearly impossible to make a theoretical forecast about the cracking sensitivity of a filter cake. If a certain and not directed negative pressure is present in the cake, tensile stresses are occurring, and at the weakest point, a primary crack is forming. From this primary cracks, secondary and tertiary cracks are diverting in rectangular direction and it forms a material characteristic crack pattern (cf. Figures 8.2 and 8.98). If the typical undamaged area between the cracks is greater than the filter area of a lab-scale pressure filter cell (cf. Figure 6.12), a tendency to cake cracking may not be recognized in the filter cell and will become apparent not until the slurry is filtered on the greater filter area of a technical apparatus. A certain hint

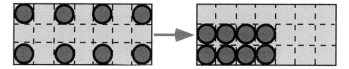

Figure 8.99 Unhindered three-dimensional shrinkage of homogeneous cakes.

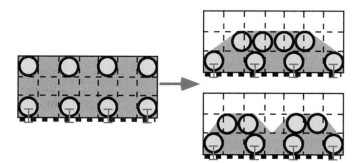

Figure 8.100 Hindered two-dimensional shrinkage of homogeneous cakes.

on a crack probability is given, if the cake is detaching a little bit from the wall of the cell. This must not happen in any case because there also exists an adhesion between particles and apparatus wall.

The shrinkage and its consequences regarding cracking depend on the respective boundary conditions. If, according to Figure 8.99, a homogeneous cake, which had been previously pressed into the equilibrium state and thus exhibits a homogeneous structure, is able to shrink unhindered in all three dimensions and is not fixed at the filter medium, the probability for cracking is minimal.

Normally, the filter cake is adhering at the filter medium and thus hindered to shrink parallel to the medium. In this case, the upper particle layers of the cake are able to shrink more easily than the lower layers. This leads to a tensile stress distribution, and depending on the adhesive forces between the particles, the cake can crack or not. This behavior is expressed in Figure 8.100 for an initially homogeneous cake structure.

If a filter cake of certain compressibility is not pressed to the equilibrium state, a porosity gradient exists, and the upper layers of the cake are able to shrink more than the lower layers. This situation is depicted in Figure 8.101 and increases the probability of shrinkage cracking.

Last but not least, the two-dimensional shrinkage of a filter cake exerts some tensile stress between the filter medium and the adhering first particle layers along the edges of the cake. If the tensile stress becomes greater than the adhesion forces, the cake is detaching from the medium and bending upward as can be seen in Figure 8.102.

This detachment of cake fragments from the filter medium is very disadvantageous because the hydraulically connected contact area between cake and filter medium is drastically reduced. Besides the reduced pressure difference due to the

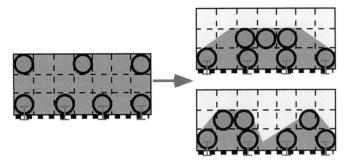

Figure 8.101 Hindered two-dimensional shrinkage of cakes with a porosity gradient.

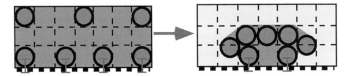

Figure 8.102 Shrinkage-induced detachment of cake from filter medium.

high gas flow through the cracks, the conditions for desaturation are additionally hindered.

8.5.3 Prevention of Shrinkage Cracks by Squeezing and Oscillatory Shear

Shrinkage cracks can occur as long as shrinkage is still possible. The characteristic curve to quantify this shrinkage behavior of filter cakes was given previously in Figure 8.94. From this diagram, one answer to the question can be derived, how to avoid shrinkage cracks during cake deliquoring with a gas differential pressure. If the filter cake would be preconsolidated mechanically by a piston, a diaphragm, a press belt, or press rollers to the pore number e_{min}, which characterizes the limit of shrinkage, no further shrinkage will be possible and thus no crack formation will occur. These correlations can be observed very good in the example of filter experiments with precipitated calcium carbonate. Filter cakes had been formed and desaturated for different gas pressure differences, whereby the pressure for cake formation and desaturation in every case was identical. If preconsolidated and nonconsolidated cakes are compared, the results can be seen in Figure 8.103 (cf. [77], p. 950).

If the cake is not consolidated before desaturation, independent of the pressure difference, a porosity gradient remains after the cake formation, and the cake is able to shrink. By this reason, every cake showed shrinkage cracks during desaturation. This effect becomes stronger toward smaller pressure differences because the porosity of the cakes increases and thus the particle mobility. If the cake is preconsolidated mechanically, the pore structure becomes homogeneous and the entire porosity decreases. This increases the inner deformation resistance of the cake and the tendency to crack is reduced. As can be seen in Figure 8.103, in

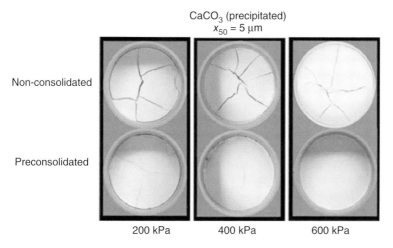

Figure 8.103 Avoiding shrinkage cracks by cake preconsolidation.

this special case, a preconsolidation pressure of 600 kPa is sufficient to reach the minimal pore number and to avoid crack formation completely during the subsequent desaturation of the cake. Unfortunately, the required pressure to reach the limit of shrinkage amounts for many products to the range between 1 and 2 MPa or even greater. Such pressures cannot be realized for continuously operating drum, disc, or belt filters. If crack formation makes the preferred application of continuously operating filters impossible, batchwise operating press filters or filter centrifuges, such as peeler centrifuges, hence are a promising alternative. Nevertheless, it would be advantageous if a crack-free filtration with continuously operating vacuum or pressure filters would be possible because these apparatuses are able to provide the comparatively greatest specific throughputs and often the lowest residual cake moisture after desaturation. A principal strategy to enable crack-free cake desaturation on drum or belt filters could be the installation of a press belt or another pressing device directly after the end of the cake formation zone and before desaturation to stabilize the cake. Unfortunately, the previously mentioned minimum pressures of often more than 1 MPa cannot be realized by design on drum or belt filters because of mechanical restrictions. This leads to the question how the cake could be consolidated alternatively without such great uniaxial press forces. The principal answer to this question is given in Section 8.4.1 (cf. Figure 8.78). If horizontal shear forces are induced into the cake structure additionally to the vertical press forces, the consolidation of the structure should become much easier and faster than with uniaxial pressing alone. Inducing additional oscillation with small amplitude and high frequency into the cake structure can generate such shear forces. This vibration can be realized by a press belt in the case of a moving filter cake on drum filters and belt filters with permanent belt movement or a piston-like pressing device in the case of belt filters with intermittent belt transport. The pressing device in any case must additionally oscillate parallel to the filter cake surface. Important is to start the preconsolidation directly after the end of cake formation to avoid any intrusion of gas into the cake pores and thus the initiation of shrinkage crack formation. Figure 8.104

Figure 8.104 Drum filter with vibrating press belt.

gives a rough impression of the idea to install a device for oscillatory shear and consolidation on a drum filter.

The rotating press belt, which is guided between the vibrator plates and the cake, is not included in this sketch for the reason of clarity. It is important here to decouple the vibratory unit from the filter itself and to provide a vibration damping between vibration unit and floor. The realization of this principle on drum filters is interesting for the installation in a pressurized vessel to realize overpressure filtration for more difficult to filter products, which tend to frequently crack formation during desaturation. Alternatively, such an installation could be realized more easily on vacuum belt filters with intermittent belt transport.

It could be experimentally demonstrated that only a very low uniaxial pressure in combination with vibration can lead to lower cake moisture contents or porosities, respectively, than exclusive uniaxial squeezing at very high pressure. As can be seen from experimental results in Figure 8.105, about 900 kPa would be necessary in the case of pure uniaxial squeezing to reach the critical crack-avoiding cake moisture of about 32.5% in the equilibrium state, which means after rather long time [78].

On the other hand, the critical cake moisture is reached much faster in less than one minute with vibration and only very slight vertical pressure of 20 kPa. Besides the drastically reduced vertical pressure, the fast consolidation is very important here because the squeezing/vibration zone on a continuously operating filter is limited. If the squeezing pressure is increased to 80 kPa, as shown in Figure 8.106 (cf. [78], p. 72), the critical moisture can be reached at lower frequencies than documented in Figure 8.105.

Extensive investigations have shown that consolidation by vibration is not in every case as effective as demonstrated above for the precipitated calcium

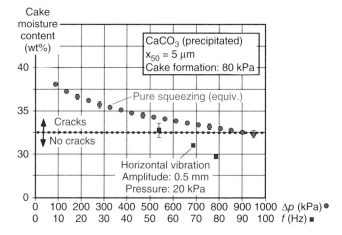

Figure 8.105 Comparison of uniaxial squeezing and vibration/squeezing.

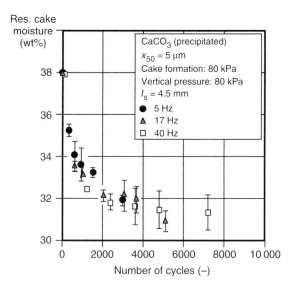

Figure 8.106 Vibration/squeezing.

carbonate, which exhibits a relatively cubical particle shape. For very compressible filter cakes from flaky china clay particles, the advantage of additional vibration in comparison to uniaxial squeezing was much less distinct [79].

First estimations clearly show that the additional necessary energy for the preconsolidation is much less in comparison to the energy for compressing the huge amount of gas, which would flow through cracks in the cake [80]. In many cases, there is no choice between preconsolidation or allowed cracking because the compressor would not be able at all to maintain the filtration pressure in the case of cracked cakes.

8.6 Electrically Enhanced Press Filtration

During several decades in the past, electrokinetic and electro-osmotic effects have been tried to implement into press filters for enhancing the process. The idea is to use the phenomenon that particles in liquid are normally charged and hydrated ions are present in the liquid. If an electric field is applied to the process room, the particles and hydrated ions are traveling to the countercharged electrodes. Figure 8.107 explains the principle effect.

If the particles are negatively charged and the negative electrode is located beneath the filter medium, the particles are moving against the filtration direction and a delayed cake filtration and simultaneous slurry concentration results. Thus, the filtrate can be removed faster and the cake formation time is reduced. This effect has been investigated and evaluated very intensively, for example, by Weber [81, 82]. It could be demonstrated that for the correct boundary conditions, an electrical field could influence the cake formation positively. Weber has developed a model to describe the acceleration of the cake formation process by an extension of Yukawas model [83]. The Yukawa model is derived from the basic equation of filter cake formation and extends this equation by the effects of electro-osmosis and electrokinesis. Weber formulated in Eq. (8.111) the cake formation equation including the electro-osmotic pressure Δp_{eo} and a critical electrical field strength E_{cr} in accordance with the basic equation, which was given in Section 6.3.2 (cf. Eq. (6.27))

$$\frac{t}{V_L} = \frac{\eta_L \cdot r_c \cdot \kappa \cdot \left[\frac{E_{cr}-E}{E_{cr}}\right]}{2 \cdot (\Delta p + \Delta p_{eo}) \cdot A^2} \cdot V_L + \frac{\eta_L \cdot R_m}{(\Delta p + \Delta p_{eo}) \cdot A} \qquad (8.111)$$

Δp_{eo} and E_{cr} can be quantified by filtration experiments. The critical electrical field strength characterizes the electrical field strength, at which the filtration velocity is equal to the contrary directed electrophoretic velocity. This electro-enhanced process has been proved beneficial especially for the separation of biopolymers such as xanthan gum. Such kind of slurries exhibits a remarkable electrical resistance, which means high voltage and low current [84, 85].

The reduction of the moisture content of compressible filter cakes by electrical means could not be confirmed clearly exclusively as an effect of the electro-osmotic pressure. As additional effects, besides electrokinesis and

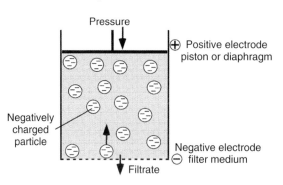

Figure 8.107 Electrically enhanced press filtration.

electro-osmosis, have to be considered, electrolysis and thermally induced viscosity decrease. An observed certain reduction of cake moisture content can be attributed mainly to gas bubbles from electrolysis at the electrodes, which are captured in the cake structure by capillarity. The development of hydrogen and oxygen due to electrolysis or the electrolytic disintegration of the electrodes and liberation of respective ions have to be considered and to be managed. In addition, a certain increase of squeezing velocity because of the thermally induced viscosity decrease could be observed. A reduced equilibrium cake moisture due to faster squeezing could not been verified. In summary, it can be stated that electrical enhancement of press filtration sometimes could be very interesting for acceleration of the filtration process in several special cases, but less for the deliquoring of the filter cake. In many cases, it makes more sense to increase the mechanical pressure than to invest in a more complicated and costly electrification of a filter press.

References

1 VDI-Guideline 2762, Part 3 (2017). Mechanical solid–liquid-separation by cake filtration – Mechanical deliquoring of incompressible filter cakes by undersaturation using a gas pressure difference, Beuth, Berlin.
2 Schubert, H. (1982). *Kapillarität in porösen Feststoffsystemen*. Berlin: Springer. ISBN: 3-540-11835-7.
3 Gibbs, J.W. (1960). *Thermodynamics, Scientific Papers*, vol. 1. New York: Dover Publications Inc (reprint of the original edition of 1906).
4 Marchand, A., Weijs, J., Snoeijer, J., and Andreotti, B. (2011). Why is surface tension a force parallel to the interface? *American Journal of Physics* 79 (10): 999–1008.
5 Myers, D. (1988). *Surfactant Science and Technology*. Weinheim: Wiley.
6 Rosen, M.J. (1988). *Surfactants and Interfacial Phenomena*, 2e. New York: Wiley.
7 Miller, R. and Lunkenheimer, K. (1986). Adsorption kinetics measurements of some nonionic surfactants. *Colloid and Polymer Science* 79 (1): 357–361.
8 Tanaka, T., Kon-No, T., and Kadoya, T. (1989). Bubble volume method for the measurement of dynamic surface tensions using the maximum bubble pressure method. *Journal of Colloid and Interface Science* 132 (1): 139–143.
9 Stroh, G. (1994). Die Wirkung von Tensiden auf die Entfeuchtung von Filterkuchen durch einen aufgeprägten Gasdruck, Dissertation, Universität Karlsruhe (TH), Karlsruhe.
10 Adamson, A.W. (1967). *Physical Chemistry of Surfaces*. New York: Wiley.
11 Weser, C. (1980). Die Messung von Grenz- und Oberflächenspannung von Flüssigkeiten – eine Gesaamtdarstellung für den Praktiker. *GIT Fachzeitschrift für das Laboratorium* 24: 642–648. and 734–142.
12 Butt, H.-J., Graf, K., and Kappl, M. (2006). *Physics and Chemistry of Interfaces*. Wiley. ISBN: 9783527406296.
13 Buff, F.P. and Saltsburg, H. (1957). Curved fluid interfaces, II. The generalized Neumann formula. *Journal of Chemical Physics* 26: 23–31.

14 Young, T. (1805). An essay on the cohesion of fluids. *Philosophical Transactions of the Royal Society of London* 95: 65–87.

15 Lester, G.R. (1967). Contact angles on deformable solids, Society of Chem. Industry, Monograph No. 25, London.

16 Bikerman, J.J. (1970). *Physical Surfaces*. New York, London: Academic Press.

17 Schwartz, A.M. (1975). The dynamics of contact angle phenomena. *Advances in Colloid and Interface Science* 4 (4): 349–374.

18 Neumann, A.W. (1974). Contact angle and their temperature dependence: thermodynamic status, measurement, interpretation and application. *Advances in Colloid and Interface Science* 4: 105–191.

19 Wenzel, R.N. (1936). Resistance of solid surfaces to wetting by water. *Industrial and Engineering Chemistry* 28 (8): 988–994.

20 Barthlott, W., Schimmel, T., Wiersch, S. et al. (2010). The Salvinia paradox: superhydrophobic surfaces with hydrophilic pins for air-retention under water. *Advanced Materials* 22: 2325–2328. https://doi.org/10.1002/adma.200904411.

21 Morrow, N.R. (1970). Physics and thermodynamics of capillary action in porous media. *Industrial and Engineering Chemistry* 62: 32–56.

22 Adam, N.K. and Jessop, G. (1925). Angles of contact and polarity of solid surfaces. *Journal of the Chemical Society* 127: 1863.

23 Heertjes, P.M. and Kossen, N.W.F. (1967). Measuring the contact angles of powder–liquid systems. *Powder Technology* 1 (1): 33–42.

24 Bröckel, U. and Löffler, F. (1991). A technique for measuring contact angles at particles. *Particle and Particle Systems Characterization* 8 (1–4): 215–221. https://doi.org/10.1002/ppsc.19910080139.

25 Laplace, P.S. (1806). Méchanique céleste, On capillary attraction, suppl. to Vol. 10, Paris

26 Thomson, W. (Baron Kelvin) (1871). LX: on the equilibrium of vapour at a curved surface of liquid. *The London, Edinburgh, and Dublin Philosophical Magazine and Journal of Science*, Series 4 42 (282): 448–452.

27 Morrow, N.R. and Harris, C.C. (1965). Capillary equilibrium in porous materials. *Society of Petroleum Engineers Journal* 5 (15): 15–24.

28 Schubert, H. (1972). Untersuchungen zur Ermittlung von Kapillardruck und Zugfestigkeit von feuchten Haufwerken aus körnigen Stoffen, Dissertation, Universität Karlsruhe, Karlsruhe.

29 Schubert, H. (1973). Kapillardruck und Zugfestigkeit von feuchten Haufwerken aus körnigen Stoffen. *Chemie Ingenieur Technik* 46 (6): 396–401.

30 Anlauf, H. and Sorrentino, J. (2004). The influence of particle collective characteristics on cake filtration results. *Chemical Engineering and Technology* 27 (10): 1080–1084. https://doi.org/10.1002/ceat.200403252.

31 DIN 66145 (1976). *Graphical Representation of Particle Size Distributions; RRSB-grid*. Berlin: Beuth Publishing DIN.

32 Stieß, M. (1995). *Mechanische Verfahrenstechnik 1*, 2e, 46–50. Berlin, Heidelberg, New York: Springer. ISBN: 3-540-59413-2.

33 Anlauf, H. (1989). Kombination von Eindicker und kontinuierlichem Druckfilter zur Verbesserung von Fest/Flüssig-Trennprozessen. *Chemie Ingenieur Technik* 61 (9): 686–693.

34 Wyckoff, R.D. and Botset, H.G. (1936). The flow of gas-liquid mixtures through unconsolidated sands. *Physics* 7: 325–345.
35 Brutsaert, W. (1963). On pore size distribution and relative permeabilities of porous media. *Journal of Geophysical Research* 68: 2233–2235.
36 Brutsaert, W. (1967). Some methods of calculating unsaturated permeability. *Transactions of the American Society of Agricultural Engineers* 10: 400–404.
37 Anlauf, H. (1986). Entfeuchtung von Filterkuchen bei der Vakuum-, Druck- und Druck/Vakuum-Filtration, Fortschrittsberichte VDI, Reihe 3, No.114, VDI, Düsseldorf, ISBN: 3-18-141403-4
38 Anlauf, H., Bott, R., and Stahl, W. (1986). Über die Kinetik desEntfeuchtungsvorganges bei der Druckfiltration feinstkörniger Suspensionen. *Chemie Ingenieur Technik* 58 (3): 232–233.
39 Nicolaou, I. (1999). Fortschritte in Theorie und Praxis der Filterkuchenbildung und -entfeuchtung durch Gasdruckdifferenz, Fortschrittsberichte VDI, Reihe 3, No. 583, VDI, Düsseldorf, ISBN: 3-18-358303-8
40 Bott, R. (1985). Zur Kontinuierlichen Druckfiltration, Dissertation, Universität Karlsruhe (TH), Karlsruhe.
41 Stahl, W., Bott, R., and Anlauf, H. (1983). Pressure filtration of iron ore slurries. *Aufbereitungstechnik/Journal for Preparation and Processing* 24 (5): 243–252.
42 Bott, R., Anlauf, H., and Stahl, W. (1984). Continuous pressure filtration of very fine coal concentrates. *Aufbereitungstechnik/Journal for Preparation and Processing* 25 (5): 245–258.
43 Anlauf, H. (2016). Gas consumption, solids throughput and residual cake moisture – relation between operating expenses and process results of rotary filters. In: *Global Guide of the Filtration and Separation Industry* (ed. H. Lyko and S. Ripperger), 182–191. Rödermark: VDL. ISBN: 978-3-00-052832-3.
44 Korger, V. and Stahl, W. (1993). Vapour pressure dewatering – a new technique in the field of mechanical and thermal solid–liquid-separation. *Aubereitungstechnik/Mineral Processing* 34 (11): 555–563.
45 Korger, V. (1995). Dampfdruckentwässerung - Ein neues kombiniertes Entfeuchtungs-verfahren, Fortschrittsberichte VDI, Reihe 3, No.403, VDI, Düsseldorf, ISBN: 3-18-340303-X.
46 Gerl, S., Korger, V., Stahl, W., and Krumrey, T. (1994). Dampf-Druckfiltration-Ein Verfahren zur kombinierten mechanisch/thermischen Entfeuchtung von Filterkuchen. *Aufbereitungstechnik/Mineral Processing.* 35 (11): 563–572.
47 Peuker, U. and Stahl, W. (2000). Dewatering and washing flue gas gypsum with steam. *Filtration, Filtration and Separation* 37 (8): 28–30.
48 Peuker, U. and Stahl, W. (2001). Steam pressure filtration: mechanical-thermal dewatering process. *Drying Technology* 19 (5): 807–848.
49 Peuker, U. (2002). *Über die kombinierte Dampfdruck- und Zentrifugalentfeuchtung von Filterkuchen.* Aachen: Shaker 3-8322-0817-8.
50 Korger, V. and Stahl, W. (1994). Visualisierung der Dampfdruckentwässerung durch MR-Tomographie. *Chemie Ingenieur Technik* 66 (9): 1264–1265.
51 Gerl, S. (1999) Dampf-Druckfiltration – Eine kombinierte mechanisch/thermische Differenzdruckentfeuchtung von Filterkuchen, Fortschrittsberichte VDI, Reihe 3, No.604, VDI, Düsseldorf, 94, ISBN: 3-18-360403-5.

52 Peuker, U. and Stahl, W. (1999). Scale up of steam pressure filtration. *Chemical Engineering and Processing* 38: 611–619.

53 BOKELA GmbH (2005). Hi-Bar Filtration – Continuous Pressure & Steam Pressure Filtration brochure 08/05, Karlsruhe, 28.

54 Perry, R.H. (1950). *Perry's Chemical Engineering Handbook*. New York: McGraw-Hill.

55 Mersmann, A. (1971). Beispiele dimensionsloser Kennzahlen in der mechanischen Verfahrenstechnik. *Verfahrenstechnik* 5 (1): 23–28.

56 Mersmann, A. (1972). Restflüssigkeit in Schüttungen. *Verfahrenstechnik* 6 (6): 203–206.

57 Batel, W. (1961). Menge und Verhalten der Zwischenraumflüssigkeit in körnigen Stoffen. *Chemie Ingenieur Technik* 33 (8): 541–547.

58 Bender, W. (1971). Zur Berechnung des Durchsatzes von Schälschleudern, Dissertation, Universität Karlsruhe, Karlsruhe.

59 Mayer, G. (1986). Die Beschreibung des Entfeuchtungsverhaltens von körnigen Haufwerken im Fliehkraftfeld, Dissertation, Universität Karlsruhe (TH), Karlsruhe.

60 Reif, F. (1990). Transport und Entfeuchtung körniger Schüttgüter in Schneckenzentrifugen, Fortschrittsberichte VDI, Reihe 3, No.202, VDI, Düsseldorf, ISBN: 3-18-140203-6.

61 Zeitsch, K. (1981). Eine neue Theorie der Zentrifugalentfeuchtung. *Chemische Technik* 33 (9): 456–461.

62 Stadager, C. (1995) Die Entfeuchtung von Filterkuchen durch die Kombination von Zentrifugal- und Gasdruckkraft, Fortschrittsberichte VDI, Reihe 3, No.415, VDI, Düsseldorf, ISBN: 3-18-341503-8.

63 Bikerman, J. (1956). Drainage of liquid from surfaces of different rugosities. *Journal of Colloid Science* 11 (4/5): 299–307.

64 Reif, F., Wünsch, M., König, H., and Stahl, W. (1990). Dewatering kinetics of finely granulated products in a centrifugal field. *Aufbereitungstechnik/Mineral Processing* 31 (3): 117–125.

65 Alles, C.M. (2000) Prozessstrategien für die Filtration mit kompressiblen Kuchen, Dissertation, Universität Karlsruhe (TH), Karlsruhe.

66 Shirato, M., Murase, T., and Iwata, M. (1986). Deliquoring by expression – theory and practice. In: *Progress in Filtration and Separation*, vol. 4 (ed. R.J. Wakeman), 181–288. Amsterdam: Elsevier.

67 Leclerc, D. and Rebouillat, S. (1985). Dewatering by compression. In: *Mathematical Models and Design Methods in Solid–Liquid Separation*, NATO ASI Series (ed. A. Rushton), 356–391. Dordrecht, NL: Nijhoff.

68 Tiller, F.M., Yeh, C.S., and Leu, W.F. (1987). Compressibility of particulate structures in relation to thickening, filtration and expression – a review. *Separation Science and Technology* 22 (2/3): 1037–1063.

69 Terzaghi, K. and Peck, R. (1961). *Soil Mechanics in Engineering Practice*. Berlin: Springer.

70 Train, D. (1956). An investigation into the compaction of powders. *Journal of Pharmacy and Pharmacology* 8 (1): 745–761. https://doi.org/10.1111/j.2042-7158.1956.tb12206.x.

71 Böhmkes, J. (1986). Untersuchungen zum Auspressen von Filterkuchen unter ein- und zweidimensionaler Hochdruckbeanspruchung, Dissertation, Universität-Gesamthoch-schule Essen, Essen.

72 Rumpf, H. (1970). Zur Theorie der Zugfestigkeit von Agglomeraten bei Kraftübertragung an Kontaktpunkten. *Chemie Ingenieur Technik* 42 (8): 538–540.

73 Schubert, H. (1974). Haftung zwischen Feststoffteilchen aufgrund von Flüssigkeitsbrücken. *Chemie Ingenieur Technik* 46 (8): 333–334.

74 Schubert, H., Herrmann, W., and Rumpf, H. (1975). Deformation behaviour of agglomerates under tensile stress. *Powder Technology* 11: 121–131.

75 Wiedemann, T. and Stahl, W. (1996). Experimental investigation of the shrinkage and cracking behaviour of fine particulate filter cakes. *Chemical Engineering and Processing* 35 (1): 35–42.

76 Wiedemann, T. (1996) Das Schrumpfungs- und Rissbildungsverhalten von Filterkuchen, Fortschrittsberichte VDI, Reihe 3, No.453, VDI, Düsseldorf, ISBN: 3-18-3-345303-7.

77 Illies, S., Anlauf, H., and Nirschl, H. (2016). Avoiding filter cake cracking: influence of consolidation on desaturation characteristics. *Drying Technology* 34 (8): 944–952. https://doi.org/10.1080/07373937.2015.1087023.

78 Illies, S., Pfinder, J., Anlauf, H., and Nirschl, H. (2017). Filter cake compaction by oscillatory shear. *Drying Technology* 35 (1): 66–75. https://doi.org/10.1080/07373937.2016.1159576.

79 Illies, S. (2017). Darstellungen zur Entfeuchtung von zu Rissbildung neigenden Filterkuchen, Dissertation, Karlsruher Institut für Technologie, Karlsruhe.

80 Illies, S., Anlauf, H., and Nirschl, H. (2017). Vibration-enhanced compaction of filter cakes and ist influence on filter cake cracking. *Separation Science and Technology* 52 (18): 2795–2803. https://doi.org/10.1080/01496395.2017.1304416.

81 Weber, K. (2002) Untersuchungen zum Einfluss eines elektrischen Feldes auf die kuchenbildende Pressfiltration, Fortschrittsberichte VDI, Reihe3, No. 733, VDI, Düsseldorf, ISBN: 3-18-373303-X.

82 Weber, K. and Stahl, W. (2003). Influence of an electric field on filtration in a filter press. *Chemical Engineering & Technology* 26 (1): 44–48.

83 Yukawa, H., Kobayashi, K., Tsukui, Y. et al. (1976). Analysis of batch Electrokinetic filtration. *Journal of Chemical Engineering of Japan* 9 (5): 396–401.

84 Vorobiev, E. and Lebovka, N. (eds.) (2008). *Electrotechnologies for Extraction from Food Plants and Biomaterials*. NewYork: Springer. ISBN: 978-0-387-79374-0.

85 Hofmann, R. (2005) Prozesstechnische Entwicklung der Presselektrofiltration als innovatives Verfahren zur Abtrennung von Biopolymeren, Fortschrittsberichte VDI, Reihe3, No. 835, VDI, Düsseldorf, ISBN: 3-18-383503-7.

9

Selected Aspects of Filter Media for Cake Filtration

9.1 Introduction and Overview

The success of a filtration process depends not only on the filter medium itself but also on the apparatus design and the process conditions. The filter medium acts as the decisive interface between slurry, filter apparatus, and operational conditions. Only if these three aspects are well adjusted to each other, a filtration process will work properly. As Figure 9.1 illustrates, the filter medium interacts with slurry properties, such as particle concentration, particle size distribution, temperature, or pH.

Further, it interacts with design parameters of the filter apparatus, such as filter type, batch, or continuous operation, size and shape of the filter area, the kind of filter cake discharge, stress by pressure, drag, or friction. Last but not least, it interacts with operational parameters, such as vacuum or gas overpressure, relative velocity between filter element and slurry, or cycle time of the filtration operation.

Particle retention and flow resistance are representing the main properties of a filter medium, which are relevant for the filtration process. The cut size should be as small as possible to improve the filtrate clarity, and the permeability of the filter medium should be as great as possible to improve the throughput. This seems to be a contradiction, but as discussed previously in Section 6.2, not exclusively the filter media structure is decisive for the cake filtration process. In addition, the structure-intruded particles and especially the first bridge forming particle layer are influencing the filter medium flow resistance. As a consequence, in many cases, the first bridge forming particle layer on the filter medium surface is dominating its flow resistance and the kind of filter media used is of second order. This enables to apply theoretically microporous membranes in many cases for cake filtration without restriction of throughput in comparison to conventional woven filter fabrics (cf. Section 9.4). However, it should be avoided to combine filter media and particle size distributions of the same average pore and particle size, respectively, because in such cases, the danger of clogging becomes maximal. If the filter media structure itself should be minimized in flow resistance, high porosity, great pore diameters, small span of pore size distribution, and small thickness of the filter media are beneficial, as it is the case for the filter cake or any other porous structure.

Wet Cake Filtration: Fundamentals, Equipment, and Strategies, First Edition. Harald Anlauf.
© 2019 Wiley-VCH Verlag GmbH & Co. KGaA. Published 2019 by Wiley-VCH Verlag GmbH & Co. KGaA.

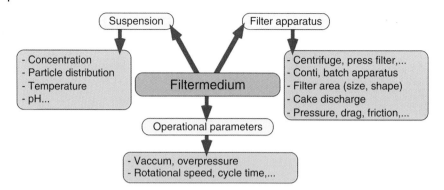

Figure 9.1 Function of filter media as the interface.

In cake filtration applications, particle retention means not exclusively the cut size as the maximal particle size, which is able to penetrate the filter medium. In many cases, particle retention means particle separation efficiency because the pore size of filter media for cake filtration is often chosen greater than the particles themselves, to avoid clogging. Only the particle bridges across the filter medium pores determine the particle retention in those cases. As discussed in Section 6.2 in more detail, slurry prethickening or particle agglomeration can reduce the filtrate pollution effectively.

Regarding washing, not the filter medium alone but the complete drainage system behind the filter cake influences dispersion effects and thus the effectiveness of washing. As has been discussed in Section 7.6, the drainage of the wash filtrate must not be hindered, for example, by the knobs, which are forming the filtrate drainage system in filter plates of filter presses or by the solid areas between the bore holes of a filter centrifuge rotor. This means that either the filter medium itself must provide enough drainage capacity in horizontal direction to the next filtrate drainage channel or an additional drainage medium must be installed between the filter medium and the support structure.

During filter cake deliquoring, the filter media normally are not the bottleneck for filtrate drainage, if the cake formation was not restricted previously. Eventually, in cases of filter cake discharge by blowback, an influence of filter media could be observed in the case that the filter cell had not been emptied completely, some filtrate is blown back and the filter cake is rewetted. The more open the filter medium becomes, the less barrier effect it can provide.

Beside the separation characteristics, the service life of filter media is an important and sometimes decisive issue for the selection of well-suited filter media. The service life directly influences the operational costs of filter apparatus and thus the efficiency of the separation process. Each shutdown to change filter media reduces the production rate and disturbs the entire process. Here, an apparatus design, which enables a quick and easy change of filter media, is of course advantageous. The filter media have to be manufactured and tailored precisely and should fit perfectly to the filter element for a reliable, safe, and optimal filter operation. The seams have to be made of strong threads of the right material. No visible needle holes should be present. The filter medium must exhibit sufficient mechanical, physical, and chemical stability. In different cake filter apparatuses, the filter media are exposed to different kinds of mechanical stress. On belt filters, the filter media are pulled through the process zone. In double-wire presses, the media is pulled, pressed, and sheared. On disc filters, the filter media must withstand the gas blowback to get rid of the filter cake, or in continuous filter centrifuges, the filter cake is transported across the filter media while it is pressed to the media by centrifugal forces.

Beside these mechanical impacts, the filter media have to resist physical stresses, such as higher temperatures or radiation, like ultraviolet radiation. In addition, the media must be stable against pH or specific chemicals.

A further very important point is the possibility for regeneration of filter media. It cannot be avoided completely that particles are intruding into the structure of the porous media. This makes it necessary to regenerate the media more or less frequently by back-pulsing with filtrate, by spraying the surface with liquid, by dissolving intruded particles, or other measures. Obviously, three-dimensional nonwovens, such as fleeces or needle felts, are much more difficult to regenerate than monofilaments of a plain weave structure.

The manifold requirements to filter media not only for cake filtration but also for filtration processes in general necessitate an appropriate variety of different types of filter media. In Figure 9.2, exemplary main types of filter media are shown.

As discussed in Chapter 1, filtration processes can be subdivided principally in depth and surface filtration processes. Depth filter media in any case consist of three-dimensional porous structures, in which the particles are intruding and deposited. The particles must be much smaller than the pores of the filter media and normally exhibit diameters of about 1 μm and less. The solid concentration of the liquid must be very low to avoid bridge formation at the filter media surface. Such filter media are formed by tailored filter elements, such as coiled yarn filter candles, resin-bound filter sheets, nonwoven fleeces and felts, or particle layers in the form of packed beds or precoat layers on cake filters. For particle layers,

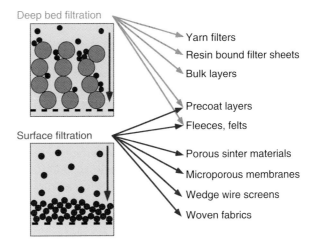

Figure 9.2 Types of filter media.

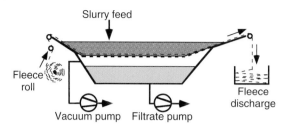

Figure 9.3 Vacuum belt filter with single-use filter media.

manifold materials such as gravel, sand, coal, activated carbon, diatomaceous earth, perlite, or different qualities of cellulose and others are utilized.

In a certain transition zone between depth and surface filtration, again particle layers in the form of a precoat or fibrous media in the form of fleeces and felts are used. As previously described in Section 5.5, a precoat, made of a relatively coarse-grained filter aid material, makes it possible to separate very small particles in low concentration on continuously operating vacuum drum filters, which are typical cake filters. Fleeces and felts are not typical media for cake filtration but are applied nevertheless in several cases because they are relatively cost-efficient and provide a certain depth filter function, which can improve filtrate clarity. In any case, the depth filter component should be avoided as far as possible because it is relatively difficult to regenerate such filter media because of their irregular pore structure. One exception is given by very simple vacuum or hydrostatic belt filters to regenerate, for example, cooling lubricants from the metal working industry. As can be seen in Figure 9.3, the fleece is used here as a single use media and is not regenerated but discharged after one cycle together with the separated solid matter.

The third group of filter media, which is shown in Figure 9.2, consists of typical surface filters, whereby microporous membranes made of sintered ceramics or polymers are mainly and typically applied in cross-flow filtration or dead end micro- and ultrafiltration. The structures, shown in Figure 9.3, exhibit real pores, which allow a convective flow of liquid and normally solid particles up to diameters of about 10 µm are separated. The pore size of ultrafiltration membranes is limited downward at about 0.01 µm. Here, not only solid particles but also dissolved macromolecules and colloids can be separated. Below 0.01 µm, membranes for nanofiltration and reverse osmosis are denominated as nonporous. The transport mechanism through these membrane structures is not more convection, but diffusion.

In principle, microporous membranes would be attractive for cake filtration applications because of the possibility to produce particle-free filtrates. There are some approaches to realize such processes. A vacuum disc filter with ceramic filter media is mentioned in Section 6.3.1 [1]. Figure 9.4 shows the principle structure of filter sectors made of such material.

On both sides of the filter segment, microporous filter plates of about 1 µm pore size are acting as the filter media. The filtrate drainage system between the filter plates is made of coarse sinter ceramic granules. These filter materials exhibit a comparably high flow resistance because of the limited porosity of sintered materials and a certain thickness of the plates for sufficient stability against breakage.

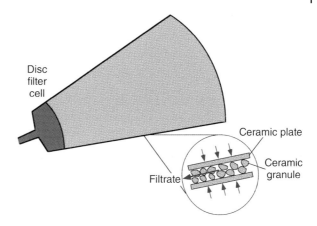

Figure 9.4 Ceramic filter element for disc filters.

On the other hand, these microporous media are delivering a particle-free filtrate and behave semipermeable against liquid and gas, if they are applied as vacuum filter. Liquid can pass and gas is stopped at the filter media surface. This saves a lot of energy for gas exhaust by the vacuum pump. The reason is the capillary pressure of the hydrophilic ceramic material, which is in the case of vacuum filtration greater than the gas differential pressure. Because of the atmospheric pressure and the vapor pressure of liquids, only pressure differences of less than 100 kPa can be realized with vacuum filters. The microporous ceramics with pore sizes of about 1 μm exhibit, in contrast to that, capillary entry pressures of about 100 kPa. An alternative would be the application of much thinner and flexible polymeric membranes. The filter media resistance of such membranes is remarkably lower than for sintered ceramics and thus the solid throughput would be remarkably higher. On the other hand, polymeric membranes do not exhibit mechanical strength, such as ceramic materials. This concept will be discussed in more detail in Section 9.4. The first concept of microporous ceramic filter elements is already commercially available, whereas the second concept of microporous polymeric membranes is still under development.

For continuous centrifugal cake filters, such as pusher or worm screen centrifuges, particularly robust filter media are necessary to withstand the friction between media and cake during solid transport, while centrifugal acceleration presses the cake toward the media. Likewise, the clogging tendency must be minimized to guarantee full filterability of the media because the residence time of the product in the centrifuge rotor is short and increased filter media resistances must not hinder the cake formation velocity. These correlations are discussed in more detail in Section 6.5.2.

Last but not least, woven filter media are mainly used for cake filtration processes. Because of the fact that it is possible to vary the design features of fabrics in a very broad range and thus the separation properties and the mechanical behavior of fabrics can be influenced with high variability, this type of filter media can be adapted with very high flexibility to the needs of slurry, separation apparatus, and operational conditions.

9.2 Woven Filter Media for Cake Filtration

Woven filter media are constructed structures. Warp and weft threads are crossing each other in a 90° angle. All threads running lengthwise of the fabric as woven are named warp threads, and all threads running crosswise of the fabric as woven are named weft threads. In this way, manifold configurations with different properties regarding particle retention and mechanical stability can be produced. In Figure 9.5, three selected basic weave types are illustrated, to give some examples.

The most simple weave structure is given by the plain weave. Each warp thread crosses one weft thread. Square-shaped open pores are the consequence. If one weft thread crosses two warp threads, a twilled weave originates. If warp and weft threads exhibit different diameters, a dutch weave is produced. A plain weave can be transformed in this way into a single plain dutch weave. In the same manner, twilled dutch weaves can be realized. The pore channels of dutch weaves are wound through the structure and exhibit a triangular shape at the narrowest position. The pore geometry at this position determines, according to Figure 9.6, the maximal size of a sphere, which is able to penetrate the weave structure (cf. Section 9.3).

The pore sizes of filter fabrics depend on the weave structure and the thickness of the threads. In general, it can be stated that the thinner the threads become, the smaller pores can be formed. Depending on the thread material, the thickness of the wires is limited. Polymers and metals exhibit a certain crystalline structure, which becomes unstable below a certain wire thickness. Metallic or polymeric

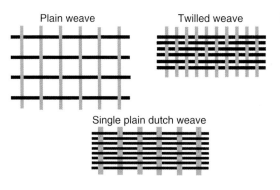

Figure 9.5 Weave structures.

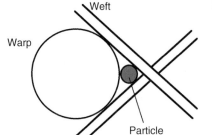

Figure 9.6 Maximal cut size of a dutch weave.

wires are available down to diameters of about 20 μm, which is less than half the thickness of human hair. As a consequence, the minimal pore sizes of woven fabrics are definitely greater than 1 μm. As a second consequence follows that the mechanical stability of fabrics decreases with decreasing thread diameters. Recent developments in weaving technology partly overcome this disadvantageous correlation.

One solution is demonstrated in Figure 9.7, which uses the principle of function separation, as it is known for asymmetric microporous membrane structures [2].

This type of weave structure can be adjusted very specifically to the needs regarding mechanical stability and particle retention. The fine porous upper side of the fabric has the job to retain the particles, whereby the coarse porous lower side is made of strong threads and realizes the necessary mechanical stability of the entire fabric. Both layers are woven together on the same weaving loom. This leads to one interconnected fabric and not to two separate layers, which would cause problems such as blockage or wrinkling.

Another strategy to get small pores, high permeability, and sufficient mechanical stability consists in a three-dimensional weave technology of dutch weaves, as shown in Figure 9.8 [2].

Although pore sizes of down to about 5 μm are possible, this type of fabrics can be produced from comparable thick threads of materials such as Avesta,

Figure 9.7 Double-weave structure.

Figure 9.8 Three-dimensional weave structure RPD-HIFLOW©. Source: Courtesy of HAVER & BOECKER OHG.

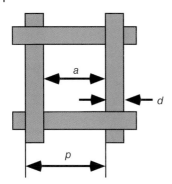

Figure 9.9 Plain weave pattern.

Hastelloy, Inconel, or titanium. Such materials could not be produced in such thin threads, as would be needed to get the mentioned fine pore sizes with conventional weave structures.

To describe the geometric properties of weave structures, different parameters are used. The repeating weave pattern describes the principle structure of a fabric by a basic element of warp and weft binding, which is repeated in the fabric. It includes two threads in the case of a plain weave and, for example, at least three wires for twilled weave. Figure 9.9 shows as an example for the simplest basic element a plain weave pattern.

The aperture opening a is in this case defined as the distance between two adjacent warp or weft wires of diameter d, measured in the projected plane at the mid positions. The pitch p is defined as the distance between the midpoints of two adjacent wires, as is expressed in Eq. (9.1)

$$p = a + d \tag{9.1}$$

In Anglo-American countries, the aperture opening often is given in mesh. This means the number of apertures per inch counted in a row one behind the other and divided by the pitch because the wires have to be considered. There is no explicit correlation between mesh and particle size especially for more complex weave structures than a plain weave. A more differentiated view to the pore size and its definition is given in Section 9.3. Beside the aperture with the open screening area or the porosity, respectively, is a decisive factor for the permeability or the pressure loss of a fabric. The open screening area results from the sum of all aperture area compared to the total area of the fabric. The porosity relates the void volume with the total volume of the fabric.

To select a well-suited fabric, the interaction with slurry, filter apparatus, and operational conditions have to be respected. This leads, beside specific geometrical properties, as discussed before, to several additional requirements to the filter fabric properties, such as elasticity, tensile strength, abrasion stability, weight, swelling properties, temperature stability, thermal or electrical conductivity, or light resistance. Very important of course is the stability against chemicals, such as mineral or organic acids, bases, salts, oxidizing agents, organic solvents, or special chemicals, such as formaldehyde, glycol, glycerin, mineral oils, or phenol. Figure 9.10 illustrates the thermal and chemical stability of different polymers.

Polymer	Temperature (°C)	Polymer
PTFE	240	PTFE
PEEK	220	PEEK
PPS	190	PPS
PCTA	160	
PETP$_{dry}$	150	
PVDF	120	PVDF
PETP$_{wet}$	100	PA6/6.6/11
PP		PP/PA12
Acid	Neutral	Alkaline

PA Polyamide
PCTA Polycyclohexylendi-
         methylenterephtalate
PEEK Polyetheretherketone
PETP polyethylene terephthalate
PP Polypropylene
PPS Polyphenylene sulfide
PTFE Polytetrafluoroethylene

Figure 9.10 Thermal and chemical stability of polymers for woven filter fabrics.

Figure 9.11 Thread structures.

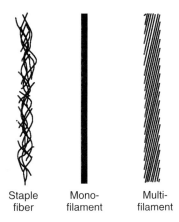

Staple fiber   Mono-filament   Multi-filament

As can be seen clearly, the thermal stability of thermoplastic polymers is limited, but if higher temperatures have to be managed, other materials for the fabrics such as metals are available.

After the choice of the material, the kind of threads has to be selected. According to Figure 9.11, different options can be chosen.

From the viewpoint of history, the staple fiber is the oldest thread modification. In former times, threads had been produced from natural fibers of limited length, such as wool or cotton. These fibers had been spun to yarns of relatively rough surface. The rough surface is caused on the one hand by the fibers themselves, which are spun to a thread and on the other hand by the ends of the fibers, which are sticking out of the threads surface. This roughness is disadvantageous for the filter cake discharge because the adhesion between cake and cloth is increased.

Because of the described properties of the threads, the pore geometry is relatively imprecise. A further consequence consists in a certain depth filter component, which may be positive to avoid filtrate pollution in the initial phase of cake formation but is subject for pore blockage on the other hand.

In contrast to staple fibers, monofilaments consist of endless wires with a smooth surface. This allows producing the most exact pore geometry. Fabrics, which are made of monofilaments, exhibit a much smoother surface than fabrics made of staple fibers. This makes the cake discharge and the media cleaning more comfortable. One deficiency may be that in the case of a wire breakage immediately, a defect is present in the fabric. This can be of importance, if the filter cloth is stressed by abrasive ambient conditions or permanent alternating stress.

The multifilament thread represents a compromise between monofilament and staple fiber. In this case, a certain number of endless monofilament wires are spun to a yarn, which exhibits a relatively precise pore geometry of relatively smooth surface, but is more stable than monofilaments because the breakage of some wires not immediately leads to a destruction of the thread.

Monofilaments and multifilaments are often combined in a fabric. The material, kind of threads, and weave pattern already allow a very flexible adjustment of the filter media to slurry, apparatus, and apparatus operation, but there are additional measures after the weaving process to influence the fabric properties.

During the weaving process, the threads are wound around each other. This causes tensions in the fabric, which can deform the structure and interfere the following manufacturing steps. To release such tensions, a so-called thermosetting can be undertaken. This temperature treatment leads to stable mesh geometry and mechanical dimension stability.

If special surface properties, such as wetting behavior or electrical conductivity are needed and not provided by the material of the threads itself, fabrics can be impregnated with respective substances, although one has to be aware of the limited durability of such coatings. Today, the wettability of polymers can be influenced directly by modifications of the thread material with a special plasma treatment [3, 4].

As discussed previously, woven filter media exhibit a more or less rough surface, which influences the adhesion between cloth and cake. In the case of staple fibers, a scorching process can be used to smooth the medium surface because the ends of the fibers, which stick out of the threads surface, are reduced or eliminated in that way. Beside staple fibers, multifilaments and monofilaments lead to a certain roughness of the fabric surface. In these cases, the fabric surface can be smoothened by a calendering process. The fabric is pressed between hot rollers and the unevennesses become weak and are smoothed out. This is shown schematically in Figure 9.12 on the example of a polymeric monofilament plain weave.

If the calendering process is too strong, the pore geometry will be affected and the pore size will be reduced. As a consequence, the calendering has to be carried out in every case carefully. On the other hand, as had been shown in Figure 6.4, the main contribution to the effective filter media resistance in most cases is caused by the first bridge forming particle layer across the meshes of the filter fabric and

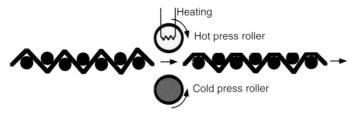

Figure 9.12 Calendering to smoothen fabric surfaces.

Figure 9.13 Filter bag on a disc filter segment.

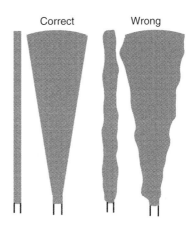

not by the pore size of the media. In addition, the calendering leads to a kind of asymmetric pore structure, which could be beneficial to reduce blockage.

After all treatment procedures to optimize the properties of the fabrics, the filter elements of the filter apparatuses must be equipped with them. Decisive for optimal operation is a perfect fit of the media on the filter element. Figure 9.13 demonstrates on the example of a disc filter segment the correct and wrong fit of the filter bag.

To get an optimal filtration performance, the filter media must fit perfectly on the filter elements. This has been discussed on the example of filter cake discharge from disc filters in Figure 6.39. Furthermore, no visible needle holes are allowed. If the needle was chosen too thick in comparison to the pore size of the fabric, needle holes originate, which constitute defects in the pore structure. The seams must be precise, straight, and the threads must be made of the same material as the fabric itself.

Beside the conventional way to equip filter apparatuses, a change to alternative and innovative methods can improve the overall performance of filter processes. This should be discussed on the example of continuously operating vacuum or pressure drum filters. A change of damaged or blocked filter clothes is sensitively influencing the effectiveness and performance of drum filter installations. Traditionally, the filter cloth consists of one large single piece of fabric, which is wound around the drum and fixed by strings in dovetail grooves and with a spiral span wire around the drum. The change of the complete filter cloth and fixing the new cloth by traditional caulking is very hard work and needs for large units from many hours to more than one day. This means high costs for maintenance and

Figure 9.14 Exchangeable drum filter cells. Source: Courtesy of BOKELA GmbH.

downtime for the process. To enhance this situation, two effective developments enable to make the change of the filter cloth on drum filters much easier and faster [5].

In the case of not too big drum filters, single cells, as can be seen in Figure 9.14, have been developed, which can be exchanged quickly and similar to the filter sectors of a disc filter.

For large drum filters, this technique is not more applicable, but also for these cases, a special method had been developed to equip each single filter cell with a separate cloth, as Figure 9.15 illustrates.

The so-called "FrameTrak" system enables the quick change of the filter cloth for single cells and especially for large drum filters. An elastomer profile with cloth inserts is fixed in special metal profiles by a locking strip of elastomer with the help of a little manually operated tool. For smaller filter units of less than 10 m^2 of filter area, single cells can be changed completely, as in the case of disc filters. Cells, which are covered with a new filter cloth, can be stored beside the filter and changed very fast, when needed. The removed cells can be regenerated afterward.

9.3 Porometry – Using Capillarity to Analyze Pore Sizes of Filter Media

9.3.1 Introduction

Particle retention and flow resistance of a filter medium are determined by pore size or pore size distribution, respectively, media structure, and interaction with the slurry. Especially, the interaction between filter media and slurry leads to the practically relevant filter media resistance. Not only the particles, which are intruding into the media structure, but primarily the first bridge forming particle layer determines the filter media resistance (cf. Section 6.3.2). The definition of pore sizes in filter media unfortunately is in most cases not as easy as

Figure 9.15 "FrameTrak" system for changing filter cloth on single drum or pan filter cells. Source: Courtesy of BOKELA GmbH.

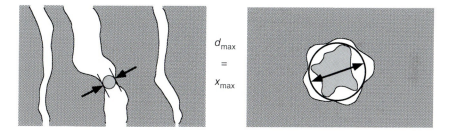

Figure 9.16 Definition of the maximal pore size.

shown in Figure 9.9 for a square-shaped pore. Thus, a more general definition is needed, what the pore size really means. Figure 9.16 exemplifies the problems schematically.

The maximal pore diameter is characterized by d_{max} and the maximal particle diameter by x_{max}. In the case of the cut size, both parameters are identical. The pores in a three-dimensional filter medium structure are normally not straight and cylindrical but wound and show constrictions and extensions. There does exist a certain pore size distribution and often no pore looks exactly like the others. Things become easier, for example, in the case of etched microsieves or monofilament fabrics, where the pore geometry is well defined. However, it makes sense to define the pore size also for pore systems, such as membranes, fleeces, or sinter materials, for which the pore geometry is much more complicated. If one looks to the pore size, the largest relevant pore cross section for particle separation can be found as the largest cross section of all constrictions of pore channels, which are connecting upper and lower surface of the media. This means that this pore cross section determines the largest particle, which can penetrate the filter medium. If the pore cross section is not circular and the particle is not spherical, one has to state more precisely that the largest pore cross section is given by the largest circle, which can be drawn into this largest cross section of all constrictions. The diameter of this circle equals the smallest circle around the projection area of the largest particle, which can penetrate this pore. This particle diameter is not easy to determine because each physically different particle size analyzer gives a different equivalent particle size for the same particle. The equivalent particle diameter describes the diameter of a sphere of the same physical property as the measured real and not spherical particle. For example, the result of a sedimentation analysis gives the diameter of a sphere, which would settle with the same velocity, as the investigated real particle (cf. Section 2.3.2).

This discussion makes clear that the practical relevant pore size and particle size are not easy to determine. For practical use and corresponding to Figure 9.16, the maximal pore size should be defined here as the diameter of the largest sphere, which is able to penetrate the filter medium.

Beside the cut size of a filter medium, the flow resistance is an important issue because it determines the filter throughput. Similar to the discussion about the pore size also for the flow resistance, the separated slurry and the particle system, respectively, have to be considered because the relevant flow resistance of a filter medium originates from the interaction between the filter medium and first separated particles.

Finally, process-relevant phenomena, such as filtrate pollution by not separated fine particles or detachment of a filter cake from the filter cloth, are influenced not only by the filter medium but also by design parameters of the apparatus and operational parameters of the process. As illustrated in Figure 9.1, all these parameters have to be coordinated to get an optimal final result.

9.3.2 Methods of Pore Size Determination

The pore size can be measured directly or indirectly. Direct measuring methods consist in the microscopic image analysis or the sieving with spherical particles, such as glass beads (challenge testing). Indirect measuring methods, such as the capillary flow porometry or the bubble point test, provide hydraulic pore diameters, which can be interpreted analogous to the equivalent particle diameter. The measured pore diameter corresponds to the diameter of a circular pore of the same capillary pressure, as the investigated pore.

The conventional microscopic analysis is relatively fast and the largest pore of the investigated sample can be identified clearly, if pores throughout the filter media exist, which are not constricted. Much more extensive but more informative is the X-ray tomography or μ-CT method, which allows to get a three-dimensional insight into the media structure and any cross section can be analyzed [6, 7].

Image analysis generally offers the possibility not only to measure the greatest pore but the entire pore size distribution. There is no limit of pore sizes, which can be measured by this technique. On the other hand, normally it is not practicable to analyze complete filter cartridges with microscopic methods, but only samples of the filter media. The technique of image analysis is only applicable for open-meshed filter media with pore channels perpendicular to the filter area, which can be analyzed with transmitted light. Nontransparent media, such as dutch weaves, cannot be analyzed in that way.

As a second direct pore size measurement technique, a screening test with spheres, such as glass beads, can be applied to any structure of the filter media [8, 9]. This test is relatively work intensive and can detect the largest pore approximately, but not the pore size distribution. A lower limit for wet sieving is given at about 5 µm because van der Waals attractive forces are becoming more and more important for decreasing particle diameters in relation to the mass forces. This leads to particle/particle and particle/media adhesion and impedes the passage of particles through the medium pores. The first precondition to find the largest pore is the existence of a sphere exactly of this diameter present in the sample. Secondly, this sphere must find the largest pore. A larger sphere must not block before this pore, and last but not least, the decisive sphere has to be found again in the sieve underflow. To get the best results as possible, special standard fractions of glass beads and special sieves for the analysis are available.

An indirect pore size measurement is given by the capillary flow porometry [10–12] or the bubble point test. The procedure for the bubble point test is described in the standards ASTM F316, ASTM E1294, ISO 2942, DIN BS 3321, and DIN EN 14898. This method is applicable for any pore system, is relatively fast, the largest pore can be detected explicitly, the pore size distribution can be measured, and complete filter cartridges can be analyzed. The physical background of this method is the overcoming of the capillary pressure p_{cap} in wetted pores by an externally applied gas pressure difference Δp. This capillary pressure is correlated with the hydraulic pore diameter d_h, the surface tension of the liquid γ_L, and the cosine of the wetting angle δ by the Young–Laplace equation. Solving this equation for d_h leads to Eq. (9.2)

$$d_h = \frac{4 \cdot \gamma_L \cdot \cos \delta}{p_{cap}} \frac{4 \cdot \gamma_L \cdot \cos \delta}{\Delta p - \rho_L \cdot g \cdot h_L} \tag{9.2}$$

As defined previously, the hydraulic diameter of a pore represents the diameter of a circular pore of the same capillary pressure, as the investigated pore of any shape. The method is limited to great pores of about 500 µm diameter because the precise measurement of very low capillary pressures is limited toward very low capillary pressures in the range of some pascal. On the other hand, the method is limited toward pore sizes in the nanometer range because a gas flow is then caused only by diffusion at high gas pressures in the megapascal range and not more by overcoming of the capillary pressure and convective gas flow. This leads to a misinterpretation of the observed gas bubbles.

The principle arrangement of a bubble point analyzer is shown in Figure 9.17.

The filter medium sample of a few square centimeter is fixed in a measuring cell. A liquid quantity of known surface tension, which must wet the sample perfectly, is given into the cell on top of the sample. The liquid layer height h_L should be adjusted to about 1 cm but must be known exactly to calculate its

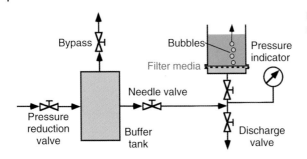

Figure 9.17 Bubble point measurement apparatus.

hydrostatic pressure. The perfect wetting is necessary for having a contact angle of 0° between solid and liquid. In this case, $\cos \delta = 1$, which makes Eq. (9.2) more simple. Now, a gas pressure is applied from the backside of the filter medium sample, which is increased very carefully by a needle valve. If the gas pressure overcomes the lowest capillary pressure of the largest pore, gas flows in the form of bubbles through this pore. The corresponding gas pressure behind the filter medium sample and the atmospheric pressure has to be measured to know the exact gas pressure difference. This value has to be corrected by the hydrostatic pressure of the liquid layer h_L on top of the filter medium sample. Measuring the corresponding gas pressure, the pore diameter can be calculated. It is important to know exactly the value of the surface tension of the wetting liquid, which depends on the temperature. Eventually, the surface tension should be measured for control purposes in parallel.

Beside the exclusive measurement of the bubble point of the largest pore in a filter medium, the entire pore size distribution can be measured by the same physical principle. The measurement is similar to the measurement of the capillary pressure distribution of a filter cake (cf. Section 8.3.4.4) by gas pressure method. After overcoming the bubble point, the pressure difference between both sides of the sample is stepwise increased, until no more pores are desaturated. The pressure difference at each step corresponds to the capillary pressure of a certain pore size. This pore size can be calculated, as discussed above according to Eq. (9.2). Simultaneously with the applied pressure difference, the corresponding gas flow is measured to quantify the respective amount of pores. Unfortunately, the gas volume flow increases after a pressure increase due to two reasons. Additionally opened pores are the first reason. The second reason is given by the increased gas flow due to the increased pressure. To be able to separate these two components and to consider exclusively the gas flow due to the opening of the pores, one has to realize first a so-called wet run to desaturate the pores and afterward a so-called dry run through the already opened pores. Figure 9.18 illustrates this procedure schematically.

In practice, sometimes originate deviations from the ideal behavior, if a small part of the liquid is evaporated during the permeation or if the sample is not completely wetted [13, 14]. Such measurements are carried out today with different types of automated porometers, which are working either top-down with a wetted sample or bottom-up with a wetted and liquid superposed sample, as shown in Figure 9.17.

Figure 9.18 Wet and dry run to determine the pore size distribution of filter media.

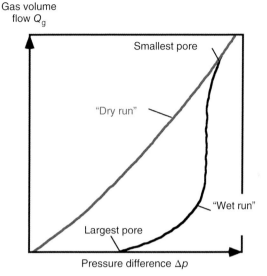

9.3.3 Theoretical Approach to Correlate Bubble Point and Largest Penetrating Sphere

The question is now how the indirectly measured hydraulic pore diameter d_h and the relevant pore diameter d_{max} of the largest sphere, which can penetrate the pore, are correlating with each other. It can be assumed that there will exist a certain correlation factor k, which allows, according to Eq. (9.3), calculating d_{max} from the measured d_h

$$d_{max} = k \cdot d_h \tag{9.3}$$

Beside methods of numerical simulation [15], a relative simple approach to determine k consists in the formulation of a force balance vertical to the pore cross section. At first, this should be carried out for a circular pore, as shown in Figure 9.19.

The real pore diameter d and the diameter of the sphere d_{max} obviously are the same. The mentioned force balance can also validate this, as can be seen from Eq. (9.4)

$$p_{cap} \cdot \frac{\pi}{4} \cdot d^2 = \gamma_L \cdot \pi \cdot d \Rightarrow p_{cap} = \gamma_L \cdot \frac{4}{d} = \gamma_L \cdot \frac{4}{d_h} \tag{9.4}$$

From Figure 9.19 it becomes clear which are the sizes of the hydraulic diameter d_h (Eq. (9.5)) and the largest penetrating sphere (Eq. (9.6))

$$d_h = d \tag{9.5}$$

$$d_{max} = d \tag{9.6}$$

The resulting correlation factor k is given in Eq. (9.7) as a ratio of d_h and d_{max}

$$\frac{d_h}{d_{max}} = 1 \Rightarrow k = 1 \tag{9.7}$$

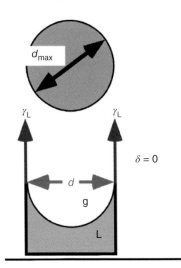

Figure 9.19 Circular pore.

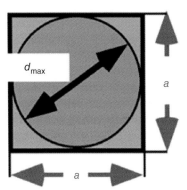

Figure 9.20 Square-shaped pore.

The calculated result correlates with the expectation. In a second step, a square-shaped pore, as depicted in Figure 9.20, should be investigated.

Analogous to Eq. (9.4), the force balance has to be formulated for this case in Eq. (9.8).

$$p_{cap} \cdot a^2 = \gamma_L \cdot 4 \cdot a \Rightarrow p_{cap} = \gamma_L \cdot \frac{4}{a} = \gamma_L \cdot \frac{4}{d_h} \tag{9.8}$$

As described in Eqs. (9.9)–(9.11), the hydraulic pore diameter and maximal diameter of the sphere correspond to the side length of the square and thus correlation factor again results in $k = 1$.

$$d_h = a \tag{9.9}$$

$$d_{max} = a \tag{9.10}$$

$$\frac{d_h}{d_{max}} = 1 \Rightarrow k = 1 \tag{9.11}$$

Also in this case, the bubble point diameter is identical with the largest sphere, which can penetrate this pore.

Figure 9.21 Triangular-shaped pore (equal sided).

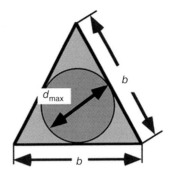

A third theoretical experiment is made with a triangular-shaped pore (equal sided), as shown in Figure 9.21.

The calculation of k (Eqs. (9.12)–(9.15)) leads to the same result as in the cases before

$$p_{cap} \cdot \frac{\sqrt{3}}{4} \cdot b^2 = \gamma_L \cdot 3 \cdot b \Rightarrow p_{cap} = \gamma_L \cdot \frac{12}{b \cdot \sqrt{3}} = \gamma_L \cdot \frac{4}{d_h} \quad (9.12)$$

$$d_h = \frac{\sqrt{3}}{3} \cdot b \quad (9.13)$$

$$d_{max} = \frac{\sqrt{3}}{3} \cdot b \quad (9.14)$$

$$\frac{d_h}{d_{max}} = 1 \Rightarrow k = 1 \quad (9.15)$$

As an example for a nonsymmetric geometry, a pore in the form of a slot, according to Figure 9.22, can be investigated.

Now, the correlation factor k (Eqs. (9.16)–(9.19)) depends on the ratio of length d and width c of the slot

$$p_{cap} \cdot c \cdot d = \gamma_L \cdot 2 \cdot (c + d) \Rightarrow p_{cap} = \gamma_L \cdot \frac{2 \cdot (c + d)}{c \cdot d} = \gamma_L \cdot \frac{4}{d_h} \quad (9.16)$$

$$d_h = \frac{2 \cdot c \cdot d}{(c + d)} \quad (9.17)$$

$$d_{max} = c \quad (9.18)$$

$$\frac{d_h}{d_{max}} = \frac{\frac{2 \cdot c}{\frac{c}{d} + 1}}{c} \Rightarrow k = \frac{\frac{2 \cdot c}{\frac{c}{d} + 1}}{c} \quad (9.19)$$

Figure 9.22 Slotted pore.

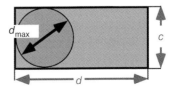

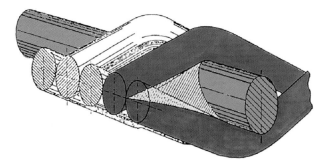

Figure 9.23 Single plain dutch weave pore.

If $d = c$, the pore is a square again and $k = 1$. If $d \to \infty$, $k = 2$. This means that in the extreme case of a slot of infinite length, the bubble point diameter is twice as large as the largest sphere, which can penetrate the slot. If c and d are known, k can be calculated exactly.

After two-dimensional pore geometries, which can be realized in the form of etched screens, three-dimensional weave structures of fabrics, which in practice are used for cake filtration processes, have to be included in the considerations. One example is shown in Figure 9.23 in the form of a single plain dutch weave.

The cut size is given here for the smallest cross section of the greatest pore. From this follows that one have to look for the smallest boundary surface, which can be placed into this cross section, because this determines the capillary pressure, which has to be overcome to desaturate the pore. This cross section is given by the shortest connection between the contact points of warp and weft threads. The result is a triangle with slightly inward bent sides. The force balance in Eq. (9.20) leads to correlation factors k, which are near to 1 depending on the curvature

$$\frac{d_h}{d_{max}} \approx 0.95 \Rightarrow k \approx 0.95 \tag{9.20}$$

Because of the fact that a fabric of monofilament wires is a geometrically well-defined object, the exact geometry can be determined by knowing the weave parameters in every case [16, 17]. The results for a multifilament fabric may differ a little bit from those of a monofilament.

9.3.4 Experimental Validation of the Theoretical Findings

Different types of etched sieves and woven fabrics had been investigated with microscopic pore size analysis, sieving with glass beads, and bubble point measurement [18]. The results had been compared to each other and to the theoretical expectations.

In Figure 9.24, bubble point and microscopic measurements for circular and square-shaped pores of etched screens are plotted against each other.

As can be seen clearly, all results are located with high accuracy on a 45° straight line. This means that bubble point measurements and image analysis give the same values for d_{max}. For each pore size, several pores had been measured and

Figure 9.24 Bubble point and microscopic analysis of circular and square-shaped pores.

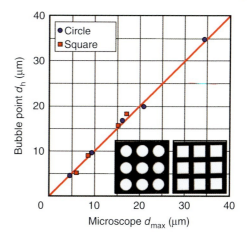

Figure 9.25 Bubble point and microscopic analysis of slotted screens.

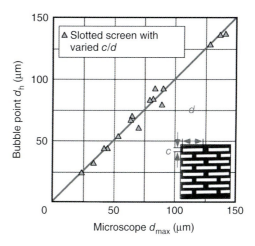

the averaged data had been used. The standard deviation had been very small because of the fact that here high-precision screens for particle size analysis had been used.

The same can be observed for slotted screens with a different ratio of length d and width c in Figure 9.25.

Again, the results from the bubble point measurement and the microscopic analysis are equivalent, if the factor k was calculated according to Eq. (9.19).

After the validation of the theoretical findings for cylindrical pores of different shapes (circle, square, and slot), three-dimensional monofilament fabrics must be investigated. In Figure 9.26, the results of bubble point, microscopic, and glass bead sieving tests for plain weaves of different pore size are plotted. All the methods are leading in principle to the same results. The results for sieving are shifted very slightly to smaller pore sizes in comparison to the bubble point and microscopic measurements. This is a hint that as discussed before, sieving can only approximate the largest pore.

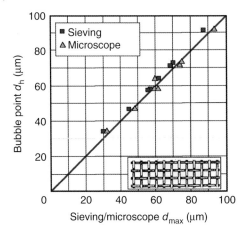

Figure 9.26 Bubble point, microscopic, and sieving analysis of plain weaved fabrics.

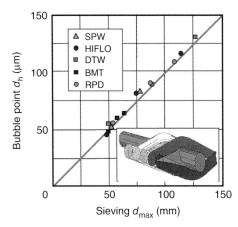

Figure 9.27 Bubble point and sieving analysis of lace-weaved fabrics.

The same result can be observed for bubble point and sieving experiments with various dutch-weaved fabrics of different pore size. Figure 9.27 documents that it is indeed possible to determine the largest sphere, which can penetrate the fabric, precisely by the bubble point measurement.

As can be seen from the theoretical approach and the experimental results, it is possible to determine the diameter of the largest penetrating sphere with the bubble point method, if the geometry of the pore cross-sectional area is well known. This is relatively easy in the case of fabrics because there exists a defined weave structure. More difficult is the determination of the pore size for fleeces and felts because the pore structure is much more complicated and formed by chance. As a consequence, different geometries of pore cross sections will be present in such structures, which lead to different k-factors. Here, modern simulation tools can help to generate nonwoven structures and to determine different pore cross sections by connecting all contact points between the fibers, which form a pore cross section. For safety, the greatest found value for k should be taken.

9.4 Semipermeable Filter Media – Gas Pressure Filtration Without Gas Flow

9.4.1 Introduction

Continuously operating vacuum filters, such as drum or disc filters, need a suction pump to generate the necessary gas pressure difference. If the vacuum overcomes the capillarity in the pores of the filter cake, the liquid is displaced and a gas flow originates through the emptied pore channels. This gas flow must be exhausted to maintain the filtration pressure. The separation of the gas–filtrate mixture after the filter apparatus needs filtrate receivers, from which the gas has to be exhausted by a vacuum pump. In the case of incompressible filter cakes with moderate capillary pressure, the gas flow can be normally managed without problems by the vacuum pump of the filter. If filter cakes show moderate compressibility, shrinkage cracks can occur in the cake during desaturation. In that case, a huge gas flow normally leads to a breakdown of the filtration pressure, and vacuum filtration is not more possible. In any case, the vacuum pump is consuming the main part of the energy, necessary to run the filtration process. By using semipermeable filter media, which are exclusively permeable for the liquid, a complete renouncement of vacuum pumps could be theoretically achieved. For that, the filter media have to be microporous and hydrophilic, if the liquid molecules exhibit polar properties, as in the case of aqueous slurries. If in that case the capillary pressure in the membranes pores is greater than the applied gas pressure difference, the filtrate can pass the membrane, but gas would be stopped. This would not only be interesting to save energy, but also to avoid gas cleaning after the vacuum pump in the case of present solvent vapor or to avoid gas recirculation in the case of using costly gases, such as nitrogen. A further advantage to use microporous filter media would be the production of particle-free filtrates. Filtrate pollution in the very first moment of cake formation, as in the case of using conventional filter clothes, would not occur. There is only one filter apparatus commercially available today, which is operating on this principle, but alternative variants are still under development. The already existing technology in the market is based on the principle of a vacuum disc filter, but instead of the conventional sectors, covered by a filter bag of woven fabrics, now microporous sintered ceramic sectors are installed. Figure 9.28 illustrates the principle filter design [19, 20].

The filter media surface is robust and even, which makes it possible to discharge the filter cake by scrapers without gas blowback. The regeneration of the filter media is possible via back-flushing, ultrasonic, or chemical treatment. Because of the micropores of about 1 µm diameter and the excellent wetting behavior of the ceramic material, the capillary pressure allows nearly no gas flow into the filtrate system during vacuum filtration. Low energy consumption and particle-free filtrate are the consequence. Because of the fact that the microporous ceramic plates must have a certain thickness because of the stability requirements, the filter medium resistance cannot be neglected, and in the case of relative permeable filter cakes, the filter media resistance can dominate

Figure 9.28 Vacuum disc filter with microporous ceramic filter media. Source: Courtesy of OUTOTEC Oyj.

the total flow resistance of the system medium/cake. This influences the cake formation equation (cf. Eq. (6.24)).

To minimize the filter media resistance under the condition of semipermeability and to make the specific solid throughput directly comparable to conventional vacuum filters, thin polymeric membrane filter clothes could be a promising approach. Such media are highly flexible and not in danger to break, as brittle ceramics. On the other hand, such media are not as robust as ceramics. This alternative process to suppress the gas throughput during cake desaturation has been validated already in a lab-scale pressure filter cell and a pilot-scale drum filter but is not yet available commercially. There are still some open questions and not solved problems regarding the filter design and the development of membrane filter clothes, which are really stable enough to overcome a sufficient long service life, is not yet finished [21].

9.4.2 Concept of Gasless Filtration on Vacuum Drum Filters and Physical Background

In Figure 9.29, a conventional vacuum drum filter including the necessary peripherals is shown schematically.

The circumferential surface of the filter drum is divided into rectangular filter cells, which are conventionally covered by a woven filter fabric. This filter fabric can be normally penetrated by liquid and by gas. If the filter cake can be desaturated and the pores of the filter media exhibit at least the pore size of the filter cake, the gas is able to overcome the capillary pressure in the medium pores too and gas can flow through. The filter cells are connected via filtrate pipes and the control head of the filter with the vacuum system (cf. Figures 6.26 and 6.27). The drum rotates continuously through a stirred filter trough, in which the slurry to be separated is fed in. The formed filter cake can be optionally washed after emerging from the slurry and is desaturated afterward because of the applied gas pressure difference. Gas flows through the emptied pores of the filter cake into the vacuum system. To maintain the filtration pressure difference, this gas must be exhausted

Figure 9.29 Conventional vacuum drum filter installation.

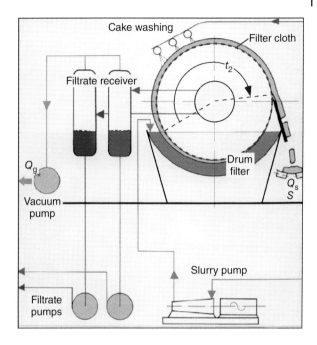

by the vacuum pump. In the cake discharge zone, the filter cake is finally removed by a supporting slight gas blowback and taken over by a scraper blade. The filtrate is separated from the gas in filtrate receivers. It must be withdrawn from these against the negative pressure by filtrate pumps. Vacuum pump and filtrate pumps are consuming most of the electrical energy to operate the filter apparatus. The more permeable the filter cake is, the more gas must be exhausted, and the more energy is consumed (cf. Section 8.3.4.6).

From the view of process technology and economics, it would be desirable to prevent the gas flow and to discharge the liquid without pumps from the vacuum system. One first step in this direction results with accordingly suitable architectural conditions in the installation of a so-called barometric leg, as sketched in Figure 9.30.

If in the filtrate receiver the absolute pressure amounts 20 kPa and the external atmospheric pressure amounts 100 kPa, a water column of 8 m heights generates a hydrostatic pressure of 80 kPa. As a consequence, additional separated liquid from the filter is able to flow unhindered out of the drainage pipe outlet. Therefore, the filtrate pumps become unnecessary. Nevertheless, the vacuum pump to generate the required negative pressure is still needed to exhaust the gas flow, which is caused by the cake desaturation.

If the filter medium would have semipermeable properties and would only let the liquid pass through, no gas flow could occur, that needs to be sucked off. In the optimal case, exclusively the 8 m high water column of the barometric leg could generate the necessary vacuum behind the filter medium. In such case also, the vacuum pump would become principally superfluous, as shown in Figure 9.31.

As cake a discharge device, alternatively a scraper or a roller can be used, depending on the handling properties of the filter cake. Alternatively to the

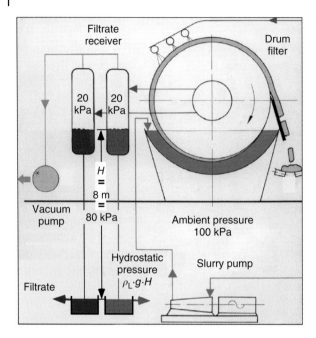

Figure 9.30 Barometric leg.

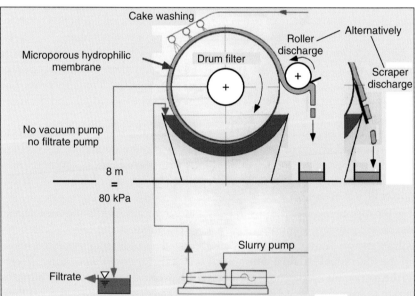

Figure 9.31 Gasless drum filter process.

filtration of easy to filter materials, this technology could be an alternative to dead-end microfiltration or centrifugal sedimentation of very small and thus hard to separate particles. If particles become smaller, the specific cake flow resistance rises up and only a very thin cake will form during the limited cake forming time on a rotary filter. Secondly, the capillary pressure rises up in such

Figure 9.32 Principle prerequisites for semipermeable membranes.

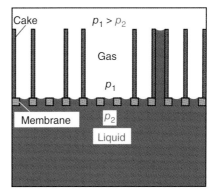

cases and eventually the cake cannot be desaturated properly anymore. Then, the cake properties change from brittle to pasty and sticky. Although such cakes cannot be desaturated well, the danger of shrinkage cracks exists and the microporous membrane in any case makes sure that no gas breakthrough occurs and the filtrate remains particle free. A roller discharge has proven as a safe method to detach such thin and pasty particle layers from the filter medium.

The semipermeability of the filter medium can be regulated via its pore size and wetting behavior because the capillary pressure depends on the pore size, surface tension of the liquid, and wetting angle (cf. Figure 8.33 and Eq. (8.38)). To withdraw liquid from the pore of the filter medium, a pressure difference must be applied from outside, which is greater than the acting capillary pressure. If the pores in the filter cake are greater than in the pores of the filter medium, the capillary pressure in the cake is smaller than in the filter medium. If the filtration pressure difference is chosen larger than the capillary pressure of the cake, but smaller than the capillary pressure in the filter medium, the gas can penetrate the cake and displace the pore liquid, but the filter medium remains fully saturated. Figure 9.32 illustrates these interrelations schematically.

Rain-protective clothing uses this effect inversely. An extreme hydrophobic polytetrafluoroethylene (PTFE)-membrane lets air and vapor pass through as the wetting fluid, whereby the capillary pressure of the micropores is so high that the impinging water droplets are not able to generate enough pressure to penetrate the membrane. As result of these considerations follows that from the theoretical point of view, a gasless filtration seems to be possible.

9.4.3 Realization of the Process in Lab and Pilot Scale

In the first step, suitable membranes are needed for the process. Suitable means that they have at first to be hydrophilic and must exhibit a capillary entry pressure or bubble point of at least 80 kPa for application on vacuum filters. Secondly, they have to be mechanically stable enough for the tests in the pressure filter cell and at the end on a rotary filter. For these purposes, hydrophilic membranes of different pore size and material had been tested in the lab scale with particle-free water regarding their capillary entry pressure. This pressure corresponds to the largest pore with the smallest capillary pressure and determines the pressure, up

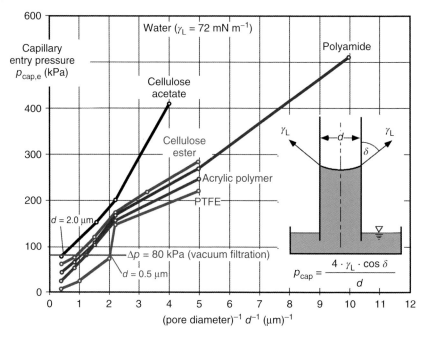

Figure 9.33 Membrane bubble point.

to which the membrane is able to hold back the gas completely. Water was used because aqueous slurries were taken in the focus. In Figure 9.33, the capillary entry pressure is plotted against the reciprocal value of pore diameter, given by the membrane manufacturers.

As expected for each membrane material, a more or less linear relationship between bubble point and inverse pore diameter could be observed. The different slope of the straight lines can be interpreted as a result of differences in the membrane wetting behavior and pore geometry. PTFE membranes, as typically hydrophobic materials, were included after making them hydrophilic with ethanol. Nevertheless, PTFE had shown the poorest results. From Figure 9.33, it can be derived that for a safe gas sealing against an external pressure of 80 kPa, pore sizes of less than 2 μm should be chosen for cellulose as best wetting material and less than 0.5 μm for hydrophilic made PTFE. According to the Young–Laplace equation for water, 25 °C, complete wetting ($\delta = 0$), and 80 kPa, theoretically a circular pore diameter of $d = 3.6$ μm would result.

In the next step, the question must be answered, whether the flow resistance of principally well-suited membranes may eventually represent a constraint in comparison to the filtration performance of conventional woven filter fabrics. For this purpose, the filter medium resistance of membranes and filter fabrics with different pore size were measured with particle-free water and various slurries. The results of these measurements already are shown in Figure 6.4. It could be demonstrated that there is of course a big difference between small flow rates for microporous membranes of 0.2 μm pore size and woven fabrics with pore sizes of more than 10 μm, if only pure particle-free water is filtered. If slurry is

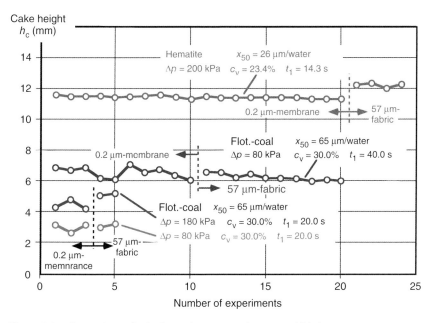

Figure 9.34 Comparison of cake formation on membranes and fabrics.

separated, in most cases, the flow resistance of the filter media itself plays only a subordinate role in comparison to the first bridge-building particle layer on the filter medium surface, which has to be assigned additionally to the filter medium resistance in practical cake filtration processes. The dominating flow resistance is generated by the bottleneck of very small pores between the particles, which are forming the bridges across the pores of the filter medium. If the micropores of the membrane are smaller than those in the cake, the bottleneck is of course located on the membrane itself, but after very short time, the absolute resistance of the cake is much greater and the membrane resistance again can be neglected.

These results find their expression in the real cake formation performance, which is documented exemplarily in Figure 9.34.

Membranes of 0.2 m pore size are compared with woven fabrics of 57 μm pore size for different materials and different pressure differences. The cake height for comparable conditions in all cases had been nearly the same. Some experiments had been repeated several times without special membrane cleaning and no spontaneous blockage of the membranes could be observed. These results are an important prerequisite for an economical cake filtration with microporous membranes because they represented in the investigated cases no constraint regarding the filter throughput in comparison to conventional filtration.

Because the filtrate flow during the cake-deliquoring phase is significantly smaller than during the cake formation, the membranes should also not hinder this process step. However, the question arises whether the air, which migrates toward the membrane and is stopped there, could have a negative impact on the filtrate flow. Figure 9.35 answers this question.

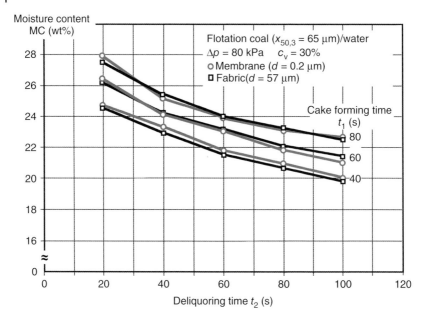

Figure 9.35 Comparison of cake deliquoring on membranes and fabrics.

It turns out that the deliquoring kinetics is behaving quite analogously for both filter media. From these results on the one hand, the proof is made that the microporous membrane is not impeding the deliquoring process, and on the other hand, it is demonstrated that a gas flow through the filter cake is not supporting the deliquoring. The only relevant mechanism to empty the pores is the overcoming of the capillary pressure by the externally applied gas pressure difference. This prerequisite has been made already previously in Section 8.3.4.2, when the desaturation kinetics of filter cakes has been discussed.

After the principle proof of the process in the lab scale, comparative experiments in pilot scale were tackled on a drum filter with a filter area of $0.7\,m^2$. This drum filter was installed in a pressure vessel to enable not only vacuum filter experiments but also an operation of pressure differences up to 180 kPa. It was accepted that the membranes had not been completely gas tight at the great pressure, but it was interesting to see how the membrane would react on greater pressure. For pressure differences of 80 kPa, the capillary entry pressure of the membranes had not been overcome. In this connection, the difficulty appeared that the available and in the lab scale with $20\,cm^2$ easy-to-handle membranes did not show sufficient mechanical stability in the pilot scale with the larger filter area of $0.7\,m^2$ and stronger mechanical impact during operation. As mentioned previously, enough robust polymeric filter media for such applications are not yet available on the market. For this reason, it was necessary to fix membranes on conventional filter fabrics in single manufacture and to generate by this measure a filter medium, which exhibited both the desired microporous structure and the required mechanical stability.

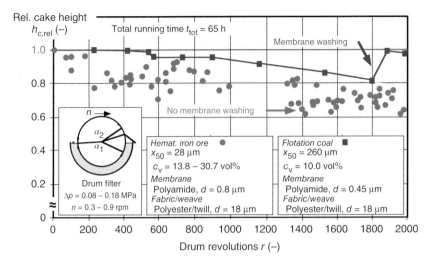

Figure 9.36 Comparison of cake formation for continuous filtration.

Figure 9.36 shows in the summarized form the results of the cake formation for 2000 drum revolutions during a one-week campaign for two different slurries and membrane filter media in comparison to a conventional woven filter fabric.

The operational parameters of the filter had varied several times during the test campaign. During the shutdown times, the membranes were protected from drying out to avoid crystallization of soluble substances in the membranes pores.

On the ordinate, the quotients of filtration with membrane filter cloth and conventional filtration with pure filter fabric are drawn, and on the abscissa, the number of drum revolutions realized during the test campaign. At a value of 1, the results of both filter media correspond exactly to each other. When the results located below 1, the cake thickness on the membrane is comparatively lower. For the flotation coal, the filtration results on the membrane got worse relatively slowly but reached after a membrane cleaning again the original full value. For the iron ore, the membrane blockage took place more quickly, and without washing of the membrane after long time, a stable result of about 70% of the conventional filter cloth was reached. From these results, it becomes clear that for the application of microporous membranes in the field of cake filtration processes, of course analogous phenomena can be observed, as well known from the cross-flow filtration. Membranes have to be cleaned periodically to maintain their filtration capacity. In the example, which is shown here, a combination of 250 μm mean particle size for coal and 0.45 μm membrane pore diameter is of course much more favorable concerning the danger of membrane blockage than the combination of 28 μm mean particle size for iron oxide and 0.8 μm membrane pore diameter.

Beside the solid throughput, the accessible residual moisture content of the filter cake is interesting as process result. The results for the cake moisture content during the test campaign, which is described above, are given in Figure 9.37.

The relative cake moisture content here again exhibits the value 1, if the results of membrane filter cloth and conventional fabric are identical. If the value becomes smaller than 1, the cake moisture for the membrane experiments

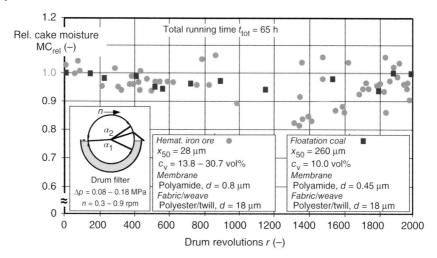

Figure 9.37 Comparison of cake moisture content for continuous filtration.

is lower than that of the cake on the conventional fabric. This is mainly caused by the different cake height. If the cake height is reduced, the moisture content of the cake is lower for the same desaturation time. The slightly increased membrane flow resistance due to some pore blockage is not the dominating factor for the filtrate flow.

For the flotation coal, a curve resulted, which corresponded with the results for cake formation. Identical moisture content resulted for identical cake height. Slightly decreasing moisture content could be observed for increased membrane resistance and thus decreasing cake height. Identical results were achieved again after membrane cleaning at 1800 revolutions of the drum.

For iron oxide, the cake moisture content could be held more or less constant. An interpretation for this phenomenon could be that here the increasing filter media resistance in combination with decreasing cake thickness has led to a compensation of negative and positive effects. There had been over the entire operating time some cakes, which had lower cake moisture on the membrane than on the conventional fabric. This may be caused by additional fluctuations from the filter operation such as variations of the slurry concentration, eventually rewetting of the cake during cake discharge or similar effects. However, in principle, the results obtained in the laboratory could be found and confirmed in the pilot scale for continuous filtration. Nevertheless, beside the optimization of the filter design, suitable membrane materials are needed in the future for durable filter operation.

References

1 Smith, J. (2015) Assessment of ceramic filtration for a metallurgical process, Dissertation, University of the Witwatersrand, Johannesburg.

2 Anlauf, H. (2015). Fest/Flüssig-Trennung auf der ACHEMA. *Chemie Ingenieur Technik* 85 (9): 1134–1145. https://doi.org/10.1002/cite.201500102.
3 Sun, D. and Stylios, G.K. (2006). Fabric surface properties affected by low temperature plasma treatment. *Journal of Materials Processing Technology* 173 (2): 172–177. https://doi.org/10.1016/j.jmatprotec.2005.11.022.
4 Morent, R., De Geyter, N., Verschuren, J. et al. (2008). Non-thermal plasma treatment of textiles. *Surface and Coatings Technology* 202 (14): 3427–3449. https://doi.org/10.1016/j.surfcoat.2007.12.027.
5 Anlauf, H. (2006). Fest/Flüssig-Trennung. *Chemie Ingenieur Technik* 78 (10): 1492–1499.
6 Anovitz, L.M. and Cole, D.R. (2015). Characterization and analysis of porosity and pore structures. *Reviews in Mineralogy & Geochemistry* 80: 61–164. http://dx.doi.org/10.2138/rmg.2015.80.04.
7 Landis, E.N. and Keane, D.T. (2010). X-ray microtomography. *Materials Characterization* 61 (12): 1305–1316. https://doi.org/10.1016/j.matchar.2010.09.012.
8 Rideal, G.R. and Sorey, J. (2002). A new high precision method of calibrating filters. *Journal of the Filtration Society* 2 (3): 18–20.
9 Rideal, G.R. (2006). Analysis and monitoring: how to improve precision pore size measurement. *Filtration & Separation* 43 (3): 28–29.
10 Gupta, J.K. (2002). Characterization of pore structure of filtration media. *Fluid/Particle Separation Journal* 3: 227–241.
11 ASTM F316-3 (2003). *Standard Test Method for Pore Size Characteristics of Membrane Filters by Bubble Point and Mean Flow Pore Test*. West Conshohocken, PA: ASTM International.
12 Li, D., Frey, M.F., and Joo, Y.L. (2006). Characterization of nanofibrous membranes with capillary flow porometry. *Journal of Membrane Science* 286 (1–2): 104–114.
13 Hernández, A., Calvo, J.I., Prádanos, P., and Tejerina, F. (1996). Pore size distributions in microporous membranes. A critical analysis of the bubble point extended method. *Journal of Membrane Science* 112 (1): 1–12.
14 Kolb, H.E., Schmitt, R., Dittler, A., and Kasper, G. (2018). On the accuracy of capillary flow porometry for fibrous filter media. *Separation and Purification Technology* 199: 198–205.
15 Herper, D. (2016). Verbesserung des Bubble-Point-Messverfahrens an Drahtgeweben mit Hilfe numerischer Berechnungsmodelle. *Filtrieren und Separieren* 30 (6): 389–392.
16 Tittel, R. and Berndt, R. (1973). Zur Bestimmung er Trennteilchengröße von Filtergeweben. *Faserforschung und Textiltechnik* 24 (12): 505–510.
17 Heidenreich, E., Tittel, R., and Nerndt, R. (1977). Zur Charakterisierung von Tressengeweben als Filtermitel. *Aufbereitungstechnik* 18 (7): 353–357.
18 Anlauf, H. (1996). Bestimmung der größten Pore in Filtermedien unterschiedlicher Struktur durch Messung des kapillaren Eintrittsdrucks. *Chemie Ingenieur Technik* 68 (11): 1476–1479.
19 Ekberg, B. and Woitkowitz, S. (1988). Cake-forming wet filtration without gas throughput with the new Ecosuc disc filters. *Aufbereitungstechnik/Mineral Processing* 29 (4): 193–196.

20 Woitkowitz, S. (1988). Ceramics fill a filter vacuum – theory and practice concerning cake-forming filtration without gas throughput. *Aufbereitungstechnik/Mineral Processing* 29 (10): 594–602.

21 Anlauf, H. (1990). Vakuum- und Druckfilter ohne Gasverbrauch. *F&S Filtrieren und Separieren* 4 (3): 135–145.

Nomenclature

Unless otherwise defined in the text, the symbols have the following meanings:

Latin Letters

a	acceleration	m s^{-2}
a	length	m
A	area	m^2
b	length	m
B	width	m
B	liquid–solid volume ratio	—
Bo	Bond number	—
c	concentration	—
c	specific heat	J kg^{-1} K^{-1}
c	length	m
c_w	drag coefficient	—
C	centrifugal number	—
CL	consolidation level	—
d	diameter, length	m
D	diffusion constant	m^2 s^{-1}
$D_{3,2}$	Sauter mean diameter (no spheres)	m
DS	dry substance	—
e	pore number	—
e	elementary charge	C
E	elasticity modulus	N m^{-2}
E	electrical field strength	V m^{-1}
Eu	Euler number	—
f	relative quantity of fine particles	

Wet Cake Filtration: Fundamentals, Equipment, and Strategies, First Edition. Harald Anlauf.
© 2019 Wiley-VCH Verlag GmbH & Co. KGaA. Published 2019 by Wiley-VCH Verlag GmbH & Co. KGaA.

F	quantity of misplaced particles	
F	free energy	J
F	force	N
Fr	Froud number	
g	gravity acceleration	m s^{-2}
g	relative quantity of coarse particles	—
h	height, length	m
h	enthalpy	J
H	Hamaker constant	J
i	number of filter areas	—
I	imperfection	—
l	length	m
L	length	m
k	number of contact points	—
k	floc factor	—
k	Boltzmann constant	J K^{-1}
k	correlation factor	—
K	kinetic parameter (gas pressure)	—
m	mass	kg
M	mass	kg
MC	moisture content	—
n	rotation speed	s^{-1}
n	number of molecules	—
Ne	Newton number	—
Q	cumulative particle distribution	—
Q	volume flow	m^3 s^{-1}
q	specific volume flow	m^3 m^{-2} s^{-1}
q	frequency distribution	m^{-1}
p	pressure	Pa
p	length	m
p	specific permeability (height related)	m^2
q_v	specific volume flow	m^3 m^{-2} s^{-1}

Q_v	volume flow	m³ s⁻¹
q_m	specific mass flow	kg m⁻² s⁻¹
Q_m	mass flow	kg s⁻¹
r	specific flow resistance (height related)	m⁻²
r	radius	m
R	radius	m
R	flow resistance	m⁻¹
R	universal gas constant	J mol⁻¹ K⁻¹
Re	Reynolds number	—
s	length	m
S	saturation	—
S	entropy	J K⁻¹
t	time	s
T	time	s
T	grade efficiency	—
T	absolute temperature	K
TC	dimensionless consolidation time	—
v	velocity	m s⁻¹
V	Volume	m³
w	velocity	m s⁻¹
w	specific energy	N
W	wash ratio	—
W	energy	N
x	particle diameter	m
$x_{av,SV}$	true Sauter mean diameter (spheres)	m
X	amount of remaining impurities	—
Y	solid–liquid mass ratio	—
z	ionic valence	
z	length	m kg⁻¹

Operators

Δ	difference operator

Greek Letters

α	specific resistance (mass related)	m kg^{-1}
α	angle	deg
β	bridge filling angle	deg
γ	angle	deg
γ	surface tension	N m^{-1}
δ	wetting angle	deg
ε	porosity	—
ε_0	absolute dielectric constant	A s V^{-1} m^{-1}
ε_r	relative dielectric constant	—
ζ	resistance coefficient	—
η	dynamic viscosity	kg m^{-1} s^{-1}
η	efficiency factor	—
ϑ	temperature	°C
κ	concentration parameter	—
κ	sharpness of cut	—
κ	isentropic coefficient	—
λ	kinetic parameter (centrifuge)	—
μ	chemical potential	J
ν	kinematical viscosity	m^2 s^{-1}
ξ	pore size distribution index	—
ρ	density	kg m^{-3}
σ_t	tensile strength	N m^{-2}
σ	standard deviation	depends on the investigated characteristic
σ^2	variance	depends on the investigated characteristic
φ	function	—
ψ	shape factor	—
ψ	electrical potential	V
ω	angular velocity	deg s^{-1}

Subscripts

ar	arithmetic
an	analytical
ag	agglomerate
ad	adhesion
av	average
A	feed material
A	area
b	buoyancy
br	break through
bs	bench scale
belt	belt
c	cake
cap	capillary
cr	critical
cent	centrifugal
cl	closed
cl	cleaning
d	drag
d	down
dr	drainage
ds	dissolved substance
D	Debye
e	entry
eq	equivalent
eo	electro-osmotic
f	feed
fi	film
F	Feret
F	force
F	fine particles
FP	filter press
g	gas
g	gravity

g	weight
g	geometric
G	coarse particles
h	hydraulic
i	dispersity characteristic index
i	inner
l	quantity index
l	line
L	liquid
m	filter medium
max	maximal
mech	mechanical
min	minimal
mod	modal
mol	mole
M	Martin
M	meniscus
n	number
N	standard conditions
o	outer, outlet
op	open
p	particle
p_i	piston
pr	pressure
P	projection
PF	pressure filter
r	remanent
r	quantity index
re	removal
rf	rotary filter
rel	relative
s	solid
s	Stern
sf	settling front

sed	sedimentation
squ	squeezing
S	surface
SV	volume-related surface
St	Stokes
t	cut point
t	tensile
therm	thermal
tot	total
v, V	volume
vac	vacuum
vdW	van der Waals
void	void
w	wash
W	Wadell
vap	vapor
0	number
0	reference value
1	length
1	cake formation process
2	area
2	cake deliquoring process
3	volume
50	median
3pcl	three-phase contact line
∞	infinite

Index

a
agglomeration   5, 8, 24, 45, 52, 53, 70–76, 78, 83, 87, 104, 107, 115, 205, 300
atmospheric pressure   79, 87, 303, 314, 323

b
barometric leg   323
belt press   142, 275, 276
bond curve   266
bond number   265
Bo number   266, 269
"Boozer"-vacuum disc filter   119
bubble method   214
bubble-point measurement apparatus   313, 314
bubble point test   312, 313
Bucher press HPX5005   140
buoyant force   213

c
cake deliquoring
  cake moisture content   250
  gas consumption   250
  on hyperbaric drum filter   249
cake desaturation
  in centrifugal field
    cake moisture   262
    capillary pressure curves   265
    equilibrium   261–267
    kinetics   267–268
    piston-film model   267
  in centrifuge beaker   262, 263
  by gas pressure difference
    capillary pressure curve for homogeneous filter cake   231
    capillary pressure curve for inhomogeneous filter cake   232
    deliquoring   235
    desaturation matrix   240
    equilibrium   231–234
    gas flow through filter cakes   240–246
    kinetics   234–240
    relative liquid and gas permeability   237, 238
    residual cake moisture   233
    slurry concentration   233
  by steam pressure filtration   257
    function   258
    principle   258
    steam pressure with combined mechanical/thermal deliquoring   261
    temperature profile   259
  on vacuum drum filter   248
cake discharge   5, 43, 84, 107, 109, 110, 113, 120, 121, 141, 151, 154, 155, 157, 159, 185, 193, 276, 282, 300, 307–309, 323, 330
cake filtration process   1, 224
  apparatuses   6, 92
  application   3
  basic principle   87
  centrifuges
    design and operation   152–169
    fundamental   146–152
  compressible filter cake formation
    fundamental   123–130

cake filtration process (contd.)
   optimization 133–136
   parameters characterization
     130–133
  effective driving pressure 87
  equipment 7
  incompressible filter cake formation
   filters design and operation
    113–123
   parameters characterization
    98–104
   principle model 94–98
   throughput of continuous vacuum
    and pressure filters 108–113
   throughput of discontinuous cake
    filters 104–108
  mechanisms 88
   characteristic value 93
   complex superposition 88
   constant solids mass 90
   deposition of particles 88
   filter medium resistance 91, 93
   flocculation 89
   flotation coal 92
   flotation coal slurry 89
   Hermans and Bredée approach 94
   linear representation 94
   phase of bridge formation 88
   pre-concentration 89
   rotary filters 90
   rotation speed 91
   solids mass 89
   special candle filter 92
  principal modeling 88
  steps 2, 87
cake moisture, conversion of parameters
   207
cake permeability 62, 65, 76, 94, 95, 97,
   98, 107, 148, 149, 239
cake structure
  during formation process 41
  ideal and real 61
  incompressible and compressible
   124
  inhomogeneous 52
  particle arrangement 49–52
  pore size 52–54

  porosity
   agglomerate 45
   cubic and rhombohedral
    arrangement 46
   elementary cell volume 47
   function 48
   irregular shaped 43
   macro-pores 45
   packing density 47
   particle 44
   pycnometric method 43
   solidosity 44
   typical values 43
   van-der-Waals forces 48
   void volume 42
   zeta-potential 48
  porous layers 41
  principle 41
  real geometry 41
cake washing 41, 87, 105, 138, 179,
   180, 192, 196, 200, 204, 206
calendering process 308
capillary entry pressure 52, 68, 71, 84,
   229, 230, 232, 233, 239, 247, 253,
   259, 266, 278, 280, 282–286, 325,
   326, 328
capillary flow porometry 41, 54, 312,
   313
capillary force 35, 83, 114, 115, 122,
   213, 235, 265, 266, 269, 280, 283
capillary method 214
capillary pressure 43, 71, 76, 83, 84,
   200, 203, 205, 214, 222–231,
   233–236, 239, 246–248, 254, 256,
   258, 262–265, 270, 303, 313, 324,
   325
capillary pressure curve 228–231, 234,
   246–248, 264, 265, 279, 282, 285
capillary pressure distribution, for ZnS
   filter cakes 256
cellulose 4, 5, 15, 81, 82, 134, 302, 326
centrifugal desaturation 268
centrifugal filtration
  acceleration 146
  applications 147
  batch centrifuges 150
  cake permeability 149

C-number 147
  design and operation 152–169
  geometrical parameters 149
  light transmission profiles 148
  pressure 146
  principle 146
  radiographic analysis 148
ceramic filter element, for disc filters 302
circular model pore 225
circular pore 312, 313, 315
co-current washing 177, 178
compressible cake structures 123, 128, 205
compressible filter cakes
  capillary pressure curve for 278
  consolidation of
    average porosity 272
    filter design and operation 274–278
    fundamental considerations 271–274
    mechanisms 273
    porosity gradient 272
    uniaxial pressing and shearing 271
  deliquoring by squeezing 205
  by gas differential pressure
    cake porosity 285
    cake shrinkage 285–288
    consolidation/desaturation 278–285
    deliquoring 282
    drum filter with vibrating press belt 290
    liquid bridge 281, 282
    mechanical and thermal cake deliquoring 283
    residual cake moisture 285
    transferable tensile strength 280
    vibration/squeezing 290
compression energy 243–245, 257
conductivity, of filtrate 181
conical disc filter cell 120
consolidation level (CL) 272, 273
continuous cake filtration process 106
continuous centrifugal cake filters 303

conventional microscopic analysis 312
counter-current washing 178, 192, 193
  drum filter for 192
critical micelle concentration (cmc) 210
cross-flow filtration 183, 302, 329
Cubic structure 46

d

Darcy equation 60, 61, 93, 103, 124, 133, 237, 240, 242, 266
Darcy law 256
Debye length 74, 76
deliquoring process
  characterization 206–208
  of compressible filter cakes by squeezing 205
density separation 3, 4, 176
desaturation 203
  kinetics 203, 234, 235, 239, 249, 262, 267, 328
desaturation, of filter cakes
  boundary surface 208
    even and bent 222
    between liquid and gas 208
    surfactant migration 210
  capillary pressure 222
    circular model pore 225
    deformation of boundary surface 223
    distribution 228
    force balance for piece of boundary surface 222
    pore channel of changing diameter 225
    positive and negative 222
    two-dimensional pore model 227
    vapor pressure 224
  contact angles 216
    for dispersed particles 221
    dynamic process 219
    hysteresis effect 216
    liquid bridges 218, 219
    particles of interest on object plate 221
    plate method 219, 220
    press force on 220

desaturation, of filter cakes (*contd.*)
  for sharp edges and corners 218
  non-wetting 216
  soil-washing 216
  surface tension 208
    bubble method 214
    capillary method 214
    of different liquids 209
    dynamic 212
    as function of surfactant
      concentration 210, 211
    plate method 213
    ring method 212
    of water 210
  surfactants
    adsorption equilibrium 211
    concentration gradient 211
    migration 210
  three-phase contact line 215
    deformation of solid body 215
    force balance 215
  wetting
    definition 216
    pore geometry 218
diafiltration, in fed batch mode 183
diaphragm filter press 275
  heat transfer plate for 275
diatomaceous earth 4, 5, 15, 80–82, 302
dilution washing 176, 182
  diafiltration in fed batch mode 183
  in hydrocyclones 183
  paddle washer 184
  in single cake filtration apparatus 185
  stirred pressure nutsche filter 186
  two-stage co-current dilution washing 185
dimension matrix 58
disc filters, ceramic filter element for 302
discontinuous cake filtration process 106
dispersion 182, 186
  in porous beds 187
displacement 80, 186–191, 214, 236, 237, 247, 258, 259, 283

dried harbor sludge, shrinkage cracks in 204
drum filter
  for counter-current washing 192
  filtrate discharge 253
dutch weaves 304
  maximal cut size 304
  three-dimensional weave technology of 305

e
electrically enhanced press filtration 292–293
energy consumption, for mechanical and thermal cake deliquoring 246–248
extraction process 176

f
Fest pressure filter 201
FEST rotary pressure filter 178, 192–194
fibrilized cellulose fibers 81
filter apparatuses 1, 6, 32, 41, 65, 76, 104, 105, 136, 178, 193, 201, 232, 251, 282, 299–301, 321, 323
filter cakes 176
  fingering effects during permeation washing 188
  gas consumption matrix 244
  homogeneous 187
  pore size distribution in 188
  shrinkage cracks in 192
  shrinkage of 198
filter cloth 92, 108, 110, 120, 123, 135, 137, 157, 159, 197, 199, 232, 309–312, 321, 322, 329
filter media
  belt filters 301
  ceramic filter element for disc filters 302
  depth 301
  in double wire presses 301
  drainage system 300
  filter cake deliquoring 300
  flow resistance 299, 310
  function 299

interactions with slurry properties 299
particle retention 299
porometry 310–320
semipermeable 321–330
surface filtration 302
types 301
vacuum belt filter 302
woven filter media 304–310
filter press cycle 138
filtration kinetics, in filter presses 140
fingering effect 188, 227, 228, 236, 258
flocculation 45, 75, 80, 89, 196
flow resistance 4, 15, 50–54, 57, 68, 91, 97, 98, 120, 125, 127, 129, 130, 132–135, 151, 152, 156, 197, 299, 310, 312, 326, 327, 330
FrameTrak system 310
friction forces 60, 197

g

gas differential pressure, incompressible cake desaturation by 203
gas pressure 4, 41, 87, 100, 159, 201, 203, 219, 226, 231, 234–236, 243, 246, 261, 270, 278, 283, 314
Gibbs equation 208
glass beads 54, 62, 92, 221, 312, 313, 318
grade efficiency 28, 31

h

Hagen–Poiseuille equation 266
HiBar drum filter 111
HiBar oyster filter 112
homogeneous filter cakes 42, 52, 103, 139, 187, 231, 232, 243
horizontal filters 32, 115, 187
horizontal peeler centrifuges 153, 154, 157, 199
horizontal vacuum belt filters 192
hydrocyclone 4, 25, 26, 31, 66, 70, 76, 77, 164, 183
particle washing in 183
hydrostatic pressure 87, 119, 142, 146, 159, 214, 224, 314, 323
hyperbaric disc filter 110, 111

hyperbaric filter 110, 111, 192, 193, 201, 249

i

ideal sharp fractionation 27
incompressible cake desaturation, by gas differential pressure 203
influencing parameters 41, 58, 60, 94, 213, 239
integral material balance 26, 27
isentropic coefficient 243

k

Kaolin particles 15, 16
Kelvin-equation 224

l

lab filtration unit 98
laminar flow 30, 60, 66, 69, 127, 187
limestone filter cake, shrinkage cracks in 204
liquid bridges 203, 218, 219, 229, 237, 266, 280, 281
liquid flow through porous particle layers
 dimension analytic approach 57–61
 empirical approach 61–63
liquid properties 13, 14
liquid saturation degree 206, 228
liquid viscosity, permeation washing 188
lotus-effect 217

m

magnetic resonance imaging (MRI) 133, 258
matrix transformation 58, 59
maximal pore size 311, 317
microporous ceramic plates 321
microporous membranes 4, 5, 90, 91, 99, 226, 248, 299, 302, 305, 325–329
mode of filtration 96, 97
moving bed washer 190

n

non-wetting 218
 definition 216

p

paddle washer 184
parallel cylindrical tubes 225
particle arrangement 41, 42, 46, 49–52
particle porosity 44
particle retention 299, 300, 304, 305, 310
particle washing
  characterization 180–182
  co-current 178
  counter-current 178
  dilution 176, 182
  limits 178–180
  permeation 176, 186
  principles 176–178
particles
  collectives
    characterization 20–24
    fractionation 24–31
  properties 14–31
  single, of characterization 16–19
peeler centrifuges 10, 151, 153–159, 199, 267, 269, 270, 289
permeation washing 176, 181
  dispersion in porous beds 187
  drum filter for counter-current washing 192
  FEST rotary pressure filter 193
  on filter presses 196
  fingering effect 188
  in horizontal peeler centrifuges 199
  on lab scale diaphragm filter press 194
  liquid viscosity on 188
  moving bed washer 190
  pusher centrifuges 194
  shrinkage cracks in filter cake 192
  shrinkage of filter cakes 198
  stagnant locations near filter media 197
  steam cabin 200
  steam pressure filtration 201
  steam washing and deliquoring 200
  vacuum belt filter 192
  vacuum pan filter 193
  in ventrical filter centrifuges 196
  washing time, influence of 189
  wash ratio, influence of 189
piston-film model 267
plate method
  contact angles 219, 220
  surface tension 213
poreless membranes 4
pore size 52, 54
  determination
    capillary flow porometry 313
    conventional microscopic analysis 312
    direct measuring methods 312, 313
    glass beads 313
    indirect measuring methods 312, 313
  distribution 311
porometry
  bubble-point and microscopic analysis
    circular pore 316
    slotted pore 317
    square shaped pore 318
  bubble-point and sieving analysis
    dutch weaved fabrics 320
    lace weaved fabrics 320
  bubble-point correlation 315–318
  bubble-point measurement apparatus 313
  circular pore 315
  maximal pore size 311
  pore size determination 312–315
  pore size distribution, wet and dry run 314
  single plain dutch weave pore 318
  slotted pore 317
  square shaped pore 316
  triangular shaped pore 317
porosimetry 41
porosity
  agglomerate 45
  for different pressure 136
  elementary cell volume 47
  function 48, 51
  irregular shaped 43
  macro-pores 45
  packing density 47

particle   44
pycnometric method   43
solidosity   44
typical values   43
van-der-Waals forces   48
void volume   42
zeta-potential   48
press force, on contact angles   220
press nips   143, 276, 277
press rollers   276
  cake shearing   277
pressure candle filter   113, 114
pressure filter cell   98–101, 109, 124, 130–132, 233, 246, 247, 249, 265, 285, 286, 322, 325
PTFE membranes   325, 326
pusher centrifuges   8, 14, 65, 160–162, 164, 165, 167, 194, 269

r

real technical fractionation   27
relative gas permeability   237, 242, 243
residual cake moisture   46, 69, 77, 135, 144, 175, 207, 233, 246, 252, 254, 255, 257, 269, 279, 282
Reynolds-number   59
rhombohedral structure   46
ring method   212
Rosin–Rammler–Sperling–Bennet (RRSB)   230
rotary filters, throughput and cake moisture of   251–257

s

scraper discharge   5, 324
screw press   134, 143, 144, 275, 277
sedimentation analysis   17, 312
semipermeable filter media
  cake deliquoring on membranes and fabrics   328
  cake formation on membranes and fabrics   327
  continuous filtration
    cake formation for   329
    cake moisture content for   329
  gasless filtration on vacuum drum filters

  barometric leg   323
  hydrophobic PTFE-membrane   325
  installation   322
  process   323
membrane bubble point   326
vacuum disc filter with microporous ceramic filter media   321
shrinkage cracks   243
  in a dried harbor sludge and limestone filter cake   204
  in filter cake   192, 286
  prevention by cake pre-consolidation   289
shrinkage, of filter cakes   198
slotted pore   317
slurry
  agglomeration
    bridging   74
    cake filtration process   71
    capillary pressure distribution   71
    coagulation   71
    electrically charged particles   72
    electrostatic repulsion energy   73
    hydroxyl group formation   72
    natural polysaccharides   74
    negatively charged particle   72
    particle size distribution   71
    polar water molecules   72
    pre-built bridges   71
    precondition   71
    single small particles   70
    water-soluble macromolecules   74
    Zeta-potential   73
  chemical conditioning   83–84
  composition and properties   13
  filter aids–body feed filtration   80–82
  fractionation   75–80
  liquid properties   14
  particle properties
    collectives, of characterization   20–24
    general aspects   14–16
    single, of characterization   16–19
  pre-treatment   65
  sampling   35–38

slurry (*contd.*)
  solids concentration   32–33
  stability   33–35
  thermal conditioning   83
  thickening
    characteristic parameter   69
    drag coefficient   67
    economical   66
    force of gravity   67
    hydrocyclone   66
    sedimentation processes   70
    single particle sedimentation and segregation   66, 68
    solids concentration   68
    Stokes law   67
slurry pre-thickening   252
soil-washing   216, 217
solid liquid separation
  physical principles   3
  processes   5, 8
solidosity   43, 44
solids concentration   32, 33
sphericity   18, 19
square shaped pore   90, 316, 318, 319
squeezing, deliquoring of compressible filter cakes   205
staple fibers   307, 308
steam cabin   200, 258
steam generator   261
steam pressure filtration   201
  cake desaturation by   257
stirred pressure nutsche filter   68, 115, 116, 185, 186
stochastic homogeneity   42
Stokes law   66, 67
sub-surface   211
surface filters   302
surface filtration   4, 301, 302
surface tension   203
  desaturation, of filter cakes   208–270
surfactants, migration   210

t
tensile strength   204, 278, 280, 282, 283, 286, 306
threads   120, 122, 304–308, 318

three-dimensional shrinkage, of homogeneous cakes   287
tower filter press PF 180   139
triangular shaped pore   317
tube diaphragm press   140
tubular pore model   53
two-dimensional pore model   227
two-dimensional shrinkage
  of cakes with a porosity gradient   288
  of homogeneous cakes   287
two-stage co-current dilution washing
two-stage co-current dilution washing, by vacuum drum filters   185

u
ultra-filtration membranes   302

v
vacuum belt filters   108, 117, 178, 192, 290, 302
vacuum disc filter, with microporous ceramic filter media   321
vacuum drum filters   82, 109
  two-stage co-current dilution washing   185
vacuum pan filters   76, 116, 178, 193, 195, 200, 257
vacuum rotary filters   179
van-der-Waals attractive forces   313
van-der-Waals forces   114, 205
vapor pressure   110, 159, 175, 224, 270, 303
ventrical filter centrifuges   196
vibrating press belt   290

w
wash liquid   46, 175, 176, 178, 179, 181–193, 196–201, 204, 286
Wenzel-equation   217
wettability   77, 216, 217, 308
wetting   218
  capillary   214
  definition   216
woven filter media
  calendering process   308
  double weave structure   305
  dutch weave   304

exchangeable drum filter cells 310
filter bag on disc filter segment 309
FrameTrak system 310
plain weave pattern 306
thermal and chemical stability of polymers 307
thread structures 307
three-dimensional weave structure 305
weave structures 304

WRING-press 141, 142

y

Young-equation 215, 217
Young-Laplace equation 222, 228, 313, 326

z

Zeta-potential 34, 35, 48, 72, 73, 87